T0305779

CANINE
ERGONOMICS
The Science of Working Dogs

CANINE
ERGONOMICS
The Science of Working Dogs

Edited by
William S. Helton

CRC Press
Taylor & Francis Group
Boca Raton London New York

CRC Press is an imprint of the
Taylor & Francis Group, an **informa** business

CRC Press
Taylor & Francis Group
6000 Broken Sound Parkway NW, Suite 300
Boca Raton, FL 33487-2742

© 2009 by Taylor & Francis Group, LLC
CRC Press is an imprint of Taylor & Francis Group, an Informa business

No claim to original U.S. Government works

ISBN 13: 978-1-4200-7991-3 (hbk)

Library of Congress Cataloging-in-Publication Data

Canine ergonomics : the science of working dogs / editor, William S. Helton.
    p. cm.
  Includes bibliographical references and index.
  ISBN 978-1-4200-7991-3
  1. Working dogs--Environmental enrichment. 2. Working dogs--Health. 3. Bioengineering. I. Helton, William S. II. Title.

SF428.2.C26 2009
636.7'0886--dc22
                                                2009008464

Visit the Taylor & Francis Web site at
http://www.taylorandfrancis.com

and the CRC Press Web site at
http://www.crcpress.com

In memory of my mother, Kit Helton.

Contents

Preface

But the poor dog, in life the firmest friend,
The first to welcome, foremost to defend,
Whose honest heart is still his master's own,
Who labors, fights, lives, breathes for him alone,
Unhonored falls, unnoticed all his worth,
Denied in heaven the soul he held on earth –
While man, vain insect! hopes to be forgiven,
And claims himself a sole exclusive heaven.

Lord Byron, **Epitaph to a Dog**

At the United Nations headquarters in Kabul, Afghanistan, 200 British, American, German, and Greek soldiers were preparing for various peacekeeping missions. They were all cautious because a German soldier had been killed recently by a suicide car bombing. Although on edge, the 200 North Atlantic Treaty Organization (NATO) soldiers were completely unaware of the imminent danger they faced. A few meters away, buried under a pile of sandbags in a nearby car park, was a bomb. Planting a second bomb is a favored tactic—playing perhaps on the common sense notion that lightning never strikes twice in the same spot. If the bomb detonated, some if not all of the 200 nearby soldiers would have been hit with shrapnel, suffering death or severe injury. Family members at home would have suffered the losses of loved ones. Fortunately, the NATO soldiers were not alone in Kabul that day. The group included Lance Corporal (L/Cpl) Karen Yardley and Sadie, a black Labrador retriever. Molecules of the bomb's odor had reached Sadie's nose and somewhere deep in her brain a pattern had been detected. Sadie's black tail began to wag excitedly as she showed keen attention to the car park. L/Cpl Yardley recognized Sadie's signal and knew an explosive device was present nearby. Yardley alerted the bomb squad and the lives of her fellow soldiers were saved by the sensitive nose of her comrade, Sadie. For her actions in Kabul, Sadie was awarded the Dicken Medal, the animal equivalent of the Victoria Cross, by Princess Alexandra of Britain. Sadie is not alone. Untold numbers of people have been saved by hard working dogs deployed in dangerous places.

Since ancient times dogs have collaborated with humans. Their status as loyal and devoted coworkers is reflected in the myths and tales we have told each other since the beginnings of human civilization. In Homer's Odyssey, the fidelity of dogs was represented in Argos, Odysseus' faithful dog, the only one who recognized Odysseus after his long delayed return from the Trojan War. Dogs are often portrayed in the important role of serving as messengers or guardians. In Hinduism, dogs are the messengers of Yama, the lord of death, and act as the guardians of heaven. In classical Greek mythology, Hades, the underworld, is guarded by Cerberus, a fierce three-headed dog. In Zoroastrianism, dogs serve as bridges to the spiritual world and are attributed with the ability to detect when the soul has left the body.

As a graduate student at the University of Cincinnati, I studied vigilance—the ability of people to pay attention over long periods. I was working in the laboratory of William Dember and Joel Warm. One day I was reading a paper about explosive detector dogs. The paper noted that dogs could search effectively only for 45 to 60 minutes, and then needed to rest. This sounded very familiar. The most ubiquitous finding in the sustained attention literature is the inability of humans to sustain their attention over long periods. If you track performance over time, it decreases progressively over the time spent on a task. People respond more slowly to signals and/or they miss increasingly more signals over time. This decrease in ability over time is called the *vigilance decrement* and it was the topic I studied with Bill Dember and Joel Warm. I looked up from the paper and there was Kiowa, a black and tan mixed-breed trained signal (hearing assistance) dog. He lay on the floor with one ear up and swiveling around searching for sounds. Kiowa, like a sonar or radar operator, was a vigilant worker, looking for relatively rare target signals among long series of irrelevant noises and sounds. Kiowa, moreover, was an expert, as he had learned to generalize his signaling to untrained but meaningful targets, such as water boiling or a bathtub filling.

The next day I went to the office of Donald Schumsky, a professor at the University of Cincinnati. I asked him whether anyone had written about nonhuman expertise. Don was a polymath and had done work on human expertise. As far as he knew, there was no such work; expertise researchers had not studied nonhumans. This seemed unbelievable at the time. Working dogs should have been of primary interest to researchers in my field of ergonomics, the science of work and workers. Working dogs perform work just as people do. My first papers advocating the ergonomic study of working dogs generated interest in the ergonomic community because the idea was then completely revolutionary. The radical perspective of seeing dogs as workers just like human workers led to an interview in the *Bulletin of the Human Factors and Ergonomics Society*, the primary news outlet for the ergonomics field and eventually led to this book.

The scientific analysis of working dogs is emerging but underdeveloped, largely because it is scattered across several disciplines: agriculture, entomology, environmental science, ethology, forensic science, medicine, psychology, and wildlife biology. As a researcher interested in dogs, I realized the need for a book that would bring working dog researchers together in a new and unique field: *Canine Ergonomics*, the science of working dogs. This book compiles the research and views of scholars who study dogs in working contexts. It is unique in drawing researchers from the different fields together with the focus on dogs as working entities.

Dogs, like people, are flexible biological workers. They are currently used in a wide variety of occupational tasks, for example, avalanche victim search, cancer detection, drug detection, explosives detection, environmental remediation, forensic tracking, hearing ear assistance, herding, livestock protection, mobility assistance, scat searching, security, seeing eye assistance, seizure detection, and termite detection, to name a few. The range of occupational tasks dogs are employed to do increases yearly. The need for working dogs, for example, is growing rapidly in a time of rising global insecurity. International events have increased interest in exploring new uses of dogs in fields such as deception detection and biochemical and nuclear weapons searches. Highly trained dogs as biological experts may also be useful as models to aid our understanding of biological factors such as genetics and early life experience that are important for analyzing human expertise and skill development and difficult or impossible to study experimentally in humans.

The scientific literature on working dogs is scattered across several nonoverlapping disciplines and, in comparison to the magnitude of its societal importance, relatively underdeveloped. The interest across disciplines and the lack of literature may be related. Because of disciplinary boundaries, discoveries in one field may be overlooked in another. In addition, the science of working dogs is not currently regarded or recognized as a specialized topic of research. This book should foster interdisciplinary dialogue and serve as a primary source of scientific information about working dogs, regardless of domain of particular application, for anyone interested in the topic, from developmental psychologists interested in using dogs as models of human experts to handlers who work with dogs in field settings.

In these pages you will find cutting edge data about working dogs. This book could serve as a supplementary text in courses on animal behavior or as a reference for people like me who have a passion for dogs. People like dogs. Amazingly, dogs have been trusted as valuable coworkers for centuries, but we are only starting to study dogs in greater detail and unravel the secrets of the very old collaboration of two morphologically disparate but compatible predators.

William S. Helton

Acknowledgments

I would like to thank my intellectual mentors at the University of Cincinnati, Joel Warm, the late William Dember, Donald Schumsky, Daniel Wheeler, Gerald Matthews, Kevin Corcoran, and Robert Frank; Anders Ericsson, Paul Feltovich, Robert Hoffman, and Nicola Hodges in the expertise community; Ian McLean, David Washburn, Reuven Dukas, Richard Caldow, Elizabeth Lonsdorf, and Robert Walker among comparative researchers; Cat Weil, Christina Sommer, and all the other dog devotees at the Queen City Dog Training Club in Cincinnati, Ohio.

I would also like to thank colleagues and students at Michigan Technological University, especially David Schaffer who helped proofread some of the chapters. I would like to especially thank Alexander Ferworn whose view of canine ergonomics is probably the most similar to my own. The dogs in my life deserve special thanks: Freedom, Yeti, Aster, Roo, Major, especially Kiowa, and Berkeley. Last but definitely not least I would like to thank my wife, Nicole Helton, who has witnessed endless iterations of the chapters, was pivotal in my pursuing research on dogs, and has been my dearest intellectual companion over the years. Without her, this book would not exist.

Editor

William S. Helton is currently a senior lecturer in the Department of Psychology at the University of Canterbury in New Zealand and an associate professor in the Department of Cognitive and Learning Sciences at Michigan Technological University in Houghton, Michigan. He earned a B.A. in philosophy and mathematics from Evergreen State College in 1995 and an M.A. and a Ph.D. in psychology from the University of Cincinnati in 1998 and 2002, respectively. His research focuses on expertise and attention in both dogs and humans. His research has been published in *Acta Psychologica, Animal Cognition, British Journal of Psychology, Journal of Experimental Psychology, Journal of Clinical and Experimental Neuropsychology* and numerous other publications.

Contributors

Craig S. Clark
U.S. Department of Agriculture
PHIS Wildlife Services
Barrigada, Guam, United States

Michael S. Davis
Department of Physiological Sciences
Oklahoma State University
Stillwater, Oklahoma, United States
msdavis@okstate.edu

Richard M. Engeman
National Wildlife Research Center
Fort Collins, Colorado, United States
richard.m.engeman@aphis.usda.gov

Paul J. Feltovich
Florida Institute for Human and
  Machine Cognition (IHMC)
Pensacola, Florida, United States
http://www.ihmc.us/

Alexander Ferworn
Department of Computer Science
Ryerson University
Toronto, Ontario, Canada
aferworn@scs.ryerson.ca

Irit Gazit
Tel Aviv University
Ramat Aviv, Israel

Allen Goldblatt
Center for Applied Animal Behavior for
  Security Purposes
Department of Zoology
Tel Aviv University
Ramat Aviv, Israel
allen.goldblatt@gmail.com

Samuel D. Gosling
Department of Psychology
University of Texas
Austin, Texas, United States
samg@mail.utexas.edu

Lindsay T. Graham
University of Texas
Austin, Texas, United States

Marc A. Hall
U.S. Department of Agriculture
PHIS Wildlife Services
Barrigada, Guam, United States

Dorit Haubenhofer
Wageningen UR
Plant Research International
Wageningen, The Netherlands
dorit.haubenhofer@wur.nl

William S. Helton
Department of Psychology
University of Canterbury
Christchurch, New Zealand
and
Department of Psychology
Michigan Technological University
Houghton, Michigan, United States
Deak_Helton@yahoo.com

Aimee Hurt
Working Dogs for Conservation
  Foundation
Missoula, Montana, United States
aimeehurt@yahoo.com

Lisa Lit
MIND Institute
Departments of Neurology and
  Psychology
University of California
Sacramento, California, United States
llit@ucdavis.edu

Nancy Hansen Merbitz
MHRC, Inc.
Pontiac, Illinois, United States

Péter Pongrácz
Department of Ethology
Eötvös Loránd University
Budapest, Hungary
peter.celeste.pongracz@gmail.com

Natalie Sachs-Ericsson
Department of Psychology
Florida State University
Department of Psychology
Tallahassee, Florida, United States
sachs@psy.fsu.edu

Michaela Schneider
Faculty of Veterinary Medicine
Ludwig Maximilians University
München, Germany
m.schneider@
  animalbehaviorandwelfare.de

Leopold Slotta-Bachmayr
Salzburg, Austria
leo@dogteam.at

Deborah A. Smith
Working Dogs for Conservation
  Foundation
Hughson, California, United States
deborahsmith1@sbcglobal.net

Joseph Terkel
Tel Aviv University
Ramat Aviv, Israel

Andrew J. Velkey, II
Department of Psychology
Christopher Newport University
Newport News, Virginia,
  United States
avelkey@cnu.edu

Daniel S. Vice
U.S. Department of Agriculture
PHIS Wildlife Services
Barrigada, Guam, United States

1 Canine Ergonomics
Introduction to the New Science of Working Dogs

William S. Helton

CONTENTS

> When you do traffic stops, it is a great feeling to know the dog is back there. That dog will save your ass more than you ever know. He will be better than another cop.
>
> **Statement of K-9 police officer cited in Sanders (2006, p. 167)**

CANINE ERGONOMICS

Since ancient times, working dogs have served as highly accurate and flexible extensions of human senses and abilities. Despite advances in artificial intelligence and modern sensor technology, machines remain unable to match the operational effectiveness of trained dogs in a number of tasks such as explosives and narcotics detection (Fjellanger et al., 2002; Furton and Myers, 2001; Gazit and Terkel, 2003; Slabbert and Rasa, 1997). Dogs fill many other occupational roles including accelerant detection, blind assistance, epilepsy detection, forensic tracking, guarding, hearing assistance, herding livestock, and detection of insect infestations and microbial growth (Brooks et al., 2003; Holland, 1994; Wells and Hepper, 2003). Their skills are presently under investigation in new arenas, for example, in the detection of cancer (Pickel et al., 2004). The study of these highly trained dogs bridges the gap between ergonomics (the study of work) and the animal sciences.

Ergonomic researchers focus on the transactional relationships between people and their environments when work is performed. Animal scientists such as comparative psychologists, veterinarians, physiologists, and animal behaviorists study

1

nonhuman animals to facilitate our understanding of animal capacities and their limitations. The study of working dogs falls between these two disciplinary traditions and offers many points of potentially fruitful exchange.

Dogs currently provide the best models for *working humans*. While studying other animals like rats or primates is informative, canines are unique in their ability to easily work beside humans. Canines, both domestic and wild, are social animals. Although the exact place and time of the divergence of wild canines and domestic dogs (*Canis lupus familiaris*) is debated, there is almost complete agreement that the genetically closest wild canines are wolves (*Canis lupus*) (Vila et al., 1997). Wolves historically were the master predators of their Northern geographic domain, spreading across Asia, Europe, and North America. Their predatory success was due to their ability to hunt and live cooperatively.

Among indigenous people of the Northern Hemisphere familiar with wolf behavior, wolves are respected for their cunning, human-like behavior. Nelson (1983), an anthropologist, living among the Koyukon of Alaska, reported an elder native hunter saying he did not like to hunt wolves: "They're too smart, too much like people." In Norse mythology Odin, the chief of the gods and the wisest, is accompanied by the cleverest indigenous animal companions, two wolves named Geri and Freki, and two ravens called Huginn and Muninn. Native American mythology is full of stories of clever, almost human, wolves and coyotes. Although people respect wolves for their cunning, human-like capabilities, wolves have also been the subject of intense fear and hatred, for example, in fairytales such as "Little Red Riding Hood." Most likely this fear arose when wolves and people shared similar ecological niches and resulted from often intense competition for resources.

Both humans and dogs descended from intelligent social predators, as research on our closest living relatives, chimpanzees (*Pantroglodytes*), suggests (Boesch and Boesch, 1999). This shared ancestry of hierarchical social hunting probably facilitated dog domestication. Human-tolerant wolves, or proto-dogs, were undoubtedly the first animals to co-exist and work with humans based on their social and mental capacities. Some researchers even suggest that this early human–dog bond may have been mutually co-evolutionary (Csanyi, 2005; Vila et al., 1997).

The exact time when domestic dogs diverged from wolves continues to be debated. The issue is complicated because dogs and wolves (and also coyotes) are similar morphologically and are subtypes of one species; they can all interbreed and produce fertile offspring. One side of the debate over the date of divergence is led by geneticists who argue that dogs and wolves diverged as early as 150,000 years ago. The other side involves paleontologists and other geneticists who argue that dogs diverged from wolves as late as 15,000 years ago (Coppinger and Coppinger, 2001; Vila et al., 1997). If the earlier date (150,000 years ago) is correct, *Homo sapiens sapiens* (modern humans) diverged from other *Homo sapiens* such as *Homo sapiens Neanderthals* around the same time. Some have argued that although these dates appear coincidental, they may instead mark a pivotal point in both human and dog evolution (Csanyi, 2005; Vila et al., 1997). Wolves and people who existed symbiotically early on, perhaps living in close proximity but not together, would have had distinct advantages over wolves or people who lived separately.

People have excellent sensory capabilities in daylight because of stereoscopic color vision, but relatively low levels of night vision, olfactory ability, and hearing put people at a distinct disadvantage in darkness. Wolves in the dark or during dawn and dusk benefit from superior night vision, exceptional olfactory abilities, and better hearing. People often rely on dogs to act as sentinels at night. Undoubtedly a pack of wolves or proto-dogs living near a human camp would have served as an early warning device, giving those people a distinct advantage over competitors without dogs. A band of warriors sneaking up on these people would be more likely to be detected because of noises made by dogs living nearby. People living without dogs in close proximity would have been easy victims of surprise night raids.

This point is used in the plot of *The 13th Warrior*, a film based on Michael Crichton's book, *Eaters of the Dead*, in which Norsemen were able to sneak into the camps of proto-humans, presumably remnant Neanderthals, at night because unlike modern humans, these proto-humans did not have dogs. The nearby proto-dogs would have benefited from access to human refuse and from the increasing ability of humans to plan for the future. Likewise, the humans living in close proximity to dogs would have gained from the ability of the proto-dogs to serve as nighttime sentinels and also by following the dogs to herds of ungulate prey and perhaps mimicking their hunting tactics. Both the dogs and people would have had to change their dispositions to become increasingly tolerant of each other in order to live symbiotically. Perhaps, and this is mere speculation, we owe as much to dogs as they do to us. Humans and dogs civilized and domesticated each other.

Wherever humans have gone, dogs have gone with them, even into outer space. As Dennett (1996) and Csanyi (2005) argue, dogs may be unique among animals in sharing culture with humans; both species obey cultural rules and ritual behavior. Recent research indicates dogs are better at reading human signals such as orientation of gaze and pointing than our closer relatives, the great apes. Likewise, dogs clearly communicate directional information to people (Agnetta et al., 2000; Miklosi et al., 1998 and 2000). These communication skills along with social natures enable dogs to become co-workers, often replacing people, for guiding the blind, herding livestock, and guarding property.

Working dogs are the best candidate models of human workers. They are, like us, the descendants of social predators. They manage to live competently in a hierarchical society as humans do. In basic psychological dimensions dogs are roughly similar in personality, emotion, cognition, and perception. Their underlying mammalian physiology is comparable to our physiology. Although they cannot use tools because they lack opposable thumbs and other manipulators and their linguistic capacities are limited (Kaminski et al., 2004), they closely resemble human workers. They live closely with us, and most importantly for ergonomic researchers, they work closely with us.

ANTHROPOMORPHISM AND ANTHROPOCENTRISM

Animal researchers are often excessively concerned about anthropomorphism—the attribution of human characteristics to nonhumans. This fear of anthropomorphism stifles research, for example, on animal cognition (Bekoff and Allen, 1997).

Especially when studying working dogs, slight anthropomorphism is indispensable and pragmatic because working dogs truly are anthropomorphic. Working dogs morph into human roles by design. They are specifically trained to work with and often replace people and may even be specifically bred to do so. Canine trainers and handlers in the working dog community actively encourage a modest anthropomorphic view when dealing with working dogs (Sanders, 2006); a strict behaviorist perspective is simply unworkable in this context.

Ergonomics (also known as human factors) is the study of human work and interaction with work systems, typically machines. Although the term *human* is used in the definition and thus may rule out consideration of nonhuman workers, I think this is a mistaken perspective based on outdated and excessive beliefs of anthropocentrism.

Humans are undoubtedly unique, but the belief that people are entirely distinct and different from the rest of nature is the result of pre-scientific, pre-Darwinian thinking. Moreover, *human* may not be a species designator in terms of ergonomics. St. Augustine in his *City of God* wrote, "Whoever is born anywhere as a human being, that is, as a rational mortal creature, however strange he may appear to our senses in bodily form or color or motion or utterance, or in any faculty, part or quality of his nature whatsoever, let no true believer have any doubt that such an individual is descended from the one man who was first created." Despite the religious wording, Augustine makes an interesting point: the term *human* may in some contexts refer to an entity with a set of abilities or human capacities, and not to species membership or exact physical form. Since Descartes' argument for the uniqueness of human cognitive capacity, our historical legacy limits ergonomics to the study of human work. This is unfortunately a fundamental error based on an outdated philosophical position that most modern scientists reject. From the perspective of modern post-Darwin science, the view that humans are disembodied rational souls and that *Homo sapiens* were created separately from the other animals in an act of special creation is no longer plausible.

Working dogs often act as human surrogates. Although they are often employed to perform tasks humans cannot perform, e.g., find bombs with their noses, they are widely employed to perform tasks humans can do, for example, guide the blind. They share many capacities with humans, in cognition and personality, for example. Ergonomics then may be considered the study of entities that share human capacities in working situations. A counter argument could be that machines now do jobs previously performed by humans and that ergonomics is not mechanical or electrical engineering. Machines, however, perform these operations and tasks in entirely different ways and modern machine architecture is very remote from human architecture. Dogs, like humans, are products of uncontrolled evolution—they were not built with a purpose in mind. Dogs, unlike machines, do jobs roughly the way humans do. They fail when humans fail, and often excel where humans excel, for example in pattern recognition. Dog architecture (physiology) is similar to human architecture. Dogs, like humans and entirely unlike machines, are flexible. No machine in existence can replicate all the tasks a dog can be trained to do.

Furthermore, society would probably not tolerate a machine with the level of autonomy a dog has. Although many working dogs work in close contact with people, many work nearly independently. Take for example dogs whose job is to defend livestock from natural predators. Although their human employers provide food and water, the animals have almost no relationships with their humans. In addition, like people, dogs cannot really be forced to work; they must be persuaded, encouraged, threatened, or enticed and the possibility of a revolt is always present. Thousands of people are admitted to emergency rooms each year for dog bites and some people are killed by dogs. Despite these occurrences, society has not demanded a recall of dogs, presumably because society accepts that dogs, like people, are individuals with distinct personalities. If a car ran someone down autonomously, the legal consequences would probably be very serious. Presumably the designer and/or builder would be held liable. An artificial agent with the capacity of a working dog, despite advances in machine intelligence, remains purely fictional. The closest model for a working dog is a working person and vice versa. Dogs and humans are more closely related than either species is to any existing machine.

This similarity is the point of the discussion of the relationship of dogs and wolves. Both dogs and humans descended from social predators with hierarchical societies. The dog, as Dennett (1996) points out, is as close to human as anything is. Dogs are almost human and are often treated as if they are human. Keep in mind, the issue is not whether dogs are truly conscious like people; that discussion is highly controversial, I am talking about their roles in society and their demonstrable cognitive and perceptual capacities.

Whether someone believes that working dogs are true human surrogates, we cannot deny that dogs are often exceptionally skilled workers. The role of working dogs in society is far greater than most people know and is likely to increase, not diminish, in the future. Mine clearance is a task primarily performed by dogs. Most other nonhuman animals are not easily trained to work in complex settings. Working dogs are unique because of the ease with which they work in complex human settings and integrate into human society. The "man's best friend" description is factual, not mythological. Dogs, of course, are far from human replicas, but working dogs are different from other animals including the great apes that are our closer relatives. Dogs are not typical subjects for study by comparative psychologists. Unlike the subjects of most comparative psychology studies, working dogs have real jobs that are completely unlike the contrived tasks of the animal laboratory. They do not work in Skinner boxes; they work on city streets, at airports, and in mine fields. The tasks they perform are more similar to the tasks of human workers than to the simplified tasks in a laboratory. Because of the scope of work performed by dogs, a new perspective that integrates animal science and ergonomics is needed. It should escape the narrow anthropocentrism and excessive fear of anthropomorphism and constitute a new science of working dogs: canine ergonomics.

This book presents the cutting edge of research in the field of canine ergonomics. It has three main themes: (1) working dogs are nonhuman models of human workers; more generally, working dogs and working people share much in common; (2) the study of working dogs can improve the science of ergonomics as the study of nonhuman models led to advances in other areas of behavioral science; and (3) an ergonomic orientation, *recognizing dogs as real workers*, can benefit working dogs.

The remainder of this introductory chapter will present two examples of issues in which ergonomics and animal scientists share a common interest: expertise development and sustained attention. Both topics will be explored in detail in other chapters in this book. The examples represent the convergence and parallel progress of two traditions, ergonomics and animal science. The examples are not definitive; myriad other examples could be selected, as will be apparent in the other chapters. Instead, these two examples are merely suggestive of the present range of common interests from which the new science of canine ergonomics will emerge from dialogues between ergonomics and animal science. These examples show that animal scientists and ergonomic professionals can learn from each other by studying working dogs.

EXPERTISE DEVELOPMENT

When regarded as models for human workers, working dogs may prove useful in resolving conflicts in contentious areas of ergonomic research. Expertise development is a case in point. Ergonomic practitioners are often asked to help individuals in the acquisition of expertise or in determining who among potential candidates actually is an expert (Weiss and Shanteau, 2003). They may also play roles in the capture of expert knowledge and assist in the subsequent deployment of that captured knowledge by an artificial expert system or a decision aid (Hoffman et al., 1995).

One area of current interest is the expert ability to detect threat items in luggage (Hancock and Hart, 2002; McCarley et al., 2004). The human expertise literature consists of many studies demonstrating qualitative and quantitative differences in performances of a variety of skills by experts and novices (Masunaga and Horn, 2000). The development of these performance differences is, however, an area of much controversy, mainly among those who lean toward learning theories of expertise development (Ericsson, 2001; Howe et al., 1998), emphasizing deliberate practice and training (Ericsson and Charness, 1994), and those who promote talent theories (Gardner, 1997) and stress personnel selection and behavioral genetics (Greenwood and Parasuraman, 2003). Researchers from both perspectives acknowledge that training and biology play roles in expertise development. The argument concerns the degree to which inherited abilities and learning determine the development of expertise.

Human studies directly testing these competing perspectives of expertise, e.g., learning–practice and genetics–talent, are nonexistent primarily because they are hopelessly confounded by participants' willingness to partake in experiments and training. Expertise in many fields requires an extensive period (up to 10 years) of skill acquisition and familiarization (Ericsson, 2001; Ericsson and Charness, 1994; Howe et al., 1998). Humans, for both ethical and practical reasons, cannot be randomly assigned to differing training groups and forced to remain in such groups for such long periods, even in controlled environments like the military. Many participants will likely remove themselves from training groups. The participants' willingness to train, unfortunately, may arise from ease of mastery, or what many call talent. The role of talent in expertise development continues to irk researchers. Additionally, humans' early experiences that may prove critical for the development of some forms of expertise cannot be controlled systematically. An exclusively human focus will leave many questions unanswerable. Humans are not, however, the only experts on this planet (Helton, 2004a and b and 2005).

Recognizing the similarities of human and nonhuman expertise would enable both ergonomics professionals and animal behaviorists to exchange information in a mutually beneficial way. Dogs are employed to do a variety of tasks, ranging from guarding and herding livestock (Holland, 1994) to detecting termites (Brooks et al., 2003). As Weiss and Shanteau (2003) note, evaluative skill is the central core of expertise. All experts, regardless of discipline, are required to evaluate situations and make judgments. Every expert dog must make judgments. For example, a live-stock guardian must decide whether an approaching animal is a harmless herbivore or an intruding predator. Judgmental competence in many arenas is hard to verify, but in other situations, for example, those involving dichotomous decisions in which one choice is objectively correct, scientists can use the signal detection metric d' as a measure of accuracy (Swets, 1986). In studies of expert dogs, d' can be calculated because dogs frequently make dichotomous judgments that are or are not objectively correct (Brooks et al., 2003; Fjellanger et al., 2002; Furton and Myers, 2001; Gazit and Terkel, 2003; Slabbert and Rasa, 1997; Pickel et al., 2004), for example, whether a substance is present or absent. Performance measures such as d' and β cited in the ergonomics literature are often applicable in studies of expert dogs and facilitate information exchanges.

Canine experts and human experts undergo long periods of formal skill training and practice (Helton, 2005). Ericsson (2001) provides a set of four behavioral determinants of deliberate practice: (1) the trainee is motivated to improve; (2) the trainee is assigned well-defined tasks; (3) the trainee receives feedback; and (4) the trainee has ample opportunity for repetition. The training of skilled dogs fits these criteria, whether the skill is narcotic detection (Slabbert and Rasa, 1997), explosive detection (Gazit and Terkel, 2003), or herding (Holland, 1994). The dogs are definitely motivated to perform a task, they have clearly defined tasks to accomplish, they receive feedback about their performance, and they train repeatedly. A trained canine engages in repeated practice to achieve a predetermined level of ability, similar to a person undergoing a prolonged period of practice. No species difference exists; both humans and dogs practice purposefully.

Many popular theories of human expertise development characterize the novice stage as the acquisition of declarative knowledge—explicit knowledge of how to perform a task (Anderson, 1983 and 1993; Dreyfus and Dreyfus, 1987; Fitts and Posner, 1967). At this novice stage, performance of the task is cognitively demanding, requiring focused attention. Gradually, over time, the declarative knowledge is proceduralized to the point of automaticity. Unlike a novice, an expert no longer needs to maintain active attention; he or she can perform the task intuitively (Dreyfus and Dreyfus, 1987). Human novices in many cases first acquire the declarative knowledge of the skill through verbal means. Dogs, although able to understand associations of words and actions or words and objects, lack human language skills. One may therefore reasonably object to using dogs as models for human expertise development in this regard; similarities of expert dogs and expert humans may be merely superficial. Humans may use mechanisms, language, and narratives unavailable to other animals; this objection is worth serious consideration.

There are, however, two potential counter-arguments to this criticism: (1) human skill development may not require the acquisition of verbally stated declarative knowledge and (2) dogs may be capable of storing declarative knowledge in images rather than via language. Individuals can acquire the ability to perform complex skills without first acquiring large amounts of declarative knowledge, and they are often unable to verbalize the rules they use to perform such tasks (Berry and Broadbent, 1988; Sun et al., 2001). This inability of human experts to fully verbalize their activities has proven problematic for engineers attempting to include knowledge acquisition in artificial expert systems (Collins, 1989). Some researchers challenge this position; Hoffman et al. (1995) argue that humans, even idiot savants, can, when pressed, declare their knowledge. The point is, however, that they must be pressed; human experts often have difficulty retrieving knowledge, implying that knowledge is not initially directly accessible to consciousness.

Humans sometimes acquire skills procedurally first and can subsequently articulate their activities; they do not always require declarative knowledge before developing procedural knowledge (Stanley et al., 1989). Some human skills are acquired initially from the bottom up (procedurally) instead of from the top down (declarative knowledge). The ability to share declarative knowledge presumably facilitates the development of human expertise, although this is not always true (Sun et al., 2001). Declarative knowledge is not always required nor is it the key to expertise. As Dreyfus and Dreyfus (1987) clearly demonstrated, relying on declarative knowledge is the sign of the novice, not the expert. Although canine experts cannot verbalize their knowledge, this in no way implies they do not have it.

Artificial expert system engineers increasingly use neural network and evolutionary learning models to help develop skilled performance (Gallant, 1993; Helton, 2004b). These adaptive models are not preprogrammed with rules and heuristics; they are generated. The rules, moreover, can be retrospectively extracted from these systems (Gallant, 1993). Hoffman et al. (1995) argue and provide supporting evidence that using verbal interviews, e.g., having human experts verbalize their knowledge, is often an inefficient method of knowledge extraction in comparison to having the experts perform contrived tasks that reveal the rules they employ. Likewise, rules can be extracted from canine experts by making them perform their skills in controlled environments (Shaffer et al., 2004). Whether the canine or the human expert is consciously aware of using a rule is irrelevant; performance ultimately reveals the presence of knowledge.

Nonhumans, of course, cannot declare their knowledge through language. Terrace and colleagues (2003) argue that language and declarative knowledge are not identical and also note that declarative knowledge predated language in the course of evolution. Encoding images analogically is an example of a nonverbal declarative knowledge system (Kosslyn, 1980; Terrace et al., 2003). Dogs, for example, appear to use encoded images to guide their behavior (Adachi et al., 2007; Topal et al., 2006). This area of research, mental imagery in dogs, however, is very underdeveloped. The dog's ability to encode declarative knowledge in images is a distinct possibility that would explain many aspects of dog behavior.

Over long courses of training, some dogs develop highly skilled behaviors. Because of their high skill levels, many are employed in operational settings where they are extremely successful. Regardless of their lack of verbal skills, these dogs are experts and investigating their skill development may be useful in understanding human expertise development. Narcotic scent detection is, for example, a complicated skill. Not only is the scent environment extremely complex, but smugglers typically attempt to hide narcotic substances and mask their scents. Dogs employed in narcotic detection must train extensively, and some become so successful that smugglers often offer large monetary rewards for their elimination. Slabbert and Rasa (1997) conducted an experiment to determine whether observations of maternal narcotic detection behavior accelerated later skill development and the acquisition of expertise by puppies. Pups from untrained and trained narcotic detection bitches were separated into two groups: those separated from their mothers at 6 weeks and those separated from their mothers at 3 months. Between the ages of 6 and 12 weeks, the pups reared by the trained bitches in the extended parental condition were allowed to observe their mothers work. The pups reared by untrained bitches in the 3-month condition presumably were subjected to the same maternal care, but were not exposed to working role models. When the groups were later tested for narcotic detection aptitude at 6 months of age, the observational group exposed to working role models performed significantly better than the nonexposed groups. Of the observational group, 85% passed the aptitude test; only 19% passed in the other groups. Early observation of a role model apparently facilitates the acquisition of narcotic detection skills.

Experiments with working dogs like that by Slabbert and Rasa (1997) directly address the talent-versus-practice debate (Helton, 2004a). Their study is one of the only true experiments, i.e., random assignment of individuals to conditions, focusing on expertise development. Dogs can be selectively bred, facilitating the work of behavioral geneticists (Schmutz and Schmutz, 1998). The ability to control the breeding of dogs and employ them in genetic studies of expertise development is further enhanced by similarities of dogs and humans in basic psychological dimensions that may have a genetic component. Canines, for example, share many personality traits with humans (Gosling and John, 1999) and may thus be useful in determining the role of individual difference variables in the acquisition of skills. Matzel and colleagues (2003) provided evidence for a general learning ability, similar to the human intelligence factor, in mice. Although it has yet to be demonstrated, presumably dogs share this factor as well. Dogs' basic cognitive abilities are similar enough to those of humans that they have become useful models for studying the effects of aging on cognition, and dogs, like humans, vary in these abilities (Adams et al., 2000). The extensive training time and associated costs required to produce reliable canine experts compel those wishing to employ expert dogs to investigate prescreening devices to determine which dogs are most likely to succeed during the training regime, i.e., personnel selection. This is an issue of direct concern to ergonomics researchers as well.

Research with skilled canines may provide insights into human expertise development useful to ergonomic practitioners. In particular, the continuing debate about the role of genetics in expertise is an issue most appropriately tackled with canine research. Although it is very unlikely that no information pertinent to human expertise would be discovered, at the very least, canine research would provide useful data for training canine experts. The widespread use of canines in settings such as explosives detection would justify the effort. Ergonomic practitioners have much to offer animal behavior specialists already engaged in studies of working canines, as will be demonstrated in the next section.

SUSTAINED ATTENTION

Sustained attention or vigilance is the ability of an observer to detect brief and unpredictable signals over time (Davies and Parasuraman, 1982; Helton et al., 2002, 2004, and 2005; Warm, 1993). This aspect of human performance is of particular interest to ergonomic specialists because of its critical role in many operational settings, especially in relation to automated human–machine systems (Warm, 1993; Helton et al., 2005). The introduction of automated systems into work settings radically transformed many tasks performed by human operators. As a result of automation, operators increasingly shift from acting as hands-on controllers to monitoring information display systems and taking action only when an event the system was not designed to handle occurs. Vigilance is thus a key element in many complex automated systems deployed in operational environments.

Systematic research focusing on vigilance did not begin until Norman Mackworth (1948 and 1950) worked on a problem encountered by the Royal Air Force (United Kingdom) during World War II. Pulse-position radio detection and ranging (radar) was a new technology employed by airborne observers to detect surfaced Axis submarines. Airborne radar reconnaissance represented a significant improvement over earlier techniques such as training seagulls to flock around U-boats (Lubow, 1977). The system, however, had a serious problem: regardless of their high level of motivation and extensive training, the airborne observers began to miss the "blips" on their radar scopes indicating the presence of submarines after only 30 minutes on watch. Unless this tendency was corrected, the result would have been increased losses of Allied vessels to U-boat attacks.

Mackworth tackled the vigilance problem experimentally by using a simulated radar display. He discovered that the ability to detect critical signals waned over time; initial levels of signal detections were high—85% detection rate—but declined by 10% after 30 minutes and continued to drop as a vigil continued. This decline in performance efficiency over time is known as *vigilance decrement* or the *decrement function*. It is the most common effect observed in vigilance research (Davies and Parasuraman, 1982; Helton et al., 2002, 2004, and 2005; Matthews et al. 2000; Temple et al., 2000; Warm, 1993). It was observed with experienced monitors such as those employed by Mackworth and with novices in both operational and simulated settings (Pigeau et al., 1995).

The vigilance decrement has also been noted in studies of nonhuman observers (Bushnell, 1998; Bushnell et al., 2003). The long history of studies of humans and other animals provides a convincing case that the central nervous system, regardless of species, cannot effectively sustain attention for an indefinite period. The focus of attention appears to be temporally limited. Bushnell et al. (2003) investigated similarities of sustained attention levels of humans and rats. Humans and rats performed homologous vigilance tasks under varying conditions, three levels of signal presentation (event) rate, and six levels of signal intensity. The study shows nearly identical results for both human and rat observers. Thus, as Dukas and Clark (1995) note, vigilance decrement is a dominant factor determining all animal behavior. A fundamental biological limitation of the nervous system to sustain focused information processing for extended periods would limit an animal's ability to detect camouflaged predators or prey. This limitation of sustained attention plays a role in determining any animal's work-and-rest cycle if work is the active search for predators and prey and rest is a period of recovery.

Some researchers propose that limited attention resources are available for information processing (Kahneman, 1973; Matthews et al., 2000). This limited resource perspective has been used extensively to explain findings related to vigilance (Davies and Parasuraman, 1982; Dukas and Clark, 1995; Helton et al., 2002, 2004, and 2005; Temple et al., 2000; Warm, 1993). This view considers resources as reservoirs of *mental energy* dedicated to the performance of a task (Hirst and Kalmar, 1987). During vigilance tasks, observers must make active, continuous signal–noise discriminations under uncertainty conditions without rest. The continuous mental requirements for vigilance do not allow for replenishment of resources. Hence, the resource or energy pool depletes over time and is reflected as a decline in efficiency. Demands upon the limited pool of resources also increase as the difficulty of a task increases. For a more demanding task, the pool of resources depletes more rapidly, thus demonstrating a steeper drop in performance efficiency.

Working canines are often assigned the task of continually monitoring the environment for relatively rare critical signals. In naturalistic terms, this is analogous to searching for cryptic prey. Whether the dogs search for explosive odors or listen for fire alarms, they are often required to be vigilant. For example, an explosive detection dog must continually sample the air and judge whether the molecular signature of an explosive substance is present or not. This is akin to a human baggage inspector's task of continuously sampling visual scenes and judging whether a restricted item is present. Many questions about the vigilance of working dogs remain unanswered. For example, how long can a dog work before needing a rest? This problem of work–rest cycles has long been an issue of ergonomics. Vigilance researchers have investigated performance recovery following rest breaks since Mackworth's seminal work in the late 1940s. Thus, the work–rest cycle is an area of common interest for both human and canine ergonomic researchers (Dukas and Clark, 1995).

The role of resource limitations during sustained attention operations, as manifested by the vigilance decrement, is another area where human ergonomics may make a practical contribution to canine ergonomics. Animal behaviorists studying canine workers are aware of potential limitations in the dogs' capacity to sustain effort over time; they, however, tend to interpret decrements in performance efficiency as motivational problems (e.g., boredom), not fundamental limitations of attention resources (Garner et al., 2000). Ergonomic practitioners can enhance this area of study by sharing a long history of research on sustained attention and methodological expertise in investigating worker duty cycles.

Ergonomic researchers and animal behaviorists, moreover, may facilitate each other's efforts by studying the underlying neurological mechanisms controlling the vigilance decrement (Hitchcock et al., 2003; Holley et al., 1995; McGaughy and Sarter, 1995). Ergonomic researchers are beginning to investigate the underlying physiology of attention resources. Studying nonhumans may facilitate the development of this neuroergonomic approach. *Neuroergonomic* is an appropriate term when applied to the neural science of working canines. As mentioned in the section on expertise development, dogs share many basic psychological traits with humans (Gosling and John, 1999). They may, therefore, be useful in determining which individual traits facilitate or impair performance of vigilance tasks. This is not a one-way street; research on human individual differences may aid canine researchers as well. Dogs have been used as models to study the effects of aging on cognition (Adams et al., 2000) and thus may also aid studies of the effects of aging on sustained attention in operational settings. Collaborations of ergonomic and animal behaviorists may prove mutually beneficial in explaining vigilance in people and dogs.

CONCLUSION

Trained dogs are employed in many occupational settings. One of our primary goals as scientists is to improve human welfare. Working dogs safeguard human life. Sandler (1996) estimated that 10,000 service (disabilities assistance) dogs were employed in the United States. This number increases every year. These service dogs protect and enrich the lives of their human companions. Children and adults with crippling disabilities, who in earlier days would have been kept in institutions, have a new lease on life through the help of their quadruped guardians. Thousands of dogs protect us from smugglers, terrorists, arsonists, and others seeking to do harm. Dogs moreover defend our environment by detecting smuggled produce potentially carrying invasive species that could wreak havoc on local agriculture. Dogs defend livestock from endangered predators and in the process defend the endangered predators from us. If we are interested in keeping society safe, we must abandon both our anthropocentric perspective and our fear of anthropomorphism and bring working dogs in from the academic doghouse.

Whether protecting livestock from predators or searching for contraband, these dogs handle serious jobs but few studies have concentrated on canine ergonomics— the performance of work by dogs. The few researchers in this field are fairly isolated both by the majority of comparative psychologists and animal behaviorists who focus on basic research with rodents and consciousness-related constructs with primates and by human ergonomic researchers with excessively anthropocentric orientations. Many animal scientists considered dogs quasi-animals, so tampered with by people, so literally anthropomorphic, that they presented little biological interest. Likewise, for ergonomic scientists, nonhuman dogs were classified as "Cartesian others," thoughtless automatons for whom little about human *thoughtful* behavior could be learned. Working dog as anthropomorphic and exceptional coworkers force us to re-evaluate our perspective. Canine ergonomics is interdisciplinary in that it bridges the gap between animal science and ergonomics.

Dogs are unique among nonhuman animals in their adaptation to human culture. Although other nonhumans such as dolphins, elephants, and rats may be employed to perform tasks, dogs are amazingly flexible. Many issues investigated in the ergonomics literature are also relevant to the study of working canines. The two issues explored in this chapter and later in this book, expertise development and sustained attention, merely demonstrate some of the common interests.

Regardless of which discipline, ergonomics or animal science, ultimately claims canine ergonomics, the need to study the operational performances of working dogs is critical. Knowing how long an explosive detection dog is capable of working, for example, is as important as knowing how long a human inspector can work when the goal is to keep us safe. The bridge is a two-way street: the study of canine workers may directly aid the wider science of ergonomics. The issue of expertise development is a case in point. An extensive experimental investigation of canine expertise development is more likely to resolve the talent–practice debate than the continued pursuit of human anecdotal and correlation studies (Helton, 2004a). Researchers may learn more by incorporating a comparative approach examining working dogs than from exclusively focusing on humans.

REFERENCES

Adachi, I., Kuwahata, H., and Fujita, K. (2007). Dogs recall their owner's face upon hearing the owner's voice. *Animal Cognition, 10,* 17–21.

Adams, B., Chan, A., Callahan, H., and Milgram, N.W. (2000). The canine as a model of human cognitive aging: recent developments. *Progress in Neuro-Psychopharmacology and Biological Psychiatry, 24,* 675–692.

Agnetta, B., Hare, B., and Tomasello, M. (2000). Cues to food location that domestic dogs (*Canis familiaris*) of different ages do and do not use. *Animal Cognition, 3,* 107–112.

Anderson, J. (1983). *The Architecture of Cognition.* Cambridge: Harvard University Press.

Anderson, J. (1993). *Rules of the Mind.* Hillsdale, NJ: Lawrence Erlbaum.

Bekoff, M. and Allen, C. (1997). Cognitive ethology: slayers, skeptics, and proponents. In Mitchell, R.W. et al., Eds. *Anthropomorphism, Anecdotes, and Animals.* Albany: State University of New York Press, 313–326.

Berry, D. and Broadbent, D. (1988). Interactive tasks and the implicit–explicit distinction. *British Journal of Psychology, 79,* 251–272.

Boesch, C. and Boesch, H. (1999). *The Chimpanzees of the Tai Forest: Behavioral Ecology and Evolution.* Oxford: Oxford University Press.

Brooks, S.E., Oi, F.M., and Koehler, P.G. (2003). Ability of canine termite detectors to locate live termites and discriminate them for non-termite material. *Journal of Economic Entomology, 96,* 1259–1266.

Bushnell, P.J. (1998). Behavioral approaches to the assessment of attention in animals. *Psychopharmacology, 138,* 231–259.

Bushnell, P.J., Benignus, V.A., and Case, M.W. (2003). Signal detection behavior in humans and rats: a comparison with matched tasks. *Behavioral Processes, 64,* 121–129.

Collins, H.M. (1989). Learning through enculturation. In Gallatly, A. et al., Eds. *Cognition and Social Worlds.* Oxford: Oxford University Press, 205–215.

Coppinger, R. and Coppinger, L. (2001). *Dogs: A Startling New Understanding of Canine Origin, Behavior, and Evolution.* New York: Scribner.

Csanyi, V. (2005). *If Dogs Could Talk: Exploring the Canine Mind.* New York: North Point Press.

Davies, D.R. and Parasuraman, R. (1982). *The Psychology of Vigilance.* London: Academic Press.

Dennett, D.C. (1996). *Kinds of Minds.* New York: Basic Books.

Dreyfus, H. and Dreyfus, S. (1987). *Mind over Machine: The Power of Human Intuition.* New York: Free Press.

Dukas, R. and Clark, C.W. (1995). Sustained vigilance and animal performance. *Animal Behaviour, 49,* 1259–1267.

Ericsson, K.A. (2001). Expertise in interpreting: an expert performance perspective. *Interpreting, 5,* 187–220.

Ericsson, K.A. and Charness, N. (1994). Expert performance: its structure and acquisition. *American Psychologist, 49,* 725–747.

Fitts, P. and Posner, M. (1967). *Human Performance.* Monterey, CA: Brooks Cole.

Fjellanger, R., Andersen, E.K., and McLean, I. (2000). A training program for filter-search mine detection dogs. *International Journal of Comparative Psychology, 15,* 277–286.

Furton, K.G. and Myers, L.J. (2001). The scientific foundation and efficacy of use of canines as chemical detectors of explosives. *Talanta, 54,* 487–500.

Gallant, S.I. (1993). *Neural Network Learning and Expert Systems.* Cambridge: MIT Press.

Gardner, H. (1997). *Extraordinary Minds: Portraits of Exceptional Individuals and an Examination of our Extraordinariness.* New York: Basic Books.

Garner, K. J., Busbee, L., Cornwell, P., Edmonds, J., Mullins, K., Rader, K., Johnston, J.M., and Williams, J. M. (2000). *Duty Cycle of the Detection Dog: A Baseline Study.* Final Report. Federal Aviation Administration, Washington, D.C.

Gazit, I. and Terkel, J. (2003). Explosives detection by sniffer dogs following strenuous physical activity. *Applied Animal Behavior Science, 81,* 149–161.

Gosling, S.D. and John, O.J. (1999). Personality dimensions in nonhuman animals: a cross-species review. *Current Directions in Psychological Science, 8,* 69–75.

Greenwood, P.M. and Parasuraman, R. (2003). Normal genetic variation, cognition, and aging. *Behavioral and Cognitive Neuroscience Reviews, 2,* 278–306.

Hancock, P.A. and Hart, S.G. (2002). Defeating terrorism: what can human factors and ergonomics offer? *Ergonomics and Design, 10,* 6–16.

Helton, W.S. (2004a). The development of expertise: animal models? *Journal of General Psychology, 131,* 86–96.

Helton, W.S. (2004b) Utilizing genetic algorithms and neural networks in expert systems: what animals teach us. *Recent Advances in Soft Computing, 5,* 177–182.

Helton, W.S. (2005). Animal expertise, conscious or not? *Animal Cognition, 8,* 67–74.

Helton, W.S., Hollander, T.D., Warm, J.S., Matthews, G., Dember, W.N., Wallart, M., Beauchamp, G., Parasuraman, R., and Hancock, P.A. (2005). Signal regularity and the mindlessness model of vigilance. *British Journal of Psychology, 96,* 249–261.

Helton, W.S., Shaw, T.H., Warm, J.S., Matthews, G., Dember, W.N., and Hancock, P.A. (2004). Workload transitions: effects on vigilance performance, and stress. In Vincenzi, D.A. et al., Eds. *Human Performance, Situation Awareness and Automation: Current Research and Trends.* Mahwah, NJ: Lawrence A. Erlbaum, 258–262.

Helton, W.S., Warm, J.S., Mathews, G., Corcoran, K., and Dember, W.N. (2002) Further tests of the abbreviated vigil: effects of signal salience and noise on performance and stress. *Proceedings of the Human Factors and Ergonomics Society, 46,* 1546–1550.

Hirst, W. and Kalmar, D. (1987). Characterizing attentional resources. *Journal of Experimental Psychology: General, 116,* 68–81.

Hitchcock, E.M., Warm, J.S., Matthews, G., Dember, W.N., Shear, P.K., Tripp, L.D., Mayleben, D.W., and Parasuraman, R. (2003). Automation cueing modulates cerebral blood flow and vigilance in a simulated air traffic control task. *Theoretical Issues in Ergonomics Science, 4,* 89–112.

Hoffman, R.R., Shadbolt, N.R., Burton, A.M., and Klein, G. (1995). Eliciting knowledge from experts: a methodological analysis. *Organizational Behavior and Human Decision Processes, 52,* 129–158.

Holland, V.S. (1994). *Herding Dogs: Progressive Training.* New York: Howell.

Holley, L.A., Turchi, J., Apple, C., and Sarter, M. (1995). Dissociation between the attentional effects of infusions of a benzodiazepine receptor agonist and an inverse agonist into the basal forebrain. *Psychopharmacology, 120,* 99–108.

Howe, M.J.A., Davidson, J.W., and Sloboda, J.A. (1998). Innate talent: reality or myth? *Behavioral and Brain Sciences, 21,* 399–422.

Kahneman, D. (1973). *Attention and Effort.* Englewood, NJ: Prentice Hall.

Kaminski, J., Call, J., and Fischer, J. (2004) Word learning in a domestic dog: evidence for fast mapping. *Science 304,* 1682–1683.

Kosslyn, S. (1980). *Image and Mind.* Cambridge: Harvard University Press.

Lubow, R.E. (1977). *The War Animals.* New York: Doubleday.

Mackworth, N.H. (1948). The breakdown of vigilance during prolonged visual search. *Quarterly Journal of Experimental Psychology, 1,* 6–21.

Mackworth, N.H. (1950). Researches on the measurement of human performance. *Medical Research Council Special Report 2680.* London: H.M.S.O. (Reprinted from *Selected Papers in the Design and Use of Control Systems,* Sinaiko, H.W., Ed. 1961, New York: Dover, 174–331).

Masunaga, H. and Horn, J. (2000) Characterizing mature human intelligence: expertise development. *Learning and Individual Differences, 12,* 5–33.

Matthews, G., Davies, D.R., Westerman, S.J., and Stammers, R.B. (2000). *Human Performance: Cognition, Stress and Individual Differences.* East Sussex: Psychology Press.

Matzel, L.D., Han, Y.R., Grossman, H., Karnik, M.S., Patel, D., Scott, N., Specht, S.M., and Gandhi, C.C. (2003). Individual differences in the expression of a "general" learning ability in mice. *Journal of Neuroscience, 23,* 6423–6433.

McCarley, J.S., Kramer, A.F., Wickens, C.D., Vidoni, E.D., and Boot, W.R. (2004). Visual skills in airport security screening. *Psychological Science, 15,* 302–306.

McGaughy, J. and Sarter, M. (1995). Behavioral vigilance in rats: task validation and effects of age, amphetamine, and benzodiazepine receptor ligands. *Psychopharmacology, 117,* 340–357.

Miklosi, A., Polgardi, R., Topal, J., and Csanyi, V. (1998). Use of experimenter-given cues in dogs. *Animal Cognition, 1,* 113–121.

Miklosi, A., Polgardi, R., Topal, J., and Csanyi, V. (2000). Intentional behavior in dog–human communication: an experimental analysis of 'showing' behavior in the dog. *Animal Cognition, 3,* 159–166.

Nelson, R.K. (1983). *Make Prayers to the Raven: A Koyukon View of the Northern Forest.* Chicago: University of Chicago Press.

Pickel, D., Manucy, G.P., Walker, D.B., Hall, S.B., and Walker, J.C. (2004). Evidence for canine olfactory detection of melanoma. *Applied Animal Behaviour Science, 89,* 107–116.

Pigeau, R.A., Angus, R.G., O'Neill, P., and Mack, I. (1995). Vigilance latencies to aircraft detection among NORAD surveillance operators. *Human Factors, 37,* 622–634.

Sanders, C.R. (2006). "The dog you deserve": ambivalence in the K-9 officer–patrol dog relationship. *Journal of Contemporary Ethnography, 35,* 148–172.

Sandler, J.L. (1996). Care and treatment of service dogs and their owners. *Journal of the American Veterinary Association, 208,* 1979–1981.

Schmutz, S.M. and Schmutz, J.K. (1998). Heritability estimates of behaviors associated with hunting in dogs. *Journal of Heredity, 89,* 233–237.

Shaffer, D.M., Krauchunas, S.M., Eddy, M., and McBeath, M.K. (2004). How dogs navigate to catch Frisbees. *Psychological Science, 15,* 437–441.

Slabbert, J.M. and Rasa, O.A.E. (1997). Observational learning of an acquired maternal behaviour pattern by working dog pups: an alternative training method. *Applied Animal Behaviour Science, 53,* 309–316.

Stanley, W., Matthews, R., Buss, R., and Kotler-Cope, S. (1989). Insight without awareness: on the interaction of verbalization, instruction and practice in a simulated process control task. *Quarterly Journal of Experimental Psychology, 41,* 533–577.

Sun, R., Merrill, E., and Peterson, T. (2001). From implicit skills to explicit knowledge: a bottom-up model of skill learning. *Cognitive Science, 25,* 203–244.

Swets, J.A. (1986). Indices of discrimination or diagnostic accuracy: their ROCs and implied models. *Psychological Bulletin, 99,* 110–117.

Temple, J.G., Warm, J.S., Dember, W.N., Jones, K.S., LaGrange, C.M., and Matthews, G. (2000). The effects of signal salience and caffeine on performance, workload and stress in an abbreviated vigilance task. *Human Factors, 42,* 183–194.

Terrace, H.S., Son, L.K., and Brannon, E.M. (2003). Serial expertise of rhesus macaques. *Psychological Science, 14,* 66–73.

Topal, J., Byrne, R.W., Miklosi, A., and Csanyi, V. (2006). Reproducing human actions and action sequences: "Do as I Do!"in a dog. *Animal Cognition, 9,* 355–367.

Vila, C., Savolainen, P., Maldonado, J.E., Amorim, I.R., Rice, J.E., Honeycutt, R.L., Crandall, K.A., Lundenberg, J., and Wayne, R.K. (1997). Multiple and ancient origins of the domestic dog. *Science, 276,* 1687–1689.

Warm, J.S. (1993). Vigilance and target detection. In Huey, B.M. et al., Eds. *Workload Transition: Implications for Individual and Team Performance.* Washington, D.C.: National Academy Press, 139–170.

Wells, D.L. and Hepper, P.G. (2003). Directional tracking in the domestic dog, *Canis familiaris. Applied Animal Behaviour Science, 84,* 297–305.

2 Skill and Expertise in Working Dogs
A Cognitive Science Perspective

William S. Helton, Paul J. Feltovich, and Andrew J. Velkey

CONTENTS

INTRODUCTION

Human expertise has been studied intensively by cognitive scientists (Ericsson et al., 2006; Feltovich et al., 2006). The cognitive science literature includes many studies demonstrating differences in performances of a variety of skills by experts and novices (Baylor, 2001; Chi et al., 1981; Lesgold et al., 1988; Masunaga and Horn, 2000; Melcher and Schooler, 1996).

Much artificial intelligence (AI) research has focused on the construction of systems capable of demonstrating some aspects of human expertise. AI researchers initially desired to construct machines capable of domain general intelligence and reasoning. This proved, however, to be a disappointing project, and they redirected their focus on building machines emulating human abilities in very narrow domains: expert systems (Buchanan et al., 2006). Cognitive science has concentrated on expertise because expertise and its development are major functions of cognition (see Sternberg, 1999, for a developmental model of expertise). The pursuit of an understanding of domain-specific expertise is more tractable than the pursuit of an understanding of general intelligence. Simon (1990) suggests that general intelligence may represent the emergence of interactive expert subsystems.

Humans may not be the only experts (Helton, 2004, 2005, 2007, and 2008; Terrace et al., 2003), but the application of the *expertise* term to nonhuman animals (and to dogs in particular) continues to be debated (Dukas, 1998; Helton, 2005; Rossano, 2003). Use of the *expertise* term in relation to dogs is contentious because of the connotation and implication nonhuman expertise has for animal consciousness, the relative contributions of nature and nurture to development, and the allocation of research funding to technology. These controversial issues will be addressed later in this chapter. Before delving into these topics, we will review definitions of expertise proposed in the cognitive science literature and provide evidence that dogs (*Canis lupus familiaris*) meet all the definitions.

DEFINITIONS OF EXPERTISE

Despite the popularity of expertise studies in cognitive science, no single, agreed-upon definition of expertise exists among cognitive scientists (Farrington-Darby and Wilson, 2006). Expertise is not simply learning, which is a universal trait of animals and even singular animal cells and intriguingly, plants (Dukas, 1998; Trewavas, 2005). Although expertise may eventually be a proven quality of most animals, it must be differentiated from learning.

Despite the lack of a single definition of expertise among cognitive scientists, they agree that expertise should be open to empirical investigation. To achieve status as a useful scientific concept, expertise must manifest itself in an observable way. In the following section we will present five candidate definitions of expertise, each of which can be subjected to empirical investigation. The first three are product-oriented definitions, and the remaining two are process-oriented. All have merits and shortcomings. Regardless of the definition advocated, we will argue that working dogs satisfy the requirements.

SOCIAL CONSTRUCTION

Expertise can be described as a social construction or a socially derived label (Sternberg and Ben-Zeev, 2001). Agnew et al. (1997) argue the minimum criterion for expertise is having a reasonably large group of people regard an individual as an expert. Expertise is a label assigned to an individual by a group of people who claim the individual is an expert in some field (Mieg, 2006). By publicly announcing or utilizing the individual's special skill, the group manifests his or her expertise in an observable way.

For example, expertise may be assessed by polling the pertinent population (Stein, 1997) or by societal indictors or markers of expertise such as awards, medals, titles, legal status, and so on (Evetts et al., 2006). As early as 945 AD, the Welsh laws of *Hywel Dda* formally declared the difference between expert dogs and untrained dogs by imposing different legal penalties for killing them (Menache, 2000); this tradition continues today. Most U.S. states have laws that impose stiff penalties for people who injure or interfere with the work of an assistance or law enforcement dog (Randolph, 1997). Dogs that are trained to guide the blind, assist the immobile, aid the hearing impaired, and predict seizures are specially protected by the Americans with Disabilities Act and the Federal Air Carriers Act, both of which guarantee these dogs access to public places. Such access is denied to untrained pet dogs. The Internal Revenue Service recognizes assistance dogs as legitimate deductible medical expenses under Treasury Regulation 1.213-1(e)(1)(iii), thus publicly acknowledging the elite or expert status of trained assistance dogs.

The legal systems of many countries also recognize dog expertise by allowing specially trained scent detection dogs to provide evidence in criminal court cases. The admission of such evidence is analogous to expert witness testimony or presentation of forensic science data (Ludingtoni, 1988). Law enforcement agencies use trained dogs to match scents from crime scene articles to suspects. This form of evidence has, like most evidence, been challenged legally, but remains admissible in many countries (Marks, 2007). While laws, because they are formal and in representative governments do not constitute the views of fringe elements of society, are probably the best indicators of the social construction of animal expertise, they are not the only indicators.

Other social markers of expertise such as honors, medals, awards, and titles, for example, are useful. One of the major public supporters of assistance dogs, the Delta Society in the U.S., honors outstanding service dogs and presents a yearly award to an exceptional dog that provides assistance to people with visual impairments, hearing impairments, or mobility impairments (Eames and Eames, 1997). By publicly recognizing the dog, the society openly declares the winner as exceptional among the exceptional, or in other words, *an expert.* Although the U.S. military currently classifies trained dogs as equipment and thus unable to receive medals, trained dogs were in the past considered personnel and often achieved rank (Thurston, 1996). Many trained dogs have been awarded military decorations for exceptional combat performance. Chips, the most decorated American dog of the Second World War, received both a Purple Heart and a Silver Star, although these awards were later rescinded under pressure from political interest groups. The British military created a special award for war animals, the Dickin medal, granted to 25 dogs since its inception in 1943. Sadie, a Labrador retriever and recent recipient of the Dickin medal, saved the lives of North Atlantic Treaty Organization soldiers in Kabul, Afghanistan by finding a hidden explosive device. Law enforcement agencies also award medals and honors for exceptional performance. Organizations such as the U.S. National Narcotic Detector Dog Association and the American Police Hall of Fame have honored outstanding police dogs. Granting awards for exceptional performance to assistance, military, and law enforcement dogs is a public declaration of the dogs' expertise by a large group of people. This public attribution of exceptionality fits the social construction view of expertise.

In an opinion survey of 78 people working at the Pacific Northwest Laboratories, a U.S. center for security research, a range of security techniques and technologies were rated for their national security benefits, accuracy, and validity (Sanquist et al., 2006). The techniques and technologies varied from airport passenger and baggage screening by human inspectors to hidden camera surveillance of individuals using gait analysis and facial recognition. The highest rated security technique was explosive detection by dogs. This judgment of high efficacy by security experts provides further evidence of the social construction of canine expertise. In sum, working dogs are often socially constructed as experts.

EXCEPTIONAL PERFORMANCE

Some expertise scientists dismiss the social constructive view of expertise as too socio-political and instead advocate behavioral realism: what can experts objectively do and do better than their more ordinary colleagues? In what fields are they objectively experts? Expertise is manifested by skilled or exceptional performance that can be produced in repeatable settings that are representative of the expertise domain (Ericsson, 2001). According to Ericsson (1996), if expertise is to be studied scientifically, it must meet these criteria: (1) it must occur reliably in clearly specified situations with distinctive observable characteristics; (2) it should be reproducible under controlled conditions; and (3) it should be operationally defined by objective measures. Ericsson (1996 and 2001) demonstrates how human expertise (defined as exceptional performance) meets these criteria. We will now explore how nonhuman performers can also meet these criteria. Extensive studies of canine skill have been conducted (Helton, 2007a and b). Dogs vary in their performance; some demonstrate exceptional performance and others perform poorly, as discussed next.

Many examples indicate performance variations. Dogs are often trained to acquire skills such as accelerant detection, assistance for the blind, epilepsy detection, explosive detection, forensic tracking, guarding, hearing assistance, herding livestock, medical diagnosis, narcotics detection, and detection of insect infestations and microbial growth (Brooks et al., 2003; Fjellanger et al., 2002; Furton and Myers, 2001; Gazit and Terkel, 2003; Holland, 1994; Marschark and Baenniger, 2002; Pickel et al., 2004; Slabbert and Rasa, 1997; Wells and Hepper, 2003). All these canine skills are subject to the behavior-based exceptional performance definition of expertise. Some dogs run faster, more carefully negotiate obstacles, and/or exhibit better substance detection accuracy. Performance is quantifiable using objective measures and reliable ratings by human judges, and the dogs can then be ranked by their performance. Exceptional dogs could be classified by performance, for example, in the upper fifth percentile.

Most dog skills, moreover, meet Ericsson's (1996) three criteria of study by scientific methods: they are reliable, reproducible, and clearly describable. Most dogs learn skills over time and their performances vary. A set of 5 or 10% exceptional performers will excel at any skill. While this may appear trivial to some theorists, cognitive scientists cannot escape the reality that some dogs are exceptionally good at domain-specific tasks (e.g., explosive detection) and some are not.

EXTRACTABLE KNOWLEDGE

An alternative product-oriented definition is to view expertise as disembodied knowledge. Expertise may be a combination of problem-solving algorithms, heuristics (rules of thumb), and a knowledge base (facts or representations) used to guide skilled behavior in specific domains. This definition is preferred by computer scientists and other investigators interested in constructing artificial experts, expert systems, and decision aids (Buchanan et al., 2006; Sternberg and Ben-Zeev, 2001; Shanteau, 1992).

Typically, from this perspective, expertise is investigated via verbal methods, for example, asking experts what they think about when performing tasks (Ericsson and Charness, 1994). Experts are asked to generate introspective reports of their reasoning while performing in their particular skill arenas. These verbal reports are then used by a knowledge engineer or expert system designer to infer the knowledge base and problem solving methods of the experts that are then used to construct an expert system or decision aid. From this perspective, expertise is about knowledge that can be communicated to a designer and then constructed (Ford and Bradshaw, 1993; Shanteau, 1992).

This reliance on verbal reports by researchers advocating this definition of expertise makes such investigation of nonhumans difficult. This perspective intermingles language and communication capacity with expertise and this conflation is unnecessary. Dogs have knowledge ("know-how"). Their movements are dictated both by physical forces acting upon them and by self-generated impulses. Dogs are actors. The execution of action requires a means of control. Despite endless debates about the architecture of these control mechanisms that hark back to concepts such as Cartesian dualism, the mechanisms clearly exist. The control of action may be considered knowledge. A dog's control mechanisms, in theory, may be translated into symbols and used to control the actions of an artificial entity, whether a physical robot or a simulated agent. This perspective led to a recent surge of interest in biomemetic robotics and artificial life. Dogs, in particular, have been used as models for developing quadruped robots, for example, Sony's Abio, a companion pet robot, and Boston Dynamic's Big and Little Dog robots designed for the military. Dogs have also served as models for the development of artificial noses, for example, Nomadic's FIDO system.

Dogs are often experts because they are extremely proficient in their responses to certain ecological challenges; for example, the dominant models for behavioral biology are based on optimization and/or maximization (see McNamara et al., 2001 for a review). In essence, dogs and other animals inherently operate as expert systems. Challenging theoretical issues about the nature of dogs' (and other animals') control mechanisms and immense technical challenges regarding the translation of biological control into artificial codes remain to be resolved. In particular, it may not be possible to make artificial codes less complicated than a modeled system. Thus, expert system designers will confront serious problems even if they do not try to recreate the full complexity of a biological expert (Forsythe, 1993).

If a control mechanism cannot be simplified, scientists have no reason to create an artificial expert since a biological one would be essentially identical. Some robotics researchers have reached this conclusion and are starting to fuse biology to electronics. For example, the Gordon mobile robot recently (2008) unveiled to the public by the University of Reading utilizes a dish of rat neurons as a central processor. Very likely, the future of machine intelligence will arise from integration of biological intelligence.

Cognitive scientists employing the knowledge-based definition of expertise cited in this section are, however, primarily interested in capturing knowledge consisting of mental representations that are easily communicable. They want to quickly extract expert knowledge, not redevelop an expert from the ground up.

Dogs are capable of acquiring knowledge consisting of mental representations and conveying this information to others. A pertinent example is communicating the location of hidden prey or food. Dogs can track the locations of objects that are not immediately perceptually present (in view). This ability has been demonstrated even in relatively simple invertebrates, for example, jumping spiders (Tarsitano and Andrew, 1999; Tarsitano and Jackson, 1997). Although multiple mechanisms may enable animals to confront this situation, one mechanism entailing the use of mental representation is active search (Timberlake, 1993), guided by the understanding that objects continue to exist even when they are not actively sensed, a phenomenon called *object permanence*. Essentially, the animal mentally represents the existence and location of a displaced object without relying upon concurrent sense data. The previously acquired location of the object, based on directly experienced sense data, must be actively maintained by the animal so that the animal can exploit the information at a later time. Dogs, for example, can maintain a representation of a hidden object's location for an extended period (Fiset et al., 2000 and 2003; West and Young, 2002). Dogs can also learn or infer the locations of hidden objects from others' gestures and can also communicate these locations to others (Agnetta et al., 2000; Miklosi et al., 1998 and 2000). The ability to locate hidden items and the communication of such information implies that dogs are capable of communicating mental representations to others. While this method for inferring mental representation may not be exactly the kind of knowledge representation that cognitive engineers seek, it indicates that dogs can use abstracted knowledge. Furthermore, they can communicate their knowledge to conspecifics (other dogs) and interspecifics (humans).

Imitation and mimicry are also considered indicators of representational thought (Uzgiris and Hunt, 1975); an observer can reproduce a previously observed action of a model (Pearce, 1997). The observer constructs a mental representation of the action that enables it to reproduce the action in the absence of direct perceptual support. Dogs have exhibited evidence of imitation and mimicry (Range et al., 2007). They appear to use mental representations and definitely coordinate their actions to accomplish difficult tasks, thus demonstrating know-how. The reliance on verbal reports by advocates of the knowledge definition of expertise may imply a close tie between ability to *verbalize* expertise and *actual* expertise, but this conflation is a fundamental categorical mistake. It confuses being an expert with being able to talk about ways in which one is an expert.

Hoffman et al. (1995) argue and provide supporting evidence that using verbal interviews, e.g., having human experts verbalize their knowledge, can be an inefficient method of knowledge extraction, in comparison to having the experts carry out contrived (but representative) tasks that reveal their methods. Likewise, methods can be elicited from expert dogs by making them perform their skills in controlled environments. Shaffer and colleagues (2004) demonstrated this approach in a study investigating the heuristics or rules used by dogs to catch Frisbees. Their research indicated that dogs use the same heuristic humans use, or at least, their behavior can be explained using the same heuristic.

Whether canine or human experts are consciously aware of using the rule is irrelevant; their performance ultimately reveals the presence of knowledge that can be formulated as heuristics. In addition, brain studies of human and animal experts will provide another line of investigation regarding potential knowledge and mental process similarities in highly skilled humans and other animals (Hill and Schneider, 2006).

OUTCOME OF PROCESS: DELIBERATE PRACTICE

Expertise may not be a product per se and may represent the outcome of a specific process. As Rossano (2003) argues, regardless of how expertise is defined, it appears in humans to result from deliberate practice (Ericsson, 2006). Cognitive scientists emphasize the cognitive or behavioral features of deliberate practice. The *cognitive features* of deliberate practice arise from the continuous self-evaluation of a trainee's current skill state in comparison to an expert's skill state via mental representations; constant focus on the elevation (not only the maintenance) of the skill; and the sustained application of voluntary control over skill production (Rossano, 2003). Unfortunately, the cognitive features of deliberate practice are heavily laden with assumptions about conscious states, and this phenomenology is problematic for assessing dogs (Helton, 2005). The *behavioral features* of deliberate practice include motivation of a trainee to improve, assignment of well-defined tasks, delivery of feedback to the trainee, and ample opportunity for repetition (Ericsson, 2001).

Consciousness-laced language is not particularly helpful when addressing dog practice. Cognitive scientists cannot ask dogs whether they are deliberately practicing. We may never determine whether dogs know why and for what ends they perform. If cognitive scientists conflate the ability to verbalize with expertise, an issue related to the definition of knowledge cited earlier, they may miss the point that learning a task or skill and excelling at it is not the same as conveying *how* one is good at a task or a skill.

Deliberateness is not so easy to ascertain even in humans; people are often unable to state the real reasons why they do what they do, a point made earlier by Freud and contemporarily by evolutionary psychologists (Buss, 2005; Cianciolo et al., 2006). By focusing instead on the behavioral characteristics of practice, dog expertise can be investigated. Dogs engage in behavior resembling deliberate practice. Animal play, for example, is a vexing problem for evolutionary biologists who continue to debate functional explanations of the activity (Caro, 1995; Burghardt, 2005). Although good alternative explanations have been advanced for play activity (Barber, 1991; Bekoff and Byers, 1981; Caro, 1995; Fagen, 1981), some animal play activity is similar to skills practice by humans. Object play in predatory species involves the use of inanimate objects, and scientists propose that predatory object play is a form of practice in various species such as cats (Caro, 1995) and raptors (Negro et al., 1996). Object play is rare among adults of wild species, but is often observed among the young

(Hall and Bradshaw, 1998; Negro et al., 1996). Captive kestrels (birds of prey), when given the option, prefer to play with objects resembling prey. Young wild harriers play with objects resembling prey (Kitowski, 2005; Negro et al., 1996). Object play by cheetahs is associated with predatory behavior; cubs engaging in more object play are more successful in contacting live prey (Caro, 1995).

Dogs also engage in object play. While they cannot be asked whether they engage in play deliberately to improve their predatory skills, predatory object play and skill at predation of live prey seem to be connected (Caro, 1995).

Unlike the relationship between object play and natural predation, the case of working dogs, whose deliberate practice is the result of specific training, seems fairly transparent. These dogs are motivated to perform. As Coppinger and Coppinger (2001) indicate, hurt and unmotivated dogs do not work. This is why the use of compulsive methods has become unpopular among most dog trainers (Fjellanger et al., 2000). The dogs are given clearly defined tasks to accomplish and receive clear feedback about their performance. Dogs repeatedly train by performing the same actions over and over. This clearly fits the behavioral features of deliberate practice (Ericsson, 2001).

While some critics may object to attributing a motivational state to a dog as anthropomorphism, working dogs are clearly motivated, whether internally or externally, to perform their tasks. For certain types of work such as herding sheep (Holland, 1994; Marschark and Baenniger, 2002), the task serves as the reward. Reinforcers such as food and praise are not needed. Access to the task (the sheep) is all that is needed to reinforce correct behaviors. Denying a dog an opportunity to work is punishment. The other behavioral features of deliberate practice are simply descriptive facts of dog training.

While a researcher cannot ask a dog whether it practices deliberately, working dogs clearly practice to achieve a future goal: skilled performance or expertise. The dog engages in repetitive practice to achieve a predetermined level of ability. An expert dog may not engage in self-conscious monitoring of skill and may not be conscious of the future, but this does not seem to be a convincing requirement. As Rossano (2003) states, "Most [human] elite performers are introduced to their future field of expertise as children in the form of play. While still in this stage, a teacher or coach is typically assigned to harness some of the child's playfulness by setting goals...." A human trainee typically starts as a playful child who is behaviorally shaped into an expert by an adult trainer. A child, like a dog, may or may not be conscious of what is happening in the ultimate sense. An internal representation of the goal state may not be a requirement for deliberate practice or subsequent expertise development. A tree does not need an internal representation of its future to achieve its full height.

Cognitive scientists studying human performance suggest a power law of learning (Anderson, 1982; Ericsson et al., 1993). Typically, when time spent practicing a skill is plotted against a metric of skill performance in a log–log format, a straight line provides a reasonably good fit. Essentially, learning continues to improve through practice, but the benefits of time spent practicing decline over time (in a related effect, diminishment of increasing skill with age appears to be conflated by reduction in deliberate practice, as much as loss of gains from practice: Krampe and

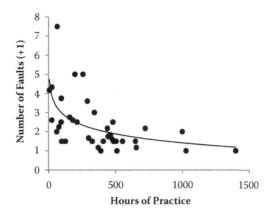

FIGURE 2.1 Number of faults (+1) plotted by number of accumulated practice hours in agility dogs (Helton, 2007).

Charness, 2006). Level of expertise may reflect the end result of this process, when performance reaches near an asymptote of ability. Helton (2007a) investigated this in dogs trained in agility and the evidence supports the application of a power law of practice to nonhuman animals (Figure 2.1). Undoubtedly, exceptional animal skill can be described as the outcome of extensive practice.

OUTCOME OF A PROCESS: LIFE-LONG LEARNING

Expertise is the outcome of a long period of learning. Most cognitive scientists studying expertise in humans agree that under normal circumstances expertise takes a relatively long time to develop—at least 7 to 10 years for humans (Ericsson and Charness, 1994). This long-term commitment differentiates expertise from more common forms of learning that can be achieved by a single trial or a brief exposure (Dickinson et al., 2004). No one becomes an expert overnight. As one of the authors proposed previously (Helton, 2008), expertise can be defined as the outcome of *sustained learning occurring over at least 10% of the species-typical lifespan.* This definition is very similar to the deliberate practice definition, but attempts to diffuse some of the angst scientists exhibit about "deliberate" actions of animals.

A growing consensus indicates that most animal foraging skills involve prolonged periods of development and require feedback (Dukas, 1998; Dukas and Visscher, 1994; Durier and Rivault, 2000; Edwards and Jackson, 1994; Keasar et al., 1996; Heiling and Herberstein, 1999; Morse, 2000). The foraging learning context fits the behavioral description of practice fairly well: the animal is highly motivated (hungry), given a well-defined task (forage under environmental and physiological restrictions), receives timely feedback, and has opportunities for extensive repetition of the skill.

Expert performance improves over the period of skill practice until some asymptote of skill is achieved: mastery (Charness et al., 2005; Ericsson and Charness, 1994). Among cognitive scientists this process is described by the power law of

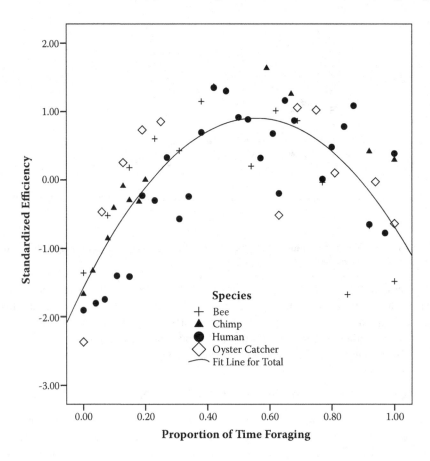

FIGURE 2.2 Standardized foraging efficiency plotted by the proportion of time foraging in honeybees (*Apis mellifera*), oystercatchers (*Haematopus ostralegus*), chimpanzees
(*Pantroglodytes*), and human hunter–gatherers (*Homo sapiens*).

practice (Anderson, 1982 and 1995). While a plateau of effective performance is
eventually achieved, at some point performance will decline due to the effects of
aging (Charness et al., 2005 and 2006). A similar quadratic relationship between age
and foraging efficiency is seen in a variety of animal species including humans when
longevity is controlled (Helton, 2008; see Figure 2.2). Regardless of the species, foraging efficiency or performance improves over time and at some point declines. All
species examined achieved a performance peak, mastery, or expertise level before
the average age of death and well after reaching physical maturity; skill development
is partially independent of physical maturation. While physical maturation may be
necessary for skill expertise, it is not sufficient; an animal must also perform the
skill for a relatively long time and receive feedback.

   This long period of development is also applicable to working dogs. For example,
running is often considered an innate ability. Improvements in *competitive* running
are primarily seen as resulting from a long history of artificial selection or to changes
in biomechanical properties due to physical maturation and growth (Heglund and

Taylor, 1988). Recent work by Mosher et al. (2007) demonstrated that faster Whippets (Class A) are more likely to carry one copy of a mutation in the myostatin gene (resulting in greater muscularity) than slower racing Whippets (Class B, C, or D) or Whippets that do not race. This perspective, however, underestimates the role experience plays in improving running; racing dogs must be specifically trained to run competitively. Moreover, many dogs initially selected for this training do not become competitive racers and only a very few race dogs become consistent winners.

As noted, the time required to attain exceptional levels of performance by humans is approximately 7 to 10 years, across a wide variety of domains (Ericsson and Charness, 1994). The exact amount of time required depends on the nature of the task. Human sprinters tend, for example, to peak earlier in life (mid-twenties) than endurance runners (mid-thirties).

Helton (unpublished manuscript) examined the performances of exceptional Greyhounds in 503-m races. Even when using a conservative metric of lifespan devoted to achieving exceptional performance, the overall mean percentage of lifespan devoted to running skill acquisition was 9.1% for Greyhound racers. This is not far from the 10% of lifespan devoted to the acquisition of expertise expected for humans (Ericsson and Charness, 1994). Greyhound 503-m track runners peak on average at 2.4 years and have approximate life spans of 8.2 years. Human sprinters peak on average at 25.0 years and have average life spans of 77.6 years (Abernethy et al., 2005). The strikingly similarities of peak ages of Greyhound track runners and human track runners relative to their average life spans support Helton's (2008) claim that the acquisition of expertise is an animal universal arising from common mechanisms. See Figure 2.3 for an example of an exceptional Greyhound's race improvement with age.

The primary difference between this process-oriented definition of expertise and the deliberate practice definition is the emphasis on deliberateness. Deliberateness evokes allusions to foresight and conscious intent. Cognitive scientists, as will be shown in the next section, continue to grapple with consciousness, and expertise is at the heart of the debate. Researchers wishing to remain agnostic about dog consciousness could instead see expertise as the outcome of an uncommon, long duration learning process that includes many of the required characteristics of deliberate practice, but without attributions of deliberate intent.

IMPLICATION OF ANIMAL EXPERTISE

CONSCIOUSNESS

Rossano (2003) argues that expertise is an indicator of consciousness because it is the outcome of deliberate practice, and deliberate practice appears to result from conscious processing (Ericsson, 1996). Any sign of expertise is, therefore, a plausible indicator of consciousness. Expertise is deeply entangled in cognitive scientists' attempts to deal scientifically with issues of consciousness. Interest in the science of consciousness has been increasing. Two scientific journals, *Consciousness and Cognition* and the *Journal of Consciousness Studies*, a society, and many popular

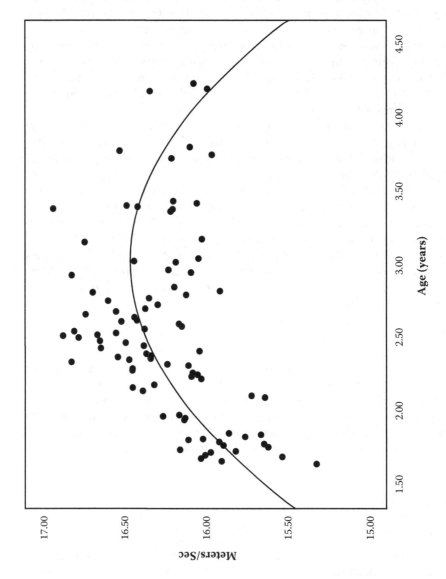

FIGURE 2.3 Running performance changes (meters per second) over time for a typical greyhound racer.

books focus on the topic (Baars, 1997). While dissenters are very skeptical of all this consciousness-related work (Uttal, 2005), consciousness is certainly a timely topic.

Block (1995 and 2001) distinguishes *phenomenal* consciousness from *access* consciousness. Phenomenal consciousness is the subjective experience of mental life or the feeling of being conscious. Of course, phenomenal consciousness is hard to operationalize in scientific terms. Access consciousness, conversely, concerns the availability of mental representations necessary for the cogent control of action and can be more formally described in information processing terminology. While the two concepts are distinct, they typically co-occur in the natural world, based on evidence from humans. Although philosophers have fantasized hypothetical creatures having access consciousness without phenomenal consciousness (zombies, super blindsighters, and intelligent but phenomenally unconscious robots), no real cases of humans having access consciousness without phenomenal consciousness have been demonstrated. The presence of access consciousness can be used as an indicator of the probable presence of phenomenal consciousness. The evidence showing humans having phenomenal consciousness without access consciousness is debatable (Block, 1995 and 2001; Rosenthal, 2002).

The development of expertise seems to be a plausible indicator of access consciousness. Deliberate practice, as previously indicated, is goal directed. According to Rossano (2003), deliberate practice requires access consciousness. From this perspective, the ambitious novice is often interpreted as striving for superior performance and the striving for superior performance is motivated, deliberate, and intentional. The intentional component is demonstrated through the process of directed attention. In human studies of skill development, changes in attention during skill acquisition are well established (Logan, 1985; Proctor and Dutta, 1995). Descriptive stage models of expertise development have been proposed (Anderson, 1995; Fitts and Posner, 1967). In an initial cognitive stage, the human trainee pays close attention to cues and feedback. Performance during this initial stage is irregular and requires active coordination of separate skill components. Skill production demands attention of the trainee. Over time and after practice, a trainee moves to an associative second stage consisting of the organization of separate skill components into larger units or chunks. This stage is characterized by a noticeable increase in skill fluidity and a decrease in the need for active attention. With more practice, the trainee shifts to a third autonomous stage in which the skill becomes essentially automatic, independent from cognitive control and attention. Skill automaticity frees the trainee's attention resources for performing other concomitant tasks and monitoring performance of the central task. When learning a complex skill, a human trainee must consciously focus on learning the skill and with ample practice, he or she can perform the task without conscious control. This is presumably why a novice learning to drive a manual transmission cannot handle additional distractions; an experienced driver can handle distractions.

In these stage models of human skill learning (Anderson, 1995), the early cognitive stage relies heavily on a declarative knowledge component. Nonhumans cannot declare their knowledge through language. Terrace and his colleagues (2003), however, argue that language and declarative knowledge are separate; declarative knowledge evolutionarily predates language. Encoding images analogically is an example of a nonverbal declarative knowledge system (Kosslyn, 1980; Terrace et al., 2003). Dogs, for example, appear to use encoded images to guide their behavior (Adachi et al., in press; Topal et al., 2006); this study of mental imagery in dogs and animals in general, however, is underdeveloped. More direct evidence supporting the application of stage models of skill learning in dogs and other animals is available, as we discuss next.

Ethological research on wild cheetahs provides some evidence supportive of a stage model of animal skill development (Caro, 1994). Cheetah mothers provide their offspring with opportunities to hunt in relatively controlled environments. They capture live prey and bring the prey back to their cubs. They then release the prey, presumably, for the cubs to chase. If the cubs are unable to capture the prey, the mother recaptures it, returns to the cubs, and again releases the prey. The cubs develop the physical skills of predation without additional cognitive burdens such as detection and decision making. Moreover, young cheetahs appear to be less capable of determining when a chase is futile, on average abandoning a chase after 18 m. An experienced adult cheetah will abandon a chase after only 2 m (Caro, 1994) and may devote more attention to detecting prey signals, because its physical skills have become automatic and demand less attention.

Recent work with dogs participating in agility sports also supports a stage model of skill development (Helton, 2007b). Agility is a new sport, developed in the 1970s. A dog runs through a series of obstacles consisting of inclined walls (A frames), hurdles, tunnels, chutes (collapsed cloth tunnels), elevated walks, weave poles, and see-saws. The dog is guided along a prescribed path through the obstacles by a handler who signals the dog via vocal or hand cues. A dog must be able to simultaneously detect handler signals and control body movement in relation to the obstacles. Essentially, the dog performs two tasks: (1) perceiving handler signals while on the move; and (2), controlling movement in a constrained setting. Agility trials allow a researcher to independently quantify motor performance and signal detections by the kinds of faults the dog initiates.

Helton (2007b) investigated signal detection and motor control precision in novice, intermediate, advanced, and expert agility dogs. Utilizing the stage model of skill development advocated by cognitive scientists (Anderson, 1995), he predicted a distinct pattern of results. Dog motor control should initially demand attention; therefore, improvements in motor control should develop early in skill development. After motor skills become sufficiently automatic, attention should be uncoupled from motor skill and facilitate an increase in signal detection precision. As seen in Figure 2.4, this pattern was detected. A statistically significant difference between novice and intermediate dogs was noted for motor precision but not for signal detection precision. A significant detection difference was found between advanced and expert dogs, but no significant motor precision difference. This pattern of results clearly fits a stage model of skill production but does not entirely rule out alternative explanations.

Limitation of attention may serve as a major constraint in animal evolution and skilled behavior, but this aspect of animal behavior has been under-studied by cognitive behavioral scientists (Dukas, 2004). Dukas and Kamil (2001) found evidence of attentional limitations in blue jay search behavior. The birds were trained to perform

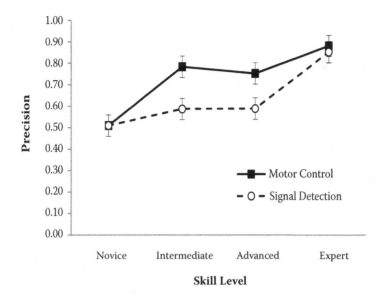

FIGURE 2.4 Mean precision measures for motor control and signal detection in agility dogs (error bars represent standard errors of mean; Helton, in press).

an artificial prey detection task by indicating the presence of prey images on a touch screen. When the birds were shown highly cryptic (low visual contrast with background) prey to locate in the center of the screen, their detection of peripheral targets decreased, as compared to the task of locating more easily detected prey in the center of the screen. As with limitations, the impact of practice and experience on the attention costs of skills in nonhumans has not been studied in detail. As Dukas (2002) indicates, "Changes in attentional requirements with skill acquisition have, to our knowledge, been studied only in humans, but they are probably relevant for other species as well."

The agility dog research (Helton, 2007b) described earlier represents an initial step employing dual-task methodology, a method commonly used to measure task automaticity in humans (Abernethy, 1988; Pashler, 1994 and 1998). As noted, automatic skills place little burden on attention, and thus performance of other concurrent tasks remains unaffected. Skills requiring active control do, however, demand attention and will interfere with performance of secondary concurrent tasks. The competition between tasks requiring limited attention leads to performance deterioration. Animal researchers could make greater use of this dual-task data derived from human expertise research (Beilock et al., 2002; Leavitt, 1979; Smith and Chamberlin, 1992) to investigate skill automaticity.

An interesting question remains: if skills in dogs become automatic, then what were they before they became automatic? We would describe this in humans as controlled or conscious. This, of course, opens the question of what *controlled* and *conscious* mean in relation to dog behavior (Helton, 2005; Rossano, 2003). The alternative possibility is that all skill learning in dogs is unconscious or implicit.

While implicit learning studies on humans have shown that contextual regularities that affect performance can be acquired independent of awareness (Chun and Jiang, 1998), human participants in such studies must actively attend to the display in order to learn the sequences (Baars, 2002). Although the participants may be unaware of learning sequences, they are certainly aware of the original perceptual inputs: they are not learning subliminally.

Recent perceptual learning studies of humans show that later visual performance can be influenced by prior exposure to subliminal stimuli (Seitz and Watanabe, 2003), but the learning occurs only when the subliminal information is presented concurrently with actively attended-to stimuli. Mere exposure is insufficient to initiate learning (Seitz and Watanabe, 2005). The near consensus among cognitive scientists is that learning complex skills requires active attention—consciousness. This issue highlights the difficulties cognitive science confronts when addressing consciousness, especially in reference to nonhumans such as dogs. Expertise is at the heart of this issue.

From a pragmatic perspective, individuals working with dogs may need to regard the dogs as conscious agents and let the philosophers deal with the nitpicky details of what consciousness means. The nuances of the differences between real consciousness and controlled attention or processes are probably unsolvable in a scientific sense and best left to philosophers and theologians who make their livings dealing with scientifically insoluble issues. Simply stated, the take-home message for people who work with dogs is to go ahead and be slightly anthropomorphic—you are going to be anyway. One inescapable reality of working with fairly intelligent but not human animals is that through many generations of close collaboration with us, the animals have acquired some human-like capacities. This idea is not new; Darwin first addressed it in 1889 in his initial work on human and animal emotional expression.

NATURE VERSUS NURTURE

One of the most controversial issues in the expertise literature concerns the relative contributions of genetics (innate ability) and practice (experience) in determining who becomes an expert. Are experts made or born? While this dichotomy is often considered simplistically, most researchers acknowledge some combination of genes and environment, or more specifically the interaction between them. However, the question cannot be sidestepped by saying that both factors contribute to expertise. This answer is not particularly useful. It is a political response to an honest debate about the relative contributions of genes and practice to skilled performance. As one of the authors previously argued (Helton, 2004), dogs, if recognized as experts, bring an immense amount of gravitas to the debate because they are the most morphologically diverse species. Their breeding (genetics) can be and has been controlled and we can regulate the amounts of practice and their early experiences (Schmutz and Schmutz, 1998; Slabbert and Rasa, 1997).

One issue emerging in the use of dogs as models for expertise is the possibility of differences in innate abilities of breeds. Breeds are "intraspecies groups that have relatively uniform physical characteristics developed under controlled conditions" by humans (Iron et al., 2003). Dogs are the most phenotypically (appearance) diverse species existing, ranging in size and shape from diminutive Chihuahuas to massive Great Danes.

Despite the recent historical origins of many breeds, advocates often suggest that breeds were developed as a result of long periods of artificial selection directed at improving occupational performance or expertise. Some breed types appear to have originated very early since they are described in ancient texts and appear in Egyptian and classical Greco–Roman art. Greyhounds and other sight hounds, for example, were selected to be quick pursuit hunters in open fields. Mastiffs were selected for combat. Interestingly, recent work reviewed by Sutter and Ostrander (2004) indicates that relatively small genetic variations have produced substantial intraspecific phenotypic variations.

Further promise in this realm is found in the recent (~2003) sequencing of the entire dog genome. Behavioral geneticists can now begin work that may quantify the amount of behavioral variation (including variations in expert performance) based on genetic factors and the amount of variance produced by other environmental factors such as training.

While this genetic approach may lead to the discovery of mechanisms for inherent ability in the development of canine expertise, Ericsson and colleagues have been the most critical of the impact of inherent ability and genetics on expertise development (Ericsson and Charness, 1994; Ericsson et al., 1993). Aside from physical size, particularly height, which may be important in areas such as athletics, they are very skeptical of the influence of genetics on human expertise development. For expertise researchers, like Ericsson, who debate the role of genetics on performance, dogs present an interesting opportunity. Breed differences based on physical characteristics are beyond debate. It is not hard to recognize the very real difference between a Chihuahua and a Great Dane. Obviously a Great Dane is better suited for bear hunting than a Chihuahua. These physical differences are clearly related to physical skill and influence running speed, jumping height, and strength (Coppinger and Coppinger, 2001; Helton, 2007a). Even the most ardent critics of the effects of genetics on expertise agree on this point.

Dogs with longer legs, for example, tend to run faster than dogs with shorter legs, and leg length is based on significant genetic influence (Helton, 2007a). This is presumably why, in canine racing sports, long-legged breeds such as Greyhounds dominate. Recent research comparing the anatomical properties of Greyhounds and Pit Bulls has shown clear differences, with trade-offs between running ability and fighting ability (Kemp et al., 2005). Pit Bull skeletons are built for grappling and fighting, whereas Greyhound skeletons are built for sprinting. These differences are to a large extent genetic and confirm common sense: in a race, bet on the Greyhound; in a fight, bet on the Pit Bull. (The authors do not advocate these activities because of serious animal welfare issues.) Coppinger and Coppinger (2001) also demonstrated how only a limited range of body sizes perform efficiently in the sport of dog sled racing, based on the thermal constraints of the task. Too large a body retains too much heat, and too small a body retains too little heat. Before the advocates of genetics and talent rejoice, a closer examination of supposed cognitive differences due to genetics should be made.

Cognitive breed differences, unlike physical differences, present more interesting "grist for the mill" and do not follow common sense. Previous researchers have reported perceived breed differences in cognitive ability, for example, trainability (Coren, 1994; Hart,1995; Rooney and Bradshaw, 2004; Serpell and Hsu, 2005). Although the sequential rankings of the breeds, from most intelligent (trainable) to least intelligent (trainable), slightly vary across studies, the rankings are fairly consistent with the abilities of certain breeds uniformly ranked high and low.

Helton (unpublished manuscript) investigated for differences in canine agility (obstacle) performance. He choose agility because speed and precision can be examined independently. Because speed is subject to physical differences (height, 2007a), precision is the more interesting trainability metric. Helton did not find any breed differences in agility precision or in the amount of training necessary to achieve mastery. "Dumb" breeds did not need more training to be as precise as "smart" breeds. Speed of course differed across breeds because body composition and shape are critical. These results agree with the conclusions of Scott and Fuller (1965) who, after a long series of cognitive tests on Basenjis, Beagles, Cocker Spaniels, Fox Terriers, and Shetland Sheepdogs, stated, "...we can conclude that all breeds show about the same average level of performance in problem solving, provided they can be adequately motivated, provided physical differences and handicaps do not affect the tests, and provided interfering emotional reactions such as fear can be eliminated."

Pongracz and colleagues (2004) also failed to find breed differences in the ability to learn a detour task from human demonstrators. In addition, previous research with drug detection dogs failed to reveal any significant links between performance and theoretically relevant genotypes (Maejima et al., in press). Breeds may not differ substantially on intelligence or cognitive ability, and this may provide strong support for Ericsson's skeptical stance. In the central nervous system, because of immense plasticity, practice may, indeed, "rule the roost."

However, before coming to premature conclusions about the role of genetics in canine cognitive abilities, two issues should be raised. First, genetics may influence neurotransmitters that are known to affect temperament. Breeds appear to differ in temperament (Scott and Fuller, 1965). Second, and perhaps more interesting, genetics exerts a strong influence on body shape, and this can influence perceptual systems. These perceptual differences can be misattributed to cognition. For example, researchers recently found a correlation between nose length and the density of ganglion cells in dog and horse visual streaks (Evans and McGreevy, 2007). High ganglion cell density is a marker of areas in the retina (light-sensitive part of the eye) that are particularly acute. In humans, ganglion cells are most dense in an area called the fovea (circular zone in the center of the retina). Dogs and horses do not have these organs. They have visual streaks—horizontal bands of high ganglion density. These are useful for animals that evolved to be sensitive to movement on the horizon (field predators and prey). Binocular focus and acuity are not as critical for dogs and horses as they are for primates like humans. Dogs with short noses, however, have fewer visual streaks and structures somewhat analogous to a fovea. Considering that canine visual systems are already more sensitive to movement than human systems, greater ganglion density in the visual streak would make these dogs even more sensitive to movement on the horizon. Long-nosed dogs should be particularly attuned to tracking moving targets along a horizon. Greyhounds and other long-nosed dogs are called sight hounds for a reason. By selecting for physical shapes, one can influence perceptual systems, and these will, in turn, influence what appears to be cognition. Sight hounds may not be more visual because of anything selected for in their brains, but in their skulls (that influence their brains). Careful studies of the relationships between shapes and cognition in dogs could lead to advances in understanding the subtle role genetics may play on cognition *indirectly*, e.g., via changes in perceptual inputs and motor outputs.

One issue to be addressed by future researchers is breed stereotyping and performance. People perceive certain breeds as more intelligent (Coren, 1994; Hart, 1995; Rooney and Bradshaw, 2004; Serpell and Hsu, 2005). The process of breed stereotyping is relatively uninvestigated. One of the authors, Helton, talked to a trainer about which breed was the best explosive detector. The response was, "depends on where you live, you see here in the United States it is the Labrador Retriever, in the United Kingdom it is the Springer Spaniel, in Belgium it is the Belgian Malinois, in Germany the German Shepherd, and in Norway, they are pushing the Norwegian Drever, a dog most people have never heard of. Personally, I think it is all goofy, pick any dog you are comfortable with." Apparently, people in the respective countries tended to push, even unconsciously, their national breed, a kind of incipient nationalism.

TECHNOLOGY

One final concern about nonhuman expertise is the allocation of effort and funding to technology development. Overlooking the expertise of nonhumans may bias funding agencies and society against biological solutions to current technical challenges. Indeed, recognizing canine expertise may lead to technical innovations. Detection dogs, for example, work in a number of applied settings. Despite steady improvements in machine intelligence and modern sensor technology, automated systems still cannot match the operational effectiveness of trained canines in a number of fields such as explosive and narcotic detection (Fjellanger et al., 2000; Furton and Myers, 2001).

However, a persistent bias against using dogs or other animals in settings of consequence like explosives detection exists, especially among engineers. In an article in *Industrial Robot*, James Trevelyan (1997) remarked, "Dogs can smell explosive vapors, but the most sensitive detectors developed so far have orders of magnitude less sensitivity. Unfortunately, dogs are not consistent and treat their job as a game; they soon become bored." Trevelyan provides no reference for his claim that dogs are prone to boredom. A study performed by the U.S. Federal Aviation Administration (Garner et al., 2000) found that trained dogs were not prone to boredom and could work continuously on detection tasks for long periods without negative consequences (see Chapter 7 for more details).

Animals, including humans, often perform better than automation devices if they have adequate training and equipment. An alternative technological perspective provided by Helton (2006) and discussed in several chapters of this volume proposes the creation of an ergonomic science of working animals, especially dogs. This research could aid the design of canine working conditions and new technologies, in particular canine–machine interfaces (Helton et al., 2007). Dismissing animal expertise may lead to inappropriate and wasteful expenditures for technological fantasies, while overlooking technology that actually works here and now. A case in point is security screening for dangerous materials. As previously described, a survey of security experts at the U.S. Pacific Northwest National Laboratory (Sanquist et al., 2006) ranked dogs as the best existing security screening technology. While canine reliability has been criticized, most of the criticism arises from poor training or misunderstanding of the role of practice in dog skill. Without rigorous practice, dogs will lapse into poor performance; the same dynamic occurs in the case of human experts. Dogs continue to serve as the gold standard for explosive detection. Funding could perhaps be allocated to better understanding how dogs develop their expertise and how humans can help them maintain their expertise in operational environments such as security screening.

CONCLUSION

Animals may exhibit expertise in certain arenas of behavior. Although many animals present genetic predispositions to develop certain skills, most animals have refined these skills over an extensive period of structured, guided practice and feedback. Regardless of the definition of expertise utilized, whether process or product oriented, some dogs qualify. The abilities of dogs present implications for cognitive science, particularly related to consciousness and language. Appropriate recognition of canine expertise may refocus the allocation of resources to promoting canine expertise in areas of skill beneficial to society.

REFERENCES

Abernethy, B. (1988). Dual-task methodology and motor skills research: some applications and methodological constraints. *Journal of Human Movement Studies, 14,* 101–132.
Abernethy, B., Hanrahan, S.J., Kippers, V., Mackinnon, L.T., and Pandy, M.G. (2005). *The Biophysical Foundations of Human Movement.* Champaign, IL: Human Kinetics.
Adachi, I., Kuwahata, H., and Fujita, K. Dogs recall their owner's face upon hearing the owner's voice. *Animal Cognition,* in press.
Agnetta, B., Hare, B., and Tomasello, M. (2000). Cues to food location that domestic dogs (*Canis familiaris*) of different ages do and do not use. *Animal Cognition, 3,* 107–112.
Agnew, N.M., Ford, K.M., and Hayes, P.J. (1997). Expertise in context: personally constructed, socially selected and reality-relevant? In Feltovich, P.J. et al., Eds. *Expertise in Context.* Menlo Park, CA: AAAI Press, 219–244.
Anderson, J.R. (1982). Acquisition of cognitive skill. *Psychological Review, 89,* 369–406.
Anderson, J.R. (1995). *Learning and Memory: An Integrated Approach.* New York: John Wiley & Sons.
Baars, B.J. (1997). *In the Theater of Consciousness.* Oxford: Oxford University Press.
Baars, B.J. (2002). The conscious access hypothesis: origins and recent evidence. *Trends in Cognitive Science, 6,* 47–52.
Barber, N. (1991). Play and energy regulation in mammals. *Quarterly Review of Biology, 66,* 129–147.

Baylor, A.L. (2001). A u-shaped model for the development of intuition by level of expertise. *New Ideas in Psychology, 19,* 237–244.

Beilock, S.L., Wierenga, S.A., and Carr, T.H. (2002). Expertise, attention, and memory in sensory motor skill execution: impact of novel constraints on dual-task performance and episodic memory. *Quarterly Journal of Experimental Psychology, 55,* 1211–1240.

Bekoff, M. and Byers, J.A. (1981). A critical reanalysis of the ontogeny and phylogeny of mammalian social and locomotor play: an ethological hornet's nest. In Immelmann, K. et al., Eds. *Behavioral Development.* Cambridge: Cambridge University Press, 296–337.

Block, N. (1995). On a confusion about the function of consciousness. *Behavioral Brain Sciences, 18,* 227–287.

Block, N. (2001). Paradox and cross purposes in recent work in consciousness. In Dehaene, S., Ed. *The Cognitive Neuroscience of Consciousness.* Cambridge: MIT Press, 197–219.

Brooks, S.E., Oi, F.M., and Koehler, P.G. (2003). Ability of canine termite detectors to locate live termites and discriminate them for non-termite material. *Journal of Economic Entomology, 96,* 1259–1266.

Buchanan, B.G., Davis, R., and Feigenbaum, E.A. (2006). Expert systems: a perspective from cognitive science. In Ericsson, K.A. et al., Eds. *The Cambridge Handbook of Expertise and Expert Performance.* Cambridge: Cambridge University Press, 87–103.

Burghardt, G.M. (2005). *The Genesis of Play: Testing the Limits.* Cambridge: MIT Press.

Buss, D.M., Ed. (2005). *Handbook of Evolutionary Psychology.* New York: John Wiley & Sons.

Caro, T.W. (1994). *Cheetahs of the Serengeti plains.* Chicago: University of Chicago Press.

Caro, T.M. (1995). Short-term costs and correlates of play in cheetahs. *Animal Behavior, 49,* 333–345.

Charness, N., Tuffiash, M., Krampe, R., Reingold, E., and Vasyujova, E. (2005). The role of deliberate practice in chess expertise. *Applied Cognitive Psychology, 19,* 151–165.

Chi, M.T.H, Feltovich, P.J., and Glaser, R. (1981). Categorization and representation of physics problems by experts and novices, *Cognitive Science, 5,* 121–152.

Chun, M.M. and Jiang, Y. (1998). Contextual cueing: implicit learning and memory of visual context guides spatial attention. *Cognitive Psychology, 36,* 28–71.

Cianciolo, A.T., Mathew, C., Sternberg, R.J., and Wagner, R.K. (2006). Tacit knowledge, practical intelligence, and expertise. In Ericsson, K.A. et al., Eds. *The Cambridge Handbook of Expertise and Expert Performance.* Cambridge: Cambridge University Press, 613–631.

Coppinger, R. and Coppinger, L. (2001). *Dogs: A Startling New Understanding of Canine Origin, Behavior, and Evolution.* New York: Scribner.

Coren, S. (1994). *The Intelligence of Dogs.* New York: Bantam.

Darwin, C. (1889). *The Expression of the Emotions in Man and Animals.* Oxford: Oxford University Press.

Dickinson, J., Weeks, D., Randall, B., and Goodman, D. (2004). One-trial motor learning. In Williams, A.M. and Hodges, N.J., Eds. *Skill Acquisition in Sport: Research, Theory, and Practice.* New York: Routledge, 63–83.

Dukas, R. (1998). Evolutionary ecology of learning. In Dukas, R., Ed. *Cognitive Ecology.* Chicago: University of Chicago Press, 129–174.

Dukas, R. (1998). Ecological relevance of associative learning in fruit fly larvae. *Behavioral Ecology and Sociobiology, 19,* 195–200.

Dukas, R. (1999). Costs of memory: ideas and predictions. *Journal of Theoretical Biology, 197,* 41–50.

Dukas, R. (2002). Behavioural and ecological consequences of limitied attention. *Philosophical Transactions of the Royal Society B, 357,* 1539–1547.

Dukas, R. (2004). Causes and consequences of limited attention. *Brain Behavior and Evolution, 63,* 197–210.

Dukas, R. and Kamil, A.C. (2001). The cost of limited attention in blue jays. *Behavioral Ecology, 11,* 502–506.

Dukas, R. and Visscher, P.K. (1994) Lifetime learning by foraging honey bees. *Animal Behavior, 48,* 1007–1012.

Durier, V. and Rivault, C. (2000). Learning and foraging efficiency in German cockroaches, *Blattella germanica* (L.), Insecta: Dictyoptera. *Animal Cognition, 3,* 139–145.

Eames, E. and Eames, T. (1997). *Partners in Independence: A Success Story of Dogs and the Disabled.* New York: Howell.

Edwards, G.B. and Jackson, R.R. (1994). The role of experience in the development of predatory behaviour in *Phidippus regius,* a jumping spider (Araneae, Salticidae) from Florida. *New Zealand Journal of Zoology, 21,* 269–277.

Ericsson, K.A. (1996). The acquisition of expert performance: an introduction to some of the issues. In Ericsson, K.A., Ed. *The Road to Excellence: The Acquisition of Expert Performance in the Arts and Sciences, Sports and Games.* Mahway, NJ: Lawrence A. Erlbaum, 1–50.

Ericsson, K.A. (2001). Expertise in interpreting: an expert performance perspective. *Interpreting, 5,* 187–220.

Ericsson, K.A. (2006). The influence of experience and deliberate practice in the development of superior expert performance. In Ericksson, K.A. et al., Eds. *The Cambridge Handbook of Expertise and Expert Performance.* Cambridge: Cambridge University Press, 683–703.

Ericsson, K.A. and Charness, N. (1994). Expert performance: its structure and acquisition. *American Psychologist, 49,* 725–747.

Ericsson, K.A., Charness, N., Feltovich, P.J., and Hoffmann, R.R., Eds. (2006). *The Cambridge Handbook of Expertise and Expert Performance.* Cambridge: Cambridge University Press.

Ericsson, K.A., Krampe, R., and Tesch-Romer, C. (1993). The role of deliberate practice in the acquisition of expert performance. *Psychological Review, 100,* 363–406.

Evans, K.E. and McGreevy, P.D. (2007). The distribution of ganglion cells in the equine retina and its relationship to skull morphology. *Anatomy, Histology, and Embryology, 36,* 151–156.

Evetts, J., Mieg, H.A., and Felt, U. (1997). Professionalization, scientific expertise, and elitism: a sociological perspective. In Ericsson, K.A. et al., Eds. *The Cambridge Handbook of Expertise and Expert Performance.* Cambridge: Cambridge University Press, 105–123.

Fagen, R. (1981). *Animal Play Behavior.* Oxford: Oxford University Press.

Farrington-Darby, T. and Wilson, J.R. (2006). The nature of expertise: a review. *Applied Ergonomics, 37,* 17–32.

Feltovich, P.J., Prietula, M.J., and Ericsson, K.A. (2006). Studies of expertise from psychological perspectives. In Ericsson, K.A. et al., Eds. *The Cambridge Handbook of Expertise and Expert Performance.* Cambridge: Cambridge University Press, 41–67.

Fiset, S., Beaulieu, C., and Landry, F. (2003). Duration of dogs' (*Canis familiaris*) working memory in search for disappearing objects. *Animal Cognition, 6,* 1–10.

Fiset, S., Gangnon, S., and Beaulieu, C. (2000). Spatial encoding of hidden objects in dogs (*Canis familiaris*). *Journal of Comparative Psychology, 114,* 315–324.

Fitts, P.M. and Posner, M.I. (1967). *Human Performance.* Belmont, CA: Brooks Cole.

Fjellanger, R., Andersen, E.K., and McLean, I. (2000). A training program for filter-search mine detection dogs. *International Journal of Comparative Psychology, 15,* 277–286.

Ford, K.M., Bradshaw, J.M., Adams-Weber, J.R., and Agnew, N.M. (1993). Knowledge acquisition as a constructive modeling activity. In Ford, K.M. and Bradshaw, J.M., Eds. *Knowledge Acquisition as Modeling.* New York: John Wiley & Sons.

Forsythe, D.E. (1993). Engineering knowledge: the construction of knowledge in artificial intelligence. *Social Studies of Science, 23,* 445–477.

Furton, K.G. and Myers, L.J. (2001). The scientific foundation and efficacy of use of canines as chemical detectors of explosives. *Talanta, 54,* 487–500.

Garner, K.J., Busbee, L., Cornwell, P., Edmonds, J., Mullins, K., Rader, K., Johnston, J.M., and Williams, J.M. (2000). Duty cycle of the detection dog: a baseline study. Final Report, Federal Aviation Administration, Washington, D.C.

Gazit, I. and Terkel, J. (2003). Explosives detection by sniffer dogs following strenuous physical activity. *Applied Animal Behaviour Science, 81,* 149–161.

Hall, S.L. and Bradshaw, J.W.S. (1998). The influence of hunger on object play by adult domestic cats. *Applied Animal Behaviour Science, 58,* 143–150.

Hart, B.L. (1995). Analyzing breed and gender differences in behaviour. In Serpell, J.A., Ed. *The Domestic Dog: Its Evolution, Behaviour and Interactions with People.* Cambridge: Cambridge University Press, 65–77.

Heglund, N.C. and Taylor, C.R., (1988). Speed, stride frequency and energy cost per stride: how they change with body size and gait. *Journal of Experimental Biology, 138,* 301–318.

Heiling, A.M. and Herberstein, M.E. (1999) The role of experience in web-building spiders (Araneidae). *Animal Cognition, 2,* 171–177.

Helton, W.S. (2004a). The development of expertise: animal models? *Journal of General Psychology, 131,* 86–96.

Helton, W.S. (2004b). Utilizing genetic algorithms and neural nets in expert systems: what animals teach us. *Recent Advances in Soft Computing, 5,* 177–182.

Helton, W.S. (2005). Animal expertise, conscious or not? *Animal Cognition, 8,* 67–74.

Helton, W.S. (2006). Canine models of expertise. *Proceedings of the Human Factors and Ergonomics Society, 50,* 875–879.

Helton, W.S. (2007a). Deliberate practice in dogs: a canine model of expertise. *Journal of General Psychology, 134,* 247–257.

Helton, W.S. (2007b). Skill in expert dogs. *Journal of Experimental Psychology: Applied, 13,* 171–178.

Helton, W.S. (2008). Expertise acquisition as sustained learning in humans and other animals: commonalities across species. *Animal Cognition, 11,* 99–107.

Helton, W.S., Begoske, S., Pastel, R., and Tan, J. (2007). A case study in canine–human factors: a remote scent sampler for land mine detection. *Proceedings of the Human Factors and Ergonomics Society, 51,* 582–586.

Hill, N.M. and Schneider, W. (2006). Brain changes in the development of expertise: neuroanatomical and neurophysiological evidence from skill based adaptations. In Ericsson, K.A. et al., Eds. *The Cambridge Handbook of Expertise and Expert Performance.* Cambridge: Cambridge University Press, 653–682.

Hoffman, R.R., Shadbolt, N.R., Burton, A.M., and Klein, G. (1995). Eliciting knowledge from experts: a methodological analysis. *Organizational Behavior and Human Decision, 52,* 129–158.

Holland, V.S. (1994). *Herding Dogs: Progressive Training.* New York: Howell.

Iron, D.N., Schaffer, A.L., Famula, R., Eggleston, M.L., Hughes, S.S., and Pedersen, N.C. (2003). Analysis of genetic variation in 28 dog breed populations with 100 microsatellite markers. *Journal of Heredity, 94,* 81–87.

Keasar, T., Motro, U., Shur, Y., and Shmida, A. (1996). Overnight retention of foraging skills in bumblebees is imperfect. *Animal Behavior, 52,* 95–104.

Kemp, T.J., Bachus, K.N., Nairn, J.A., and Carrier, D.R. (2005). Functional trade-offs in the limb bones of dogs selected for running versus fighting. *Journal of Experimental Biology, 208,* 3475–3482.

Kitowski, I. (2005). Play behaviour and active training of Montagu's harrier (*Circus pygargus*) offspring in the post-fledging period. *Journal of Ethology, 23,* 3–8.

Kosslyn, S. (1980). *Image and Mind.* Cambridge: Harvard University Press.

Krampe, R.T. and Charness, N. (2006). Aging and expertise. In Ericksson, K.A. et al., Eds. *The Cambridge Handbook of Expertise and Expert Performance*. Cambridge: Cambridge University Press, 723–742.

Leavitt, J. (1979). Cognitive demands of skating and stick handling in ice hockey. *Canadian Journal of Applied Sport Science, 4,* 46–55.

Lesgold, A., Rubinson, H., Feltovich, P., Glaser, R., and Klopfer, D. (1988). Expertise in a complex skill: diagnosing x-ray pictures. In Chi, M. et al., Eds. *The Nature of Expertise*. Hillsdale, NJ: Lawrence A. Erlbaum.

Logan, G.D. (1985). Skill automaticity: relations, implications, and future directions. *Canadian Journal of Psychology, 39,* 367–386.

Ludingtoni, J.P. (1988). Criminal law: dog scent discrimination lineups. *American Law Reports, 63,* 143–153.

Maejima, M., Inoue-Murayama, M., Tonosaki, K., Matsuura, N., Kato, S., Saito, Y., Weiss, A., Murayama, Y., and Ito, S. (2008). Traits and genotypes may predict the successful training of drug detection dogs. *Applied Animal Behaviour Science*, in press.

Marks, A. (2007). Drug detection dogs and the growth of olfactory surveillance: beyond the rule of law? *Surveillance and Society, 1,* 257–271.

Marschark, E.D. and Baenniger, R. (2002). Modification of instinctive herding dog behavior using reinforcement and punishment. *Anthrozoos, 15,* 51–68.

Masunaga, H. and Horn, J. (2000). Characterizing mature human intelligence: Expertise development. *Learning and Individual Differences, 12,* 5–33.

McNamara, J.M., Houston, A.I., and Collins, E.J. (2001). Optimality models in behavioral biology. *Society for Industrial and Applied Mathematics Review, 43,* 413–466.

Melcher, J.M. and Schooler, J.W. (1996). The misremembrance of wines past: verbal and perceptual expertise differentially mediate verbal overshadowing of taste memory. *Journal of Memory and Language, 35,* 231–245.

Menache, S. (2000). Hunting and attachment to dogs in the pre-modern period. In Podberscek, A.L. et al., Eds. *Companion Animals and Us*. Cambridge: Cambridge University Press, 42–60.

Mieg, H.A. (2006). Social and sociological factors in the development of expertise. In Ericsson, K.A. et al., Eds. *The Cambridge Handbook of Expertise and Expert Performance*. Cambridge: Cambridge University Press, 743–760.

Miklosi, A., Polgardi, R., Topal, J., and Csanyi, V. (1998). Use of experimenter-given cues in dogs. *Animal Cognition, 1,* 113–121.

Miklosi, A., Polgardi, R., Topal, J., and Csanyi, V. (2000). Intentional behavior in dog–human communication: an experimental analysis of "showing" behavior in the dog. *Animal Cognition, 3,* 159–166.

Morse, D.H. (2000). The effect of experience on the hunting success of newly emerged spiderlings. *Animal Behavior, 60,* 827–835.

Mosher, D.S., Quignon, P., Bustamante, C.D., Sutter, N.B., Mellersh, C.S., Parker, H.G., and Ostrander, E.A. (2007). A mutation in the myostatin gene increases muscle mass and enhances racing performance in heterozygote dogs. *Public Library of Science: Genetics, 5,* 779–786.

Negro, J.J., Bustamante, J., Milward, J., and Bird, D.M. (1996). Captive fledgling American kestrels prefer to play with objects resembling natural prey. *Animal Behavior, 52,* 707–714.

Sutter, N.B. and Ostrander, E.A. (2004). Dog star rising: the canine genetic system. *Nature Reviews: Genetics, 5,* 900–910.

Pashler, H. (1994). Dual-task interference in simple task: data and theory. *Psychological Bulletin, 116,* 220–244.

Pashler, H. (1998). *The Psychology of Attention*. Cambridge: MIT Press.

Pearce, J.M. (1997). *Animal Learning and Cognition: An Introduction*. Hove: Psychology Press.

Pickel, D., Manucy, G.P., Walker, D.B., Hall, S.B., and Walker, J.C. (2004). Evidence for canine olfactory detection of melanoma. *Applied Animal Behaviour Science, 89*, 107–116.

Pongracz, P., Miklosi, A., Vida, V., and Csanyi, V. (2004). The pet dog's ability for learning from a human demonstrator in a detour task is independent from breed and age. *Applied Animal Behaviour Science, 50*, 309–323.

Procter, R.W. and Dutta, A. (1995). *Skill Acquisition and Human Performance.* Thousand Oaks, CA: Sage.

Randolph, M. (1997). *Dog Law.* Berkeley: Nolo Press.

Range, F., Viranyi, Z., and Huber, L. (2007). Selective imitation in domestic dogs. *Current Biology, 17*, 1–5.

Rooney, N.J. and Bradshaw, J.W.S. (2004). Breed and sex differences in the behavioural attributes of specialist search dogs: a questionnaire survey of trainers and handlers. *Applied Animal Behaviour Science, 86*, 123–135.

Rosenthal, D.M. (2002). How many kind of consciousness? *Consciousness and Cognition, 11*, 653–665.

Rossano, M.J. (2003). Expertise and the evolution of consciousness. *Cognition, 89*, 207–236.

Sanquist, T.F., Mahy, H.A., Posse. C., and Morris, F. (2006). Psychometric survey methods for measuring attitudes toward homeland security systems and personal privacy. *Proceedings of the Human Factors and Ergonomics Society, 50*, 1808–1811.

Schmutz, S.M. and Schmutz, J.K. (1998). Heritability estimates of behaviors associated with hunting in dogs. *Journal of Heredity, 89*, 233–237.

Scott, J.P. and Fuller, J.L. (1965). *Genetics and Social Behavior of the Dog.* Chicago: University of Chicago Press.

Seitz, A. and Watanabe, T. (2003). Psychophysics: is subliminal learning really passive? *Nature, 422*, 36–46.

Seitz, A. and Watanabe, T. (2005). A unified model for perceptual learning. *Trends in Cognitive Science, 9*, 329–334.

Serprell, J.A. and Hsu, Y. (2005). Effects of breed, sex, and neuter status on trainability in dogs. *Anthrozoos, 18*, 196–207.

Shaffer, D.M., Krauchunas, S.M., Eddy, M., and McBeath, M.K. (2004). How dogs navigate to catch Frisbees. *Psychological Science, 15*, 437–441.

Shanteau, J. (1992). The psychology of experts: an alternative view. In Wright, G. et al., Eds. *Expertise and Decision Support.* New York: Plenum Press, 11–24.

Simon, H.A. (1990). Invariants of human nehavior. *Annual Review of Psychology, 41*, 1–20.

Slabbert, J.M. and Rasa, O.A.E. (1997). Observational learning of an acquired maternal behaviour pattern by working dog pups: an alternative training method. *Applied Animal Behaviour Science, 53*, 309–316.

Smith, M.D. and Chamberlin, C.J. (1992). Effect of adding cognitively demanding tasks on soccer skill performance. *Perception and Motor Skills, 75*, 955–961.

Stein, E.W. (1997). A look at expertise from a social perspective. In Feltovich, P.J. et al., Eds. *Expertise in Context.* Menlo Park, CA: AAAI/MIT Press, 179–194.

Sternberg, R.J. (1999). Intelligence as developing expertise. *Contemporary Educational Psychology, 24*, 359–375.

Sternberg, R.J. and Ben-Zeev, T. (2001). *Complex Cognition: The Psychology of Human Thought.* Oxford: Oxford University Press.

Tarsitano, M.S. and Andrew, R. (1999). Scanning and route selection in the jumping spider, *Portia labiata. Animal Behavior, 58*, 255–265.

Tarsitano, M.S. and Jackson, R.R. (1997). Araneophagic jumping spiders discriminate between detour routes that do and do not lead to prey. *Animal Behavior, 53*, 257–266.

Terrace, H.S., Son, L.K., and Brannon, E.M. (2003). Serial expertise of rhesus macaques. *Psychological Science, 14*, 66–73.

Thurston, M.E. (1996). *The Lost History of the Canine Race.* Kansas City, MO: Universal Press.

Timberlake, W. (1993). Behavior systems and reinforcement: an integrative approach. *Journal of Experimental Analysis of Behavior*, *60*, 105–128.

Topal, J., Byrne, R.W., Miklosi, A., and Csanyi, V. (2006). Reproducing human actions and action sequences: "Do as I Do!" in a dog. *Animal Cognition, 9,* 355–367.

Trevelyan, J. (1997). Robots and landmines. *Industrial Robot, 24,* 114–125.

Trewavas, A. (2005). Plant intelligence. *Naturwissenschaften, 93,* 401–413.

Uttal, W.R. (2005). *Neural Theories of the Mind.* Mahway, NJ: Lawrence A. Erlbaum.

Uzgiris, I. and Hunt, J.M. (1975). *Assessment in Infancy: Ordinal Scales of Psychological Development.* Urbana: University of Illinois Press.

Wells, D.L. and Hepper, P.G. (2003). Directional tracking in the domestic dog, *Canis familiaris. Applied Animal Behaviour Science, 84,* 297–305.

West, R.E. and Young, R.J. (2002). Do domestic dogs show any evidence of being able to count? *Animal Cognition, 5,* 183–186.

3 Social Learning in Dogs

Péter Pongrácz

CONTENTS

The bond between humans and dogs is a unique phenomenon. Although it is easy to describe a dog as one of the many domesticated animal species, our interactions with this social carnivore reach far beyond the usual extent of our relationships with other pets and livestock. In addition to domestication, several features of dogs' social behavior and mental capacity make the human–dog bond one of the most intricate interspecific co-existence relationships. The ability to learn from others by shaping their own behaviors according to a template observed from a demonstrator, in other words, the capacity for social learning, is thought to be a key factor in helping dogs fit into the human social environment.

Traditional methods in training dogs to perform various tasks are usually based on conditioning. Positive and/or negative reinforcers (reward and/or punishment) are used to mark behaviors that humans want or do not want in specific situations (usually following a command or in a specific situation). Through training, humans exploit a dog's willingness to serve, attraction to food rewards, and the close social bonds of dogs and people. This bond makes dogs willing to participate in human activities and this is why social reinforcement (verbal praise and/or scolding) works so well. However, in general, conditioning can be regarded as a special form of asocial or trial-and-error learning in which dogs perform all the action (trial) and humans signal (reward or punish) the correctness of the behavior (error).

Everyday observations and the scientific literature support the concept that dogs are more than treat consumers who react and learn only if they receive rewards. Dogs are also keen observers of others' actions. However, social learning (using a demonstrator) is seldom used as a technique to train dogs. This chapter summarizes the ethological research on social learning in dogs in the hope that it will provide clear examples of the learning capacities of dogs from an unusual point of view.

THEORY OF SOCIAL LEARNING: SHORT INTRODUCTION

Perhaps the simplest and still valid definition of social learning was put forward by Whiten and Ham (1992): "B learns some aspect of the behavioral similarity from A." Such a general description leaves plenty of room for refining the numerous subcases of social learning. Several theoretical and review papers discuss the details of social learning (Galef, 1988; Heyes, 1993; Byrne and Russon, 1998; Zentall, 2001), typically based on exactly what the observer learned from exposure to the behavior of the demonstrator. As we shall see, the extent of matching of the demonstrator's and observer's actions, the ability of the observer to incorporate its own and learned sequences into a behavioral sequence, and insights about the demonstrator's goals all assume importance when we try to subdivide social learning mechanisms.

The next section summarizes the most common definitions and subcategories of social learning. We want to emphasize that the definitions of mechanisms behind social learning are sometimes fiercely debated. It is beyond the scope of this chapter to compare existing parallel theories.

SUBCATEGORIES OF SOCIAL LEARNING

Table 3.1 summarizes the best known subcategories of social learning and characterizes them on the basis of several conditions that are important for distinguishing subcategories. Some of these conditions describe actual behavior (action) learned. Other conditions (object of the action or reward) belong to the circumstantial environment of learning. Finally, a primarily cognitive condition (understanding the demonstrator's goal) is probably the most difficult to investigate.

TABLE 3.1

Subcategories of Social Learning and Associated Factors

	Same Object	Fidelity of "Copying"	Understanding of Goal	Need for Reward	New Action	Known Action
Social Influence (No Social Learning)						
Contagion	Yes	n/a	No	No	No	Yes
Social facilitation	Yes	n/a	No	Yes	No	Yes
Simpler Social Learning Mechanisms						
Observational conditioning	Yes	High	No	Yes	Yes	No
Response facilitation	n/a	High	No	No	No	Yes
Enhancement	Yes[a] No[b]	Medium	Unnecessary	Yes	Yes	n/a
Complex Cognitive Understanding Required						
Emulation	No	Low	Yes	Yes	Yes	No
Imitation	Unnecessary	High	Yes	Yes	Yes[c]	No

n/a = not applicable.

[a] Local enhancement.

[b] Stimulus enhancement.

[c] Imitation requires (1) an action not in the repertoire of the observer or (2) a sequence of individual actions not tried earlier.

DISTINGUISHING SOCIAL LEARNING FROM SOCIAL INFLUENCE

In certain situations, behavioral congruence of observer and demonstrator is present but no real learning is involved. Zajonc (1965) called the process *social facilitation*—the mere presence of the demonstrator increased the motivational level of the observer, but the performed action was not important from the aspect of the resulting behavior. Some experiments suggest a social facilitation in behavioral synchronization among dogs. Vogel et al. (1950) and Scott and McGray (1966) found that dogs in pairs ran significantly faster than single animals to obtain food rewards. Similarly, puppies eat much more food in a group compared to feeding alone (Ross and Ross, 1949). Whiten and Ham (1992) considered the spread of barking among dogs as contagious behavior (social influence).

When the mere presence of the demonstrator is not enough for the observer to adopt similar behavior and observation of an action is also needed, some kind of social learning occurs. The demonstrator can direct the observer's attention (1) to the location or object of the action, (2) to the problem situation and solvability, and (3) to the particular form of action needed. Experimental separation of these three components may aid our understanding of the underlying cognitive processes of the animal mind. However, in most cases, it is very difficult to test these issues individually and their separation remains arbitrary.

LEARNING THROUGH SOCIAL ENHANCEMENT

When social learning does not require deeper understanding of a demonstrator's goal or the connection between the action and its consequence, the underlying mechanism is usually described as some kind of enhancement—a term referring to the effect of demonstration, which increases the proportion of the observer's behavior directed toward the location or object of the demonstrator's activity (Spence, 1937). This, in turn, produces a similar action on the part of the observer. Local enhancement (Roberts, 1941) involves attraction of an animal to a site or object by the presence of a conspecific at the site or by residues of the demonstrator's activity (e.g., odor cues) at the site. Stimulus enhancement has a more general effect, as in the cases in which an observer is attracted to all objects that appear physically similar to the one manipulated by the demonstrator.

Local and stimulus enhancements concern certain objects of the environment that were "marked" by the action of a demonstrator. In the case of response facilitation (Byrne, 1999 and 2002; Byrne and Russon, 1998), an already known, familiar motor pattern (behavior) of the observer is primed in a new context (Janik and Slater, 2000; Byrne, 2002). Some authors refer to this as mimicry (Tomasello and Call, 1997), emphasizing the lack of involvement of complex cognitive mechanisms.

OBSERVATIONAL CONDITIONING

Conditioning is a very common form of learning: it happens continually during the life of every dog and during training. The main principle is the formation of a new mental connection between a formerly indifferent action and a reinforcer (reward or punishment). For example, a dog learns that performing a particular action will produce a reward.

Observational conditioning differs in this basic case from the view of the original actor who initially served as the demonstrator. Most importantly, the observer should see that the demonstrator's action yields a reward (obtained and/or consumed by the demonstrator) in order to reinforce the demonstrator's action as a stimulus. The special feature of observational conditioning is that observing is enough for the learning of an action and the observer does not need to perform the action. This mechanism may be far more common than we thought and may be responsible for what we view as "true imitation" (Heyes, 1993).

EMULATION

Learning from a demonstrator does not mean necessarily that the resulting action of the observer will be the same as the demonstrator's because some problems or tasks have multiple solutions that lead to the same result. Instead of copying the motor pattern, an observer may recognize the problem to be solved and develop his or her own technique to solve it (Tomasello and Call, 1997). This kind of social learning is called emulation. Emulating observers are able to realize the nature of a problem and under experimental conditions may even solve tasks if the action can be demonstrated by the automatic working of a device (performed by "ghost control").

IMITATION

Thorpe (1963) provided the first refined description of imitation: "True imitation has been defined as the copying of a novel or otherwise improbable act or utterance, or some act for which there is clearly no instinctive tendency." According to Heyes (1993), "Imitation means learning something about the form of behavior through observing others, while other kinds of social learning are learning about the environment through observing others." Many others proposed that imitation was an almost uniquely human-specific capacity for cultural learning (Meltzoff, 1996; Tomasello, 1990). Tomasello and Call (1997) emphasize that imitation is a mechanism by which an observer is capable of high fidelity copying of an action and also recognizes the goal of the behavior. This capability is very difficult to prove in animals.

A possible solution may be testing whether an observer is able to generalize a demonstrated action, for example, making noise by beating two wooden sticks together instead of two metal ones (Yunger and Björklund, 2004). Their study revealed no imitation by orangutans. Another way to test imitation is by assigning a task consisting of a sequence of actions. The order is important and individual actions can be facilitated by other mechanisms. Copying all or part of the sequence would provide proof of imitational abilities (Whiten, 1998; Byrne and Russon, 1998).

DOGS AS MODELS FOR STUDYING CONSPECIFIC AND HETEROSPECIFIC SOCIAL LEARNING

Current social learning experiments were conducted typically on species whose foraging behaviors require refined and meticulous skills to manipulate small objects such as boxes or locking devices (Norway rats: Galef and Whiskin, 2001; keas: Gajdon et al., 2004; capuchin monkeys: Caldwell and Whiten, 2004; chimpanzees: Custance et al., 2001). More recently, many studies investigated social interactions between dogs and humans (Miklósi, 2007) and described rich patterns of attachment (Gácsi et al., 2001; Topál et al., 2005) and communication (Miklósi et al., 2003; Gácsi et al., 2004).

It has been argued that selection for living in a human niche changed the behavior of dogs and this led to behavioral parallels of dogs and humans. As a result, the canine species acquired the potential to model specific features of social interactions including social learning that appeared during early human evolution before the emergence of language. The behavioral dog model is based on three key features: (1) selection for highly social behavior and for living in varying environments; (2) possible effects of domestication; and (3) naturalistic socialization to humans.

CANID SOCIAL BEHAVIOR AS BACKGROUND
FOR SOCIAL LEARNING

In a review, Nel (1999) pinpoints that in canids, social learning may represent an adaption to local environmental conditions. The lack of data may be explicable by canids' hidden nocturnal lives. However, Nel cites some examples in the prey acquisition context. For example, avoidance of poisoned bait by several canid species makes it difficult to control their numbers. A laboratory study found that mates and cubs of experienced jackals learned to avoid a common cyanide gun, suggesting acquisition through social learning or even some type of animal teaching behavior (Brand and Nel, 1997).

The dog's closest relative, the wolf lives in family-based packs consisting of an unrelated pair and their offspring of various ages. Wolf packs maintain hierarchical social structures based on leadership. Resident wolf packs are territorial and aggressively defend their home range areas against other wolves. Wolf cubs are cared for by both parents and other group members and most cubs leave the pack by or before the age of 3 years. In principle, such a social (family) environment could favor the acquisition of certain skills through useful social learning, for example, communicative interactions and hunting. It is interesting to notice the similarities of the long childhoods of juvenile wolves in their original packs and the extended puppy states of companion dogs in human society. Theoretically, dogs could extend their sources of social learning from pack members to their new human companions during domestication.

The life of a family dog can be divided into two parts. A puppy spends its first 8 to 10 weeks of life with its mother and littermates, then moves to a human group. This case is special because (1) conspecific parental care is shorter in dogs than in their wild relatives and (2) humans often interfere with parental care shortly after puppies are born and become the most significant social partners later in their lives. In this mixed group, humans usually assume leading roles and are more experienced with regard to the physical and social environment. According to Coussi-Korbel and Fragaszy (1995) and Laland (2004), this would predict that dogs should be inclined to learn from humans in comparison to learning from conspecifics, even if observations are hindered by species-specific abilities in perception and motor skills.

The only known study that made a comparison of the social learning abilities of dogs and wolves was done by Frank and Frank (1985), who noted that compared to dogs, captive wolves showed enhanced insight skills because they were able to recognize mean-and-ends relationships. Similarly, they argued that social learning abilities of dogs should also be inferior. Although comparative studies conducted by this research team indeed found behavioral and performance differences between dogs and wolves, their concept about social learning was "supported" only by a short anecdotal observation. Accordingly, wolves locked in a kennel fitted with a complicated gate mechanism were able to escape after watching a human operate the gate once; dogs could not perform this task. Obviously, such remarks, although interesting, provide little support for the theory, partly because motivational differences can also explain the described species differences.

In more controlled and detailed experiments Miklósi and colleagues (2003) showed that even hand-raised, socialized wolves performed less successfully than companion dogs in such tasks when they had to "read" human visual signals. This research group found a simple reason behind the big difference in dog and wolf behavior in this situation. The dogs looked at the human and the wolves did not. Because the rearing conditions were the same in both species, the difference appears genetically based.

EXPERIMENTAL APPROACH TO SOCIAL LEARNING

Experiments are organic components of modern ethology and examining social learning is an especially fruitful field for experimental testing. A few crucial factors are considered in most social learning experiments:

1. Does social learning exist? Comparing the behaviors of observers and nonobservers constitutes the most important control process to determine effects of a demonstrator's actions. Examining the behaviors of nonobservers will provide insight to the preferred way of individual problem solving (without demonstration). The effect of seeing a demonstrator can prevail in acquiring new behaviors, but also lead to more effective (e.g., faster) performance of an old action. For practical reasons, the difficulty of the task to be learned is an important factor. An easy task leaves no room for improvement after a demonstration is observed, but a difficult task will unnecessarily raise the percentage of failing subjects. To determine whether a task is sufficient but not too difficult, pilot tests without demonstrations can be executed, or, if the demonstrator is a dog instead of a human, the speed of learning by the demonstrator will indicate the difficulty of the task.
2. Exactly how and what did the observer learn? Comparison of subjects exposed to slightly different demonstrations may be achieved via several mechanisms that govern social learning in animals. To differentiate the mechanisms, a demonstration should be carefully manipulated. Changing the quality and quantity of information given to an observer can reveal the details of the means by which the observer utilizes the observed actions.
3. Did the observer learn anything new? Especially for testing imitative social learning, the novelty of the action plays a crucial role. Naturally, novelty can relate to a stimulus, an object, a location, a problem, or a mode of action that must be implemented (the main component of imitation). More importantly, novelty is usually a relative concept (Whiten and Custance, 1996), so care is required when planning an experiment because particular tasks present different levels of novelty, even for subpopulations of dogs (as with garden-kept and apartment-kept dogs investigated by Topál et al., 1997).

DIFFICULTIES WITH SOCIAL LEARNING EXPERIMENTS

As noted above, social learning experiments have been conducted mostly on foraging species that are able to manipulate small objects (boxes, locking devices, etc.). Rats (Galef and Whiskin, 2001), parrots (Gajdon et al., 2004), monkeys (Caldwell and Whiten, 2004), and apes (Custance et al., 2001) usually feed on seeds, fruits, and other foods that involve labor. In general, obtaining a reward (usually food) is the easiest way to stimulate animals to solve tasks and observe demonstrators. However, dogs are carnivorous mammals and it is difficult to find a suitable way to test their social learning abilities if we want to follow some of the traditional methods. Dogs rarely manipulate objects and their physical abilities are not geared to operating complex locking mechanisms. Their paws are very different anatomically from human hands, creating manipulation difficulties if a human demonstrator manipulates a device with his or her hand.

A further complicating factor is the species of the demonstrator. In the case of species other than dogs, demonstrators are conspecifics. Both humans and dogs can play the demonstrator role during dog training. While humans may be preferred because they do not require training and dogs can very rapidly develop social relationships with unfamiliar humans, the utilization of a human demonstrator leads to problems with interpretation of the underlying mechanisms because of differences in species-specific aspects of behavior. Thus it seems to be optimal to test dogs in the same context both with conspecific and human demonstrators. It is possible that dogs learn different aspects of the same situation based on the species of the demonstrator.

EARLY STUDIES OF SOCIAL LEARNING

Experimental investigation of the ability of dogs to learn from conspecifics was sporadic during the twentieth century. Adler and Adler (1967) found that Dachshund puppies learned more quickly to pull a cart by means of a string after observing their trained littermates doing so. Slabbert and Rasa (1997) conducted a more natural study on German Shepherd puppies. The subjects were drug-seeking police dogs bred at a South African kennel. The puppies were raised beside their trained drug-seeking mothers for elongated periods. They were present when their mothers routinely participated in drug retrieval trials and later performed these tasks more easily and faster than puppies raised without this social experience. This experiment is often cited as evidence for social learning in dogs, but if we examine it in detail, it is easy to see that many questions remain open. We agree that the results are very important from the aspect that they illustrate the effect of early exposure to social and individual information on subsequent behavioral performance. However, we do not get a clear answer as to what exactly the puppies learned. Was the maternal behavior (similar to a demonstrator's action) the important factor or did the exposure to the drug sachets affect the puppies' subsequent preference for this kind of task? If the mother dog's behavior was the key, exactly what did the puppies learn from her? Did they learn to retrieve the target or simply manipulate it? Which activity was more important to the puppies? These questions could be answered with more experimental groups that controlled for these alternative possibilities.

CASE STUDY: LEARNING DETOUR BY OBSERVATION

Species differ to a large degree in their capacities to rapidly solve detour tasks and this is partially a reflection of their adaptation to specific niches. This difference was observed during prey hunts by wolves. Some individual wolves have been observed to leave the chase pack and follow an alternative route that placed them in front of the running prey (Mech, 1970). Obviously, such skills may be learned through individual experience, but young animals may also improve their performance by observing other pack members.

While looking at such effects under natural circumstances is nearly impossible, experimental models may provide some help. Buytendijk and Fischel (1932) found that dogs can improve their performances in consecutive trials (through trial-and-error learning) when navigating around barriers (Scott and Fuller, 1965). We chose a detour task in which dogs had to walk around a V-shaped transparent wire mesh fence to obtain rewards (favorite toys or foods; Figure 3.1). The experiments were conducted in open areas near dog schools and the participants were volunteers participating in training courses. Each dog was tested only once in a series of minute-long trials. If the dog obtained the target within 1 minute, the trial was terminated earlier. Owners who remained at the start point were allowed to encourage the dogs. Learning performance was characterized by measuring the time required to reach the target (latency).

ESTABLISHING BASELINE: INDIVIDUAL LEARNING

The first step of any new study in social learning is to establish how a task is solved by trial-and-error learning. We initially tested dogs in the detour experiments without demonstration (Pongrácz et al., 2001). The experimenter hid the target object

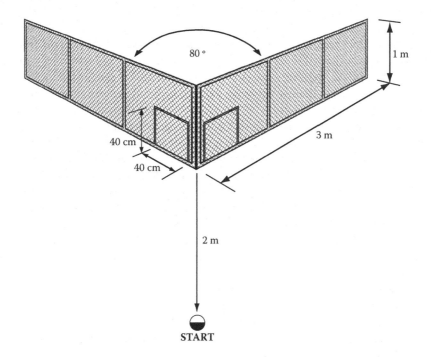

FIGURE 3.1   Fence navigation task.

without observation by the subjects (the owners used their hands to cover their dogs' eyes). After the experimenter placed the target behind a fence, each owner led his or her dog to the fence, showed the target to the dog through the wire mesh, returned to the start point, and then released the dog to locate the target. This process was repeated in each trial.

Dogs were tested in six consecutive trials without demonstrations. We found no significant differences in latencies in the first five trials; only the latency of the sixth detour was significantly shorter than the first. Thus dogs improved the speed of their detours very slowly when they had to rely on their own experiences (Pongrácz et al., 2001). The performance did not appear to be influenced by the breed (Pongrácz et al., 2005) or social status in the home pack (Pongrácz et al., 2007).

The relative slow acquisition of this task by family dogs (only 16% of the subjects were able to detour the fence in less than 30 seconds in the first trial) suggests that learning plays a role in developing skillful behavior. No current data suggest that previous experience with similar problems (e.g., that rural dogs may acquire skills in their native environments) may affect performance in a specific situation.

INCREASED PERFORMANCE THROUGH SOCIAL LEARNING

In the experiments that followed, dogs could observe a human or a dog demonstrator. The human demonstrator walked behind the fence with the target in his hand, put it down, and returned along the other wing of the fence to the dog and owner who waited at the start point. During the detour, the experimenter attracted the dog's attention by calling the dog's name and displaying the target. When a dog served as the demonstrator, an experimenter placed a target behind the fence and a trained dog demonstrator showed the detour by running for the target and returning it to the experimenter. In both cases, the observer witnessed unfamiliar demonstrators making a detour and returning to their starting locations. Each subject participated in three trials, the first without demonstration, and two subsequent trials with demonstration.

A series of experiments established that human demonstration produced a facilitating effect on the acquisition of detouring skill (Pongrácz et al., 2001). By the second trial, the latency was significantly shorter. Further experiments showed that the familiarity of the demonstrator played no significant role, that is, dogs learned equally rapidly by observing unfamiliar experimenters and owners (Pongrácz et al., 2001). Species of the demonstrator (human or dog) did not appear to affect performance. Rapid detouring can be learned from observing an unfamiliar dog or human (Pongrácz et al., 2003a). However, the social rank of the observer (dominant or a subordinate status in family in relation to other dogs) influenced social learning performance. We found that independently of social status, observer dogs learned equally well from a human demonstrator and subordinate dogs were superior in comparison to dogs of higher status if the demonstrator was an unfamiliar dog (Pongrácz et al., 2007).

These observations reveal a range of processes that influence social learning. The indiscriminate learning from conspecific or heterospecific demonstrators, regardless of familiarity, suggests that demonstrator behavior serves to direct the attention of the observer to certain parts or aspects of the physical environment (e.g., end points of the fence). Such cases are usually categorized as stimulus or local enhancement (see above).

INTERACTIONS OF INDIVIDUAL AND SOCIAL EXPERIENCES

In a series of experiments we manipulated the experience of a dog before it was exposed to the demonstrator. In one study, a dog was allowed (once) to obtain the target through an opening at a near corner of the fence (Pongrácz et al., 2003a). Next, dogs were divided into two experimental groups. One group observed a detouring demonstrator and the other group did not. Only the dogs exposed to a demonstrator were able to accomplish the detour task within the time limit set for the observation.

In a second study some dogs were provided with the experience of detouring before the demonstration. We also varied the demonstration by walking along only one side of the fence (unambiguous detour) or both sides (ambiguous detour; see above). The results showed that both aspects affected performance. Experienced dogs that witnessed a demonstrator walking along one side of the fence (unambiguous detour) surmounted the problem mainly by using the same side of the fence. In contrast, dogs that had previous experience detouring the fence or witnessed an ambiguous detour did not show such preferences (Pongrácz et al., 2003b). This was true for both the dogs that were unsuccessful in the first trial and saw the demonstration in the second and third trials, and for those who viewed the demonstration right before the first trial.

In a third study we tested the willingness of dogs to follow a demonstrated action if an undemonstrated but more straightforward solution is available. Closed doors were installed in the fence and the dogs observed a detouring demonstrator. After experiencing single or triple detours, the dogs were allowed to detour. The doors were opened, offering a simpler way to solve the problem. Most dogs continued to detour the fence even when easier direct access through door openings was presented. However, dogs that observed only a single detouring demonstration abandoned detouring sooner and chose to pursue the target through the doors (Pongrácz et al., 2003a).

IMPORTANCE OF CONTEXT IN SOCIAL LEARNING

Demonstration of a detour provides a complex set of information for an observer. The demonstrator moves around a fence and leaves a scent trail. He carries a target conspicuously and talks to the dog. We tested different experimental groups to examine the effects of demonstrator behavior on performance of the detour task (Pongrácz et al., 2004). The procedure was the same as described above. In the scent-only group, dogs were prevented from observing the demonstrator who detoured the fence before each trial on the same side, laying down an unambiguous scent trail. These dogs did not improve their detour speed when tested. In the walk-only group, the demonstrator detoured the fence but did not carry the target or talk to the dogs. Again, these demonstrations did not improve the detour latencies. Results were similar for the walk-with-target group; the demonstrator carried the target conspicuously behind the fence, but did not talk to the dogs. In the walk-and-talk group, the demonstrator detoured the fence without the target, but called the dogs to attract their attention. This activity was as effective as in the usual demonstration group (walking, talking and showing target). These results provide evidence that in the case of a human demonstrator, maintaining the dog's attention is the most important factor. Interestingly, the effect does not seem to be present in cases of conspecific demonstrators.

WHAT WE LEARNED FROM DETOUR TESTS

The fact that performance did not improve with the mere presence of a demonstrator, in the absence of visual access to the demonstration (scented paths around the fence), allows us to infer that social learning occurred. After several experiments, we can conclude that dogs are able to benefit from observing a human or dog demonstration of a spatial task. These results seem to contradict Frank's (1980) hypothesis that dogs have inferior social learning abilities. The detour task proved a useful method because it fit the dogs' motor capacities and was not solved immediately by inexperienced animals. On average, dogs needed six trials to solve the problem through trial and error, but performed a detour after a single demonstration.

Importantly, only limited data account for the mechanism controlling learning. Because the experiment did not involve learning a new motor pattern, imitation can be excluded. The main question is to what aspects of the environment the demonstrator's behavior directed the attention of the observers. Two nonexclusive hypotheses may be put forward. First, the key feature of the demonstrator's action was making the detour, that is, going around the edge of the fence. In this case the dog's behavior could be interpreted in terms of response facilitation (Byrne, 1999) by hypothesizing that detouring (navigating around physical obstacles) is a specific behavior pattern that could be primed by observation (in contrast to attempts to dig under a fence). Of course, another possibility is that detouring is a distinct canine motor pattern like running and jumping. Alternatively, we can assume that the behavior of the demonstrator exposed the edge of the fence to the dog, that is, the observer dogs realized that the fence ended and by running toward it (and arriving at the end) they spontaneously achieved a detour. In this case the social learning could be the described in terms of stimulus enhancement. Local enhancement can be ruled out because the dogs did not always detour on the same side as the demonstrator.

Obviously, further work is needed to elucidate the mechanisms more clearly. To separate stimulus from local enhancement, one must demonstrate that dogs observing a detour at one location (e.g., a fence) are able to transfer this knowledge to similar situations. It becomes more difficult to separate response facilitation from stimulus or local enhancement because it is problematic to separate detouring behavior from the location of the detour (edge of fence). The differences in actions of the human and dog demonstrators in relation to the target (the human takes the target behind the fence; the dog retrieves it from behind the fence) did not seem to influence the observers' performances. This suggests that goal emulation and observational conditioning are less likely to explain the results.

The context of the demonstration is important, especially in the case of humans. While the behavior of the unfamiliar demonstrator dog was always copied (with the exception of the dominant observers), unfamiliar humans were successful demonstrators only when they maintained an observer dog's attention. This could be explained by assuming that the behavior of the demonstrator dog exposed the observer to a more natural pattern of action, or in the case of humans, dogs have learned to attend human actions only in certain situations, e.g., when the demonstrator is accompanied by communicative cues (Erdőhegyi et al., 2007).

Previous experience and social rank also played an important role. In general, dogs attempt to solve a problem in their own individual ways, but they are also keen to follow a variety of social cues. Dogs with little or no experience of detouring were more likely to copy the human demonstrator's action, especially if it provided unambiguous information. The social status of a dog also affects its tendency to learn from other dogs. Dogs that held higher status in their domestic packs were not less able to solve a detour on their own but were less inclined to learn through observation.

OBSERVING SPECIFIC ACTIONS: MANIPULATION TESTS

Object manipulation testing usually involves obtaining a reward from a box by specific handling of a locking device (budgerigars: Heyes and Saggerson, 2002; gorillas: Stoinski et al., 2001). While the detour tests were useful for demonstrating interspecific and intraspecific social learning in dogs, including the role of context of demonstration, experience, and other factors, the tests relied on a very simple action pattern. Learning about manipulating an object could more clearly indicate the exact information learned.

ONE-ACTION TESTS

In a recent experiment, dogs were tested for preference of a demonstrated method of obtaining a ball from a box over their own individual methods (Kubinyi et al., 2003). Pushing a lever protruding from a box to the right or left allowed a ball to roll out of the box. Without demonstrations, dogs pushed the lever only accidentally; they preferred to shake and scratch the box to obtain the ball. After the owner demonstrated the pushing of the lever 10 times, the observer dogs showed a clear preference for using the lever. Interestingly, dogs also learned to use the lever if no ball emerged from the box during the demonstrations. It shows that human actions can be important to dogs even in the absence of a reward.

TWO-ACTION TESTS

In the previous experiment the subjects did not copy the direction of pushing the lever from left to right or right to left. This may indicate that high fidelity between the demonstrator's and observer's actions is not important for dogs. Using only a single simple locking device does not provide enough detail to test mechanisms of social learning. The so-called two-action tests provide a better insight because they offer two equally good solutions that allow a subject to obtain a reward, but only one solution is demonstrated. If an observer dog shows an enhanced preference for a demonstrated solution, it is a sign that the animal may find a functionally similar action in his own repertoire that matches the demonstrator's solution. This is perhaps a more difficult mental function than showing a preference for a demonstrated device as in the one-action test.

In another experiment with dogs, a horizontally suspended tube was used as a two-action device (Pongrácz et al., in preparation). If the tube was tilted, a ball rolled out. Tilting of the tube was achieved by pulling one of two ropes hung from both ends of the tube or by pushing down either end of the tube directly. In the control group, dogs preferred the latter method (pushing the tube) over pulling the rope. In the experimental groups, the human demonstrator showed one of the two possible methods.

The results have shown thus far that dogs were significantly more interested in the demonstrated action in both experimental groups. This is an important finding because it was the first time dogs were shown a motoric action instead of another available motoric solution of a task. In further investigations, we will clarify the importance of the social component of this experiment. We are not yet sure whether the dogs followed the behavior of the human demonstrator or reacted to the movement of the tube. The so-called ghost control technique can solve this problem: no demonstrator is present when an observer sees device movement.

MATCHING BEHAVIOR TO HUMAN ACTION SEQUENCES

Topál et al. (2006) recently obtained evidence indicating that dogs are capable of using a human behavior action as a cue to display a functionally similar behavior. To test this ability, they adopted the matching-to-sample paradigm ("do as I do" task) from studies of apes (Custance et al., 1995; Call, 2001) and dolphins (Herman, 2002). The procedure consists of two phases. First, the subject is trained to perform an action matched to the action of a demonstrator after a simple command ("do it"). In the second phase, the animal is tested with novel actions. A 4-year-old trained assistant Belgian Shepherd male learned to perform nine different actions after presentations by a human demonstrator in 10 weekly 20-minute training sessions (191 trials in total, 17 to 28 trials for each action). It should be noted that because of anatomical differences, human and dog actions were only partially equivalent in motor terms, but were functionally similar.

For example, a human demonstrator's jump with two feet was the equivalent of a dog standing on its two hind legs. Other trained actions included turning around the body axis, barking, jumping over a horizontal rod, placing objects into a container, carrying an object to an owner, and pushing a rod from a chair to the floor.

In the second phase, the dog was tested with complex novel action sequences. Three identical plastic bottles were placed at six predetermined places on a floor. The owner picked up one bottle from one place and transferred it to one of the other five places. After a "do it" command, the dog was able to duplicate the entire sequence of moving a bottle from one specific place to another. It seems that the dog understood the action sequence on the basis of spontaneous observation alone, in terms of the initial state, the means, and the goal. The subject had considerable ability to generalize his understanding of copying and was able to learn to use different forms of human behavior as samples against which to match his own behavior on the basis of resemblance to the demonstrated action.

THE REASONS BEHIND ACTIONS: WISE IMITATIONS?

Gergely et al. (2002) tested human toddlers in a social learning situation. The experiment involved switching on a light on a table by touching it with the forehead. The adult demonstrator and the baby sat on opposite sides of a table and the demonstrator hit the light with her forehead and consequently lit the lamp. Each baby was then subjected to one of two demonstrations and had to switch on the lamp. In one demonstration, the demonstrator put her hands beside the lamp when switching it on. In the second, she held a blanket around her shoulders with her hands. Babies imitated the demonstration in the first condition but used their hands in the second. The authors interpreted the results to indicate that young babies prefer to follow the action of a demonstrator if the situation represented a rule. Otherwise babies used their own, more obvious actions to solve the problem. When the demonstrator held the blanket with her hands, obviously she had to use her head to switch on the light; when her hands were free, using her forehead appeared to work as a rule dictating the proper way to switch on the light.

Range and her colleagues (2007) also tested dogs according to this scenario. A demonstrator dog was trained to open an elevated food container with his paw by pulling a handle at the end of a rope. In this situation naïve dogs tend to use their mouths rather than their paws. Observer dogs could witness two kinds of demonstrations: (1) when the demonstrator's mouth was empty and (2) when he held a ball in his mouth during pulling of the handle. Observers mostly used their paws when they saw the empty-mouthed demonstrator, but they used their mouths if the demonstrator worked with the ball in his mouth. The results indicate that dogs can also have insight to the goal of the demonstrator and may understand the causal connection between a particular motoric action (i.e., using paws) and the circumstances of the demonstration (the mouth of the demonstrator is occupied).

CONCLUSIONS: THE UTILITY OF SOCIAL
LEARNING IN DOG TRAINING

We can now ask whether a practicing dog trainer or service dog handler can learn from the scientific results cited in this chapter. Ethology is the science of research of natural behavior. For dogs, natural behavior means dog–dog interactions and most of their human-directed and human-related behaviors. Studying the capacity of dogs to learn from a human or from a conspecific in an observer–demonstrator situation has shown many times that dogs learn better from observing examples than by trial and error. In our opinion, one of the most important results is that dominance status of the dog affects the way it learns from a human or from another dog. Dominant dogs learn well from unfamiliar human demonstrators but fail to learn anything from unfamiliar dogs. Subordinate dogs learn almost equally well from both dog and human demonstrators (Pongrácz et al., 2007).

This information suggests that dogs, according to their dominance rank exhibit different suitabilities for particular tasks or training methods. Another interesting aspect of this study was that dog behavior is influenced by their dominant or subordinate attitudes and thus these relationships are also valid among dogs that are unfamiliar with each other. This means that dominant and subordinate dogs behave (and learn) differently in new environments and with unknown persons and dogs as well.

Other research results revealed the capacity of dogs to understand such abstract rules as "do as I do" (Topál et al., 2006) and reasonable copying (Range et al., 2007). This should make us aware that dogs are truly not only treat-centric automatons in the hands of trainers but they have natural abilities to understand and use more complicated mental abilities as well.

ACKNOWLEDGMENTS

Many of the studies cited in this chapter were supported by the Hungarian Scientific Research Fund (T043763 and PD48495). The author is thankful to Dr. Ádám Miklósi for his useful comments, to Dr. Enikő Kubinyi for her contribution to an earlier version of the chapter, and to Celeste Pongrácz for her help in proofreading the manuscript.

REFERENCES

Adler, L.L. and Adler, H.E. (1977). Ontogeny of observational learning in the dog. *Developmental Psychobiology, 10,* 267–271.
Brand, D.J. and Nel, J.A.J. (1997). Avoidance of cyanide guns by clack-backed jackal. *Applied Animal Behaviour Science, 55,* 177–182.
Buytendijk, F.J.J. and Fischel, W. (1932). Die Bedeutung der Feldkräfte und der Intentionalität für das Verhalten des Hundes. *Archiv Néderland de Physiologie, 17,* 459–494.
Byrne, R.W. (1999). Imitation without intentionality: using string parsing to copy the organization of behavior. *Animal Cognition, 2,* 63–72.
Byrne, R.W. (2002). Imitation of novel complex actions: what does the evidence from animals mean? *Advances in the Study of Behavior, 31,* 77–105.
Byrne, R.W. and Russon, A.E. (1998). Learning by imitation: a hierarchical approach. *Behavioral and Brain Sciences, 21,* 667–721.
Caldwell, C. and Whiten, A. (2004). Testing for social learning and imitation in common marmosets, *Callithrix jacchus,* using an artificial fruit. *Animal Cognition, 7,* 77–85.
Call, J., (2001). Body imitation in an enculturated orangutan (*Pongo pygmaeus*). *Cybernetic Systems, 32,* 97–119.
Coussi-Korbell, S. and Fragaszy, D. (1995). On the relationship between social dynamics and social learning. *Animal Behaviour, 50,* 1441–1453.
Custance, D.M., Whiten, A. and Bard, K.A. (1995). Can young chimpanzees (*Pantroglodytes*) imitate arbitrary actions? Hayes and Hayes (1952) revisited. *Behaviour, 132,* 837–857.
Custance, D., Whiten, A., Sambrook, T. and Galdikas, B. (2001). Testing for social learning in the "artificial fruit" processing of wildborn orangutans (*Pongo pygmaeus*), Tanjung Puting, Indonesia. *Animal Cognition, 4,* 305–313.
Erdőhegyi, Á., Topál, J., Virányi, Zs. and Miklósi, Á. (2007). Dog logic: inferential reasoning in a two-way choice task and its restricted use. *Animal Behaviour, 74,* 725–737.

Frank, H. (1980). Evolution of canine information processing under conditions of natural and artificial selection. *Zeitschrift Tierpsychologie, 59,* 389–399.

Frank, H. and Frank, M.G. (1985). Comparative manipulation test performance in ten-week-old wolves (*Canis lupus*) and Alaskan malamutes (*Canis familiaris*): a Piagetian interpretation. *Journal of Comparative Psychology, 3,* 266–274.

Gácsi, M., Miklósi, Á., Varga, O., Topál, J. and Csányi, V. (2004). Are readers of our faces readers of our minds? Dogs (*Canis familiaris*) show situation-dependent recognition of human's attention. *Animal Cognition, 7,* 144–153.

Gácsi, M., Topál, J., Miklósi, Á., Dóka, A. and Csányi, V. (2001). Attachment behavior of adult dogs (*Canis familiaris*) living at rescue centres: forming new bonds. *Journal of Comparative Psychology, 115,* 423–431.

Gajdon, G.K., Fijn, N. and Huber, L. (2004). Testing social learning in a wild mountain parrot, the kea (*Nestor notabilis*). *Learning and Behavior, 32,* 62–71.

Galef, B.G. (1988). Imitation in animals: history, definition, and interpretation of data from the psychobiological laboratory. In Zentall, T.R. and Galef, G.B., Eds. *Social Learning.* Hillsdale, NJ: Lawrence A. Erlbaum, 3–28.

Galef, B.G. and Whiskin, E.E. (2001). Interaction of social and asocial learning in food preferences of Norway rats. *Animal Behaviour, 62,* 41–46.

Gergely, Gy., Bekkering, H. and Király, I. (2002). Rational imitation in preverbal infants. *Nature, 415,* 755–758.

Herman, L.M. (2002). Vocal, social, and self-imitation by bottlenosed dolphins. In Nehaniv, C. and Dautenhahn, K., Eds. *Imitation in Animals and Artifacts.* Cambridge: MIT Press, 63–108.

Heyes, C.M. (1993). Imitation, culture and cognition. *Animal Behaviour, 46,* 999–1010.

Heyes, C.M. and Saggerson, A. (2002). Testing for imitative and non-imitative social learning in the budgerigar using a two-object/two-action test. *Animal Behaviour, 64,* 851–859.

Janik, V.M. and Slater, P.J.B. (2000). The different roles of social learning in vocal communication. *Animal Behaviour, 60,* 1–11.

Kubinyi, E., Topál, J., Miklósi, Á. and Csányi, V. (2003). The effect of human demonstrator on the acquisition of a manipulative task. *Journal of Comparative Psychology, 117,* 156–165.

Laland, K.N. (2004). Social learning strategies. *Learning and Behavior, 32,* 4–14.

Mech, L.D. (1970). *The Wolf: Ecology and Behaviour of an Endagered Species.* New York: Natural History Press.

Meltzoff, A.N. (1996). The human infant as imitative generalist: a 20-year progress report in infant imitation with implications for comparative psychology. In Galef, B.G. and Heyes, C.M., Eds. *Social Learning in Animals: The Roots of Culture.* New York: Academic Press, 347–370.

Miklósi, Á. (2007). *Dog Behavior, Evolution, and Cognition.* New York: Oxford University Press.

Miklósi, Á., Kubinyi, E., Topál, J., Gácsi, M., Virányi, Zs. and Csányi, V. (2003). A simple reason for a big difference: wolves do not look back at humans but dogs do. *Current Biology, 13,* 763–766.

Nel, J.A.J. (1999). Social learning in canids: an ecological perspective. In Box, H.O. et al., Eds. *Mammalian Social Learning.* Cambridge: Cambridge University Press, 259–277.

Pongrácz, P., Miklósi, Á., Kubinyi, E., Gurobi, K., Topál, J. and Csányi, V. (2001). Social learning in dogs: the effect of a human demonstrator on the performance of dogs in a detour task. *Animal Behaviour, 62,* 1109–1117.

Pongrácz, P., Miklósi, Á., Kubinyi, E., Topál, J. and Csányi, V. (2003a). Interaction between individual experience and social learning in dogs. *Animal Behaviour, 65,* 595–603.

Pongrácz, P., Miklósi, Á., Timár-Geng, K. and Csányi, V. (2003b). Preference for copying unambiguous demonstrations in dogs. *Journal of Comparative Psychology, 117,* 337–343.

Pongrácz, P., Miklósi, Á., Timár-Geng, K. and Csányi, V. (2004). Verbal attention getting as a key factor in social learning between dog *(Canis familiaris)* and human. *Journal of Comparative Psychology, 118,* 375–383.

Pongrácz, P. Miklósi, Á., Vida, V. and Csányi, V. (2005). The pet dogs' ability for learning from a human demonstrator in a detour task is independent from the breed and age. *Applied Animal Behaviour Science, 90,* 309–323.

Pongrácz, P., Vida, V., Bánhegyi, P. and Miklósi, Á. (2007). How does dominance rank status affect individual and social learning performance in the dog *(Canis familiaris)*? *Animal Cognition, 7,* 90–97.

Pongrácz, P., Bánhegyi, P. and Miklósi, Á. (2008). Successful application of the two-action paradigm for social learning in dogs. Submitted.

Range, F., Virányi, Z. and Huber, L. (2007). Selective imitation in domestic dogs. *Current Biology, 17,* 868-872.

Roberts, D. (1941). Imitation and suggestion in animals. *Bulletin of Animal Behaviour, 1,* 11–19.

Ross, S. and Ross, J.G. (1949). Social facilitation of feeding behavior in dogs I. Group and solitary feeding. *Journal of Genetic Psychology, 74,* 97–108.

Scott. J.P. and Fuller, J.L. (1965). *Genetics and the Social Behavior of the Dog.* Chicago: University of Chicago Press.

Scott, J.P. and McGray, C. (1967). Allelomimetic behavior in dogs: negative effects of competition on social facilitation. *Journal of Comparative Psychology, 2,* 316–319.

Slabbert, J.M., Rasa, O. and Anne, E. (1997). Observational learning of an acquired maternal behavior pattern by working dog pups: an alternative training method? *Applied Animal Behaviour Science, 53,* 309–316.

Spence, K.W. (1937). The differential response in animals to stimuli varying within a single dimension. *Psychological Review, 44,* 430–444.

Stoinski, T.S., Wrate, J.L., Ure, N. and Whiten, A. (2001). Imitative learning by captive western lowland gorillas *(Gorilla gorilla gorilla)* in a simulated food-processing task. *Journal of Comparative Psychology, 115,* 272–281.

Thorpe, W.H. (1963). *Learning and Instinct in Animals,* 2nd ed. Cambridge: Harvard University Press.

Tomasello, M. (1990). Cultural transmission in the tool use and communicatory signalling of chimpanzees? In Parker, S.T. and Gibson, K.R., Eds. *Language and Intelligence in Monkeys and Apes.* Cambridge: Cambridge University Press, 247–273.

Tomasello, M. and Call, J. (1997). *Primate Cognition.* Oxford: Oxford University Press.

Topál, J., Byrne, R.W., Miklósi, Á. and Csányi, V. (2006). Reproducing human actions and action sequences: "Do as I Do!" in a dog. *Animal Cognition, 9,* 355–367.

Topál, J., Gácsi, M., Miklósi, Á., Virányi, Z., Kubinyi, E. and Csányi, V. (2005). The effect of domestication and socialization on attachment to human: a comparative study on hand reared wolves and differently socialized dog puppies. *Animal Behaviour, 70,* 1367–1375.

Topál, J., Kubinyi, E., Gácsi, M. and Miklósi, Á. (2005). Obeying social rules: a comparative study on dogs and humans. *Journal of Cultural and Evolutionary Psychology, 3,* 213–237.

Topál, J., Miklósi, Á. and Csányi, V. (1997). Dog–human relationship affects problem solving behavior in the dog. *Anthrozoös, 10,* 214–224.

Vogel, H.H., Scott, J.P. and Marston, M. (1950). Social facilitation and allelomimetic behavior in dogs. *Behaviour, 2,* 120–133.

Whiten, A. (1998). Imitation of the sequential structure of actions by chimpanzees *(Pantroglodytes). Journal of Comparative Psychology, 112,* 270–281.

Whiten, A. and Custance, D. (1996). Studies of imitation in chimpanzees and children. In Heyes, C.M. and Galef, B.G., Eds. *Social Learning in Animals: The Roots of Culture.* San Diego: Academic Press, 291–318.

Whiten, A. and Ham, R., (1992). On the nature and evolution of imitation in the animal kingdom: reappraisal of a century of research. In Slater, P.J.B. et al., Eds. *Advances in the Study of Behavior.* New York: Academic Press, 239–283.

Yunger, J.L. and Björklund, D.F. (2004). An assessment of generalization of imitation in two enculturated orangutans (*Pongo pygmaeus*). *Journal of Comparative Psychology, 118,* 242–246.

Zajonc, R.B. (1965). Social facilitation. *Science, 149,* 269–274.

Zentall, T.R. (2001). Imitation in animals: evidence, function, and mechanisms. *Cybernetic Systems, 32,* 53–96.

4 Temperament and Personality in Working Dogs

Lindsay T. Graham and Samuel D. Gosling

CONTENTS

Practitioners, researchers, and laypersons in day-to-day contact with dogs have long been aware of personality differences among individual animals. Moreover, it is widely believed that these differences can predict subsequent behavior and work performance. However, only a handful of studies have addressed this issue empirically. In this chapter, we review the empirical evidence to date that has examined personality and temperament in working dogs.

Specifically, this review will start by examining general trends in research on dog temperament. What methods have been used, what breeds have been assessed, and what other trends can be identified? Next, we will review the studies of specific domains of temperament, identifying the temperament domains that have considerable cross-study support. Next, we examine past work on the reliability and validity of temperament tests to evaluate the effectiveness of temperament measures. Finally, we draw the findings together to offer some broad conclusions about the field and identify the major questions that remain to be addressed.

BACKGROUND

Until recently, in academic circles at least, the idea that temperament or personality could be identified in nonhuman animals was considered "goofy" at best, and scientifically irresponsible at worst. It was widely thought that trait ascriptions to animals reflected little more than anthropomorphic projections by overly sentimental humans. However, research over the past decade or so, much of it directly addressing concerns about anthropomorphism, has shown that individual differences in animal temperament do exist and can be measured as reliably as the corresponding traits in humans (Gosling, 1998, 2001, 2008; Gosling and John, 1999; Gosling, Kwan, and John, 2003; Gosling, Lilienfeld, and Marino, 2003; Gosling and Vazire, 2002; Vazire, Gosling, Dickey, and Schapiro, 2007).

The fact that temperament can be measured in animals is relevant to several applied domains. Perhaps the most promising domain is in the area of working dogs. Dogs are called upon to perform a great variety of tasks, ranging from rounding up sheep to assisting the visually impaired, that are of great importance to special economic interests or narrow segments of the population. However, because specially bred and trained working dogs are critical front-line assets in military and law enforcement functions such as suspect apprehension, explosives detection, and narcotics interdiction, they also have a much more pervasive influence on life in the developed world, especially in this dangerous age of terroristic threat and globe-spanning low intensity warfare. Working police and military dogs are employed in large numbers throughout the nations of Western Europe and North America, and they are increasingly being adopted by governmental agencies in the rest of the world including Asia, Australasia, South America, and Africa. As in the case of humans (Hogan, Hogan, and Roberts, 1996), it is becoming clear that some individuals are better suited to some tasks than are others (Murphy, 1998; Serpell and Hsu, 2001; Slabbert and Odendaal, 1999; Svartberg, 2002; Wilsson and Sundgren, 1997 and 1998).

An increasing quantity of empirical evidence in addition to an abundance of anecdotal reports from the field suggests that temperamental factors (e.g., emotional stability), not physical abilities (e.g., olfaction), are the primary determinants of effectiveness of detection dogs in the field (Maejima et al., 2006). For example, dogs that are temperamentally predisposed to fearful behavior are more likely to become anxious in the presence of loud noises, impeding their ability to work effectively as explosive detection dogs in a wide range of force protection missions in conflict zones, such as vehicle checkpoints, route clearance, and combat search and clear operations.

DEFINITIONS OF TEMPERAMENT AND PERSONALITY

Before we can begin a review of the temperament and personality literature, it is important to determine what these terms mean and what, if any, difference exists between them. As noted elsewhere the distinction between temperament and personality has not been maintained consistently in the human and animal literatures (Jones and Gosling, 2005). Given our goal to evaluate all potentially relevant studies, for the purpose of this review, we adopt a broad working definition that encompasses both constructs. Specifically, we use the *temperament* term to refer to individual differences in behavior that are relatively consistent over time and across situations.

LITERATURE SEARCH PROCEDURES

To capture as many relevant articles as possible, we employed several search strategies. First, we searched for articles cited in previous reviews of personality or temperament in dogs (Diederich and Giffroy, 2006; Jones and Gosling, 2005) or animals more broadly (Gosling, 2001). Second, we conducted key word searches for working dogs and temperament or personality in the PsychInfo, Biosis, and Web of Science databases. Third, in the Web of Science database, we searched for articles citing major papers (Svartberg, 2001; Murphy, 1998; Goddard and Beilharz, 1985). In all of our searches we focused on empirical articles; broad reviews (Diederich and Giffroy, 2006) and practical guidelines (Hart 2000) were not included. Our final review included 13 papers that are summarized in Table 4.1.

GENERAL SURVEY OF FIELD

Our first goal was to survey the general state of research on temperament in working dogs. To this end, Table 4.1 shows the basic features of the studies included in our review. To identify the major trends, we next summarize the research literature in terms of the methods of temperament assessment, the breeds examined, the purposes of the studies, the ages at which the dogs were tested, the breeding and rearing environments, and the sexual status of the animals.

METHODS OF ASSESSING TEMPERAMENT

Jones and Gosling (2005) identified four main methods used to assess temperament in dogs: (1) test batteries, (2) observational tests, (3) ratings of individual dogs, and (4) expert ratings of breed prototypes. Test batteries document dogs' reactions to specific stimuli; the tests are typically performed by presenting various, usually novel, stimuli singly to a canine subject and recording its reactions. Wilsson and Sundgren (1996) utilized this method when assessing the effectiveness of the behavior tests conducted by the Swedish Dog Training Center to select dogs for specific work and breeding. They conducted a series of seven testing situations and scored 10 behavioral characteristics.

Observational tests involve assessing relatively broad traits discernible in naturalistic environments; the tests were usually conducted in carefully selected, but

TABLE 4.1

Summary of Empirical Research on Working Dog Temperament

Study Design	Number	Breed Composition						Sex		Age of Assessment (months)					Purpose of Assessment				Population
		GSD	Lab	Gld Rtv	Lab x Gld Rtv	Other		M (Neut)	F (Spay)	1st	2nd	3rd	4th	5th	Guide	Police	Work	Other	
Test Batteries																			
Fallani, Previde & Valsecchi (2006)	109	6	53	42	8	0		34 (NR)	75 (NR)	12	30–32				109	0	0	0	Guides
Fuchs et. al. (2005)	149	149	0	0	0	0		NR (NR)	NR (NR)	20	29				0	0	0	149	Swiss German Shepherd Club breeding
Slabbert & Odendaal (1999)	167	167	0	0	0	0		NR (0)	NR (0)	1.85	2.77	3.70	6	9	0	167	0	0	Police work
Svartberg (2001)	2655	2219	0	0	0	436		1381 (NR)	1274 (NR)	12–18					0	0	2655	0	Privately owned
Weiss & Greenberg (1997)	9	0	NR	NR	0	NR		6	3	10–24					0	0	9	0	Service dog selection for physically challenged
Wilson & Sundgren (1997)	2107	1310	797	0	0	0		1073 (0)	1034 (0)	14.8–19.7					797	1310	0	0	Work/service, breeding
Ratings of Individual Dogs																			
Lefebvre et al. (2006)	303	0	0	0	0	303		NR (NR)	NR (NR)	12–36					0	0	0	303	Military

Study	N	GSD	Lab	Pure	Mixed	Unk	M (Neut)	F (Spay)	Age (mo)	Work	Pet	Guides	Type
Maejima et al. (2006)	197	0	197	0	0	0	109 (NR)	81 (NR)	12–24	197	0	0	Drug detection
Rooney et al. (2006)	26	0	26	0	0	0	26 (0)	(NR)	11–12, 12–13, 13–14, 14–15	26	0	0	Search
Serpell & Hsu (2000)[a]	1067	293	369	264	140	0	NR (NR)	NR (NR)	6, 12, 14–24	0	0	1067	Guides
Observational													
Goddard & Beilharz (1983 & 1985)[b]	102	16	16	0	0	70	51 (51)	51 (0)	2.77, 4, 6, 12, 12–18	0	0	102	Guides
Goddard & Beilharz (1982)[b]	887	0	731	43	0	113	436 (227)	451 (0)	12–18	0	0	887	Guides
Murphy (1995 & 1998)[c]	89	0	84	0	0	5	38 (NR)	51 (NR)	12	0	0	89	Guides

Notes: N, number of subjects in study; GSD, German Shepherd; Lab, Labrador Retriever; Pure, dog of specific, unmixed breeds including GSDs and Labs; Mixed, dog known to be of mixed breeding; Unk., breeds unknown or not recorded (guesses about mixed breeds not made); M, male. Neut., neutered; F, female; Spay, number spayed; Age at assessment subgrouped as 1st, 2nd, 3rd, 4th, and 5th because dogs may have been tested more than once, at different ages; Guides notation indicates that dogs were assessed for possible use as guide dogs; Police indicates assessment for police work; Work indicates assessment for use in other types of work (field work, search and rescue, tracking, protection); Pet indicates assessment for selection as pets or status as pets at time of assessment; Dogs in Other category do not fit into any previous categories; they may participate in studies of personality. NR indicates that the authors did not report the information; Yes indicates that the authors reported dogs of that type but did not report numbers or percentages; Absence of entry indicates that calculation or report of statistic is not appropriate or applicable.

a All but 10 dogs are intact.

b Authors report that not all original subjects were maintained throughout study, but do not indicate how many were maintained. Where applicable, number of dogs per breed is also uncertain; we did not know the breeds of animals that dropped out.

c All but four dogs were castrated; they were ex-show or ex-breeding dogs donated to the guide dog program and were several years older than the other dogs assessed.

not controlled, environments and involved the fortuitous presentation of naturally occurring stimuli. For example, in one series of studies, potential guide dogs were judged on cooperativeness based on all behaviors displayed during videotaped walks (Murphy, 1995 and 1998).

Ratings of individual dogs gather information about temperament using rating scales completed by informants familiar with target animals. Serpell and Hsu (2000) used this method in a study in which volunteer puppy raisers rated potential guide dogs on a series of behavioral rating scales.

Expert ratings of breed prototypes rely on informants deemed by the researchers to be dog experts (American Kennel Club judges, veterinarians, dog trainers), who describe, rank, or rate breeds of dogs as a whole rather than specific individual dogs.

As shown in Table 4.1, the most common assessment method was the test battery that was employed in 6 (46%) of the 13 studies reviewed, followed by ratings of individual dogs (31%) and observational tests (23%). Expert ratings of breed prototypes were not used in any of the studies reviewed.

These trends in the working dog literature mirror those found in the broader dog temperament literature reviewed by Jones and Gosling (2005). In their review, test batteries were the most commonly used forms of assessment, followed by ratings of individual dogs and observational tests. Only a few studies were based on expert ratings of breed prototypes.

BREEDS ASSESSED

Dogs come in an enormous variety of breeds, with as many as 150 breeds officially recognized by the American Kennel Club (AKC; Registration Statistics, 2004; http://www.akc.org/breeds/reg_stats.cfm) and many others not recognized by the AKC (Morris, 2002; Wilcox and Walkowicz, 1995). Table 4.1 shows that only a small subset of breeds were examined in the literature, reflecting the breed preferences for working dogs. Specifically, 62% of the studies focused on only three breeds. Of these, 54% studied the German Shepherd (GSD), 62% the Labrador Retriever, and 23% the Golden Retriever. In some cases other breeds were examined including Labrador–Golden Retriever mixes (Murphy, 1995 and 1998; Serpell and Hsu, 2000; Fallani, Prato-Previde, and Valsecchi, 2006), Kelpies, Boxers, and Belgian Tervurens (Goddard and Beilharz, 1984 and 1985; Svartberg, 2001).

PURPOSES OF STUDIES

Not surprisingly, the literature is dominated by studies of guide dogs for the blind (38%) and police dogs (8%) that focus on temperament and its ability to predict subsequent success in work contexts. In addition to guide and police work, the working dogs under investigation were also employed as military dogs, bomb and drug detection dogs, service dogs for the physically disabled, and search dogs.

AGES AT TESTING

The goal of the studies reviewed here was to determine the extent to which traits assessed during puppyhood predicted individual effectiveness at performing specific jobs in adulthood. As a result, most of the studies first assessed the dogs when they were very young: 62% of the studies first assessed the dogs at about 12 months of age and no studies first tested dogs older than 20 months. Some studies assessed dogs multiple times but even in these cases testing rarely continued beyond 24 months of age.

BREEDING AND REARING ENVIRONMENTS

Unsurprisingly, most studies examined canine participants drawn from programs in which dogs were specifically bred to become working animals. These programs included the Swedish Dog Training Center (SDTC), Jackson Laboratories, the Australian Guide Dog Association, and the American Guide Dog Association. In most cases, dogs began life with volunteer owners until they reached the appropriate age for assessment and training for their specified lines of work.

SEXUAL STATUS

As noted above, many of the dogs assessed came from programs seeking to breed animals suitable for specific tasks such as guiding or police work. Thus, most animals were not spayed or neutered. In many cases, the sexual status of the dogs was not addressed or even mentioned. In the few cases in which sexual status was discussed, most castration occurred just before the age of sexual maturity, at approximately 5 to 7 months of age. Overall, however, we know little about the effects of spaying and neutering on dog temperament in general, and even less about how age at castration affects temperament.

SUMMARY OF GENERAL SURVEY

To identify some basic patterns in the studies of temperament in working dogs, we summarized the literature in terms of six study parameters. Our analyses showed that: (a) most studies assess dogs using test batteries and ratings of individual dogs; (b) virtually all the dogs assessed were GSDs, Labrador Retrievers, or Golden Retrievers; (c) most of the studies aimed at predicting adult working performance from tests conducted during puppyhood; (d) most dogs were first assessed by 12 months of age; (e) most dogs were specifically bred and reared for use as working dogs; and (f) virtually all animals were left intact. All of these findings are broadly consistent with the findings of Jones and Gosling's (2005) review of the broader literature on dogs.

Which traits of working dogs were examined? To determine the traits identified in studies of nonhuman animals, Gosling and John (1999) reviewed 19 structural studies of temperament and personality in nonhuman species, ranging from chimpanzees to octopuses. They found evidence for several basic dimensions that recurred across species, with especially strong cross-species evidence for anxiety/ nervousness, sociability, and aggression. What can we learn from the present, more focused review of the temperament traits studied in working dogs? In this section, we describe the findings of a systematic analysis of the traits and behaviors examined.

Although a broad array of traits were examined in the working dog literature, a select few emerged quite frequently. To provide a framework for organizing the literature, we categorized the findings in terms of the seven categories identified by Jones and Gosling (2005): (1) reactivity/excitability, stability, (2) fearfulness, courage/confidence, (3) aggression–agreeableness, (4) sociability/friendliness, lack of interest in others, (5) responsiveness to training, (6) dominance and submission, and (7) activity level. Table 4.2 summarizes the working dog literature in terms of these seven categories.

As shown in the table, traits in the reactivity category were studied frequently, in 11 of the 13 investigations. These traits tended to refer to the ways in which dogs reacted to their environments and to novel stimuli. Studies of traits within this category used terms such as excitement and excitability (Murphy, 1995, 1998; Weiss and Greenberg, 1996), distraction (Goddard and Beilharz, 1982-1983), and hardness and nerve stability (Fuchs, Gaillard, Gebhardt-Henrich, Ruefenacht, and Steiger, 2005; Wilsson and Sundgren, 1996).

Traits related to fearfulness were examined in 12 of the 13 studies currently under investigation. Traits generally associated with fearfulness included anxiety (Goddard and Beilharz, 1982 and 1983; Maejima et al., 2006; Murphy, 1995 and 1998), boldness (Ronney et al., 2006; Svartberg, 2001), self-confidence (Fuchs et al., 2005; Goddard and Beilharz, 1985), and fear or fearfulness (Goddard and Beilharz, 1982, 1983; Serpell and Hsu, 2006; Weiss and Greenberg, 1996). In several cases, some overlap existed between traits assessed in the reactivity and fearfulness categories (Fuchs et al., 2005; Goddard and Beilharz, 1982 and 1983; Rooney et al., 2006; Slabbert and Odendaal, 1999; Svartberg, 2001). One possible reason for this overlap is that dogs may exhibit similar or indistinguishable behaviors as a result of differing emotional states (i.e., a single behavior may result from either reactivity or fear).

Aggression was also studied in 10 of the 13 studies reviewed. Aggressive behavior was sometimes divided into subcategories, based on the cause of the aggression (aggression in the service of dominance versus aggression as a result of nervousness) or on targets of aggression (stranger-directed fear and/or aggression versus owner-directed aggression versus dog-directed aggression). In studies seeking dogs able to work as police dogs, a very specific subset of aggression (sharpness) was tested; it was defined as willingness to bite a human being (Wilsson and Sundgren, 1996).

Responsiveness to training was assessed in 9 of the 13 articles reviewed. Terms associated with this category tended to deal with obedience (Lefebvre et al., 2006; Maejima et al., 2006; Rooney et al., 2006), ability and willingness to learn and work (Murphy, 1995 and 1998; Rooney et al., 2006, Serpell and Hsu, 2000; Wilsson and Sundgren, 2006), and concentration levels (Maejima et al., 2006; Murphy, 1995 and 1998; Weiss and Greenberg, 1996). In several studies, temperament was associated with responsiveness to training (Fuchs et al., 2005; Wilsson and Sundgren, 1996).

The sociability category was examined in 8 of 13 studies. Sociability was indexed by such behaviors as initiating friendly interactions with people and other dogs. This

TABLE 4.2

Traits Studied in Working Dogs: Review of Research

Study	Reactivity	Fearfulness	Activity	Sociability	Responsiveness to Training	Submission	Aggression	None/Other
Fallani, Previde & Valsecchi (2006)	Oriented to person Oriented to door Stare at puppet Exploration *Passive behavior* Oriented to chair Approach puppet	Avoid puppet	*Social play* Individual play *Passive behavior* Locomotion	Approach person Following *Social play* Greeting behavior Physical contact				Scratch door Drink Vocalization
Fuchs et al. (2005)	Nerve stability *Reaction to gunfire* *Temperament* Hardness	Self confident *Reaction to gunfire*	*Temperament*		*Temperament*		Sharpness Defensive drive	
Goddard & Beilharz (1985)		Confidence *Aggression, dominance* *(hackle biting)*				Submissiveness *Aggression–dominance*	*Aggression–dominance*	
Goddard & Beilharz (1982–1983)	Distraction Sensitivity *Fearfulness, high anxiety*	Fearfulness *Fearfulness and high anxiety*					*Nervous aggression* Aggression	
Lefebvre et al. (2006)	Level headedness			Sociability	Obedience		Aggressiveness	General performance sensitivity (body, sound)

TABLE 4.2 (CONTINUED)
Traits Studied in Working Dogs: Review of Research

Study	Reactivity	Fearfulness	Activity	Sociability	Responsiveness to Training	Submission	Aggression	None/Other
Maejima et. al. (2006)		Anxiety	General activity	Affection demand	Obedience training Concentration Interest in target		Aggression towards dogs	
Murphy (1995 & 1998)	*Low concentration Dog distraction* Excitability	Anxiety Suspicion Nervousness			*Low concentration Dog distraction* Low willingness		Pure aggression Nervous aggression Dog aggression	Low body sensitivity Immaturity
Rooney et. al. (2006)	*Boldness* Interest in toys or objects	*Boldness*	*Playfulness*	*Playfulness*	Obedience to human command Ability to learn from reward		Level of aggression toward human	Tendency to hunt by smell alone Stamina Acuity of sense of smell Motivation to retain possession of object Health Ease of adaptation to kennel

Serpell & Hsu (2000)	*[chasing]*	*Dog-directed fear/aggression* *Non-social fear* *Stranger-directed fear/aggression*	Energy level	*Attachment (one person)* *Stranger directed fear/aggression* *Attachment (one person)*	Trainability	*Stranger-directed fear/aggression* *Owner-directed aggression* *Dog-directed fear/aggression* *[Chasing]*	*Attachment (one person)* *Attachment (one person)*
Slabbert & Odendaal (1999)	*Startle test* *Gunshot test*	*Startle test* *Gunshot test*		*[Retrieval test]*	*[Retrieval test]* Obstacle test	Aggression test	*(Retrieval test)*
Svartberg (2001)	*Boldness/shyness*	*Boldness/shyness*		*Boldness/shyness*		*Boldness/shyness*	
Weiss & Greenberg (1997)	*Attention/distraction* Excitement	*Fear/submission*			*Attention/distraction* *Fear/submission* Dominance		
Wilsson & Sundgren (1997)	Nerve stability Hardness *Prey drive*	Courage	*[Temperament]*	Affability	Cooperativeness *[Temperament]*	Sharpness Defensive drive *Prey drive*	

Note: All dimension labels are those used by the authors. The study authors' definitions of temperament have been used, so we have not excluded items that would not normally be considered temperament constructs (i.e., specific behaviors). Those traits that fell into more than one category are italicized. We list in square brackets those traits that did not elicit 100% agreement among the expert judges in terms of category membership. We provide in standard brackets, where appropriate, more information about traits.

category was labeled as social play (Fallani et al., 2006), sociability (Lefebvre et al., 2006), affection demand (Maejima et al., 2006), playfulness (Rooney et al., 2006), and attachment (Serpell and Hsu, 2000).

Activity was covered in 6 of the 13 studies. Various labels were used for traits including social and individual play as well as locomotion (Fallani et al., 2006). Additional terms such as temperament (Fuchs et al., 2005; Wilsson and Sundgren, 1996), general activity (Maejima et al., 2006), playfulness (Rooney et al., 2006), and energy level (Serpell and Hsu, 2000) were used to describe activity.

Submissiveness was the least studied category, appearing in only 2 of 13 studies. In both instances the defining traits overlapped with traits that measured fearfulness (Goddard and Beilharz, 1985; Weiss and Greenberg, 1996). Goddard and Beilharz (1985) examined submissiveness and aggression/dominance. Weiss and Greenberg (1996) cited fear/submission and dominance in their assessments.

Finally, in several instances various traits examined did not fit into any of the seven categories defined by Jones and Gosling (2005). These traits were placed into a none/other category. Seven studies incorporated trait assessments assigned to this category. Terms such as "attachment" (Serpell and Hsu, 2000), "drink," "scratch door," and vocalizing" (Fallani et al., 2006) were included. Rooney et al.'s (2006) study examined several traits in this category, e.g., tendency to hunt by smell alone, acuity of smell, stamina, and motivation to retain possession of an object.

A broad range of traits have been studied to date. However, the wide range of terms used to describe rather similar behaviors suggests the need for a standard taxonomy of terms to provide a common currency to facilitate a more cumulative approach to studying personality and temperament in working dogs. Of course, with so many researchers having already invested heavily in different assessment systems, it will be a major challenge to create such a common pool of descriptors.

RELIABILITY OF WORKING DOG TEMPERAMENT ASSESSMENTS

If temperament tests are to be of any value, they must be shown to be both reliable and valid. Reliability is a prerequisite for validity, and so we review the evidence for reliability first. Our review enabled us to examine three forms of reliability: internal consistency, test–retest reliability, and inter-observer agreement. Internal consistency measures the degree to which items on a scale assess the same construct and is usually estimated using Cronbach's alpha coefficient (α) or intra-class correlations. In human personality research, α values are often computed following factor analyses to determine the internal coherence of the derived factors.

Of the reviewed articles, the investigation of Serpell and Hsu (2001) provides the only example of an internal consistency estimate. The study gathered data using questionnaires with five-point frequency (Likert) scales. Serpell and Hsu (2001) reported a mean α of 0.65 for the factors derived in their study, with a maximum α of 0.84 for the stranger-directed fear/aggression factor, and a minimum α of 0.53 for attachment. Although only one study reported internal consistency estimates, other studies of dogs have also reported strong α values (Gosling, Kwan, and John, 2003) supporting the idea that dog temperament can be assessed reliably.

Fuchs et al. (2005) estimated test–retest reliability of seven behavioral assessments measured 1 year apart. Scores were generally stable; 78% of the dogs achieved the same scores on two tests separated by year. However, some variability across traits was exhibited. The least stable assessment was for defense drive with only 58% of dogs getting the same scores on both occasions and the most stable assessment was reaction to gunfire; 94% reached the same scores on both occasions.

One of the studies reviewed estimated reliability from inter-observer agreement correlations. Goddard and Beilharz's (1982 and 1983) findings suggest that inter-observer agreement is possible, but not guaranteed: the mean agreement correlation of 0.47 was respectable but estimates ranged from 0.00 to 0.70.

Rooney et al.'s (2006) study of 26 male Labrador Retrievers trained as search dogs is also relevant to the discussion of reliability because although the characteristics she investigated were not personality traits per se (e.g., coverage of area during free search), the study used subjective rating methods to assess behavior. Dogs were videotaped performing a standardized search task and subsequently rated by scientists and experienced dog trainers. The mean Kendall coefficient of concordance across the 12 items rated was 0.65, suggesting that even ratings based on short behavioral samples can made with reasonable levels of reliability. In short, although the research evidence is rather scant, the studies to date suggest that dog temperament can be assessed reliably but reliability is by no means guaranteed.

VALIDITY OF ASSESSMENTS OF DOG TEMPERAMENT

After the reliability of a test has been established, the next step is to evaluate validity—an index of how well an instrument measures what it is designed to measure. The construct validation process involves determining how well a measure assesses a construct (e.g., fearfulness) as that construct has been conceptualized. A full conceptualization of a construct involves specifying items to which the construct should be related and also items to which the construct should be unrelated (Cronbach and Meehl, 1955).

These two components are known as convergent and discriminant validity. Convergent validity is supported when a measure correlates with other measures to which it should be related. Discriminant validity is supported when a measure is empirically unrelated to other measures that are theoretically unrelated (Campbell and Fiske, 1959). Thus, for example, the construct validity of a measure of fearfulness would be supported by strong correlations with other measures of fearfulness (i.e., convergent validity) and weak correlations with measures of theoretically unrelated traits such as sociability (i.e., discriminant validity; Devellis, 1991). To evaluate the validity of the tests in our review, we culled all potentially relevant validity data from the articles.

CONVERGENT VALIDITY

Table 4.3 displays the studies in which convergent validity was examined. Sections for four of the seven dimensions identified by Jones and Gosling (2005) are shown in the table; no studies reported evidence for convergent validity for activity, sociability, and aggression, so the table includes no section for this dimension. The lowest

TABLE 4.3

Convergent Validity: Ability of Dog Temperament Tests to Predict Future Behavior or Scores on Other Assessments

Dimension Study	Trait	Criterion Measure or Behavior	Basis for Scoring	Validity Coefficient	Number of Subjects
		Reactivity			
Weiss & Greenberg (1997)[a]	Excitement (rated by three observers)	Excitement-related behavior	Scoring methods not specified; behaviors included steady high level of jumping, pawing, barking, etc.	.36	9
Mean				.36	
		Fearfulness			
Goddard & Beilharz (1984)	General nervousness (rated by trainers)	Fear on walk (3 months)	Ratings by trainers based on combination of reactions to various stimuli including clap, noise, toy horse, gun shot, party whistle, rapid head movement, ear position, stranger entering house	.24	102
	General nervousness (rated by trainers)	Fear on walk (4 months)		.35	102
	General nervousness (rated by trainers)	Fear on walk (6 months)		.42	102
	General nervousness (rated by trainers)	Fear on walk (12 months)		.59	102
	General nervousness (rated by trainers)	Fear on walk (day 3 of final evaluation)		.59	102
		Fear on walk (day 4 of final evaluation)		.64	102
Mean				.47	

Responsiveness to Training

Study	Measure	Behaviors	Scoring	Value	N
Weiss & Greenberg (1997)[a]	Attention/distraction (rated by three observers)	Attention/distraction-related behaviors	Scoring method not specified, but behavior described: "dog's attention should be on the handler"	.00	9
Mean				.00	

Submissiveness

Study	Measure	Behaviors	Scoring	Value	N
Weiss & Greenberg (1997)[a]	Dominance (rated by three observers)	Dominance-related behaviors	Scoring method not specified; behaviors included front paw on handler, mounting, placing body above handler, growling during eye contact	.13	9
Weiss & Greenberg (1997)[a]	Fear/Submission (rated by three observers)	Fear/Submission-related behaviors	Scoring method not specified; behaviors included crouching, submissive urination, shoulder roll, prolonged startle/fear to noise etc.	1.00	9
Mean				.56	

Note: Placement of trait in these categories determined by assignments derived in Jones & Gosling, 2005

[a] These correlations are rho values from Spearman's rank analysis.

validity coefficient was associated with responsiveness to training ($r = 0.00$); however, it should be noted that only one study reported a validity coefficient for this dimension and only a single coefficient was reported.

Moreover, several other studies of dogs found evidence for the validity of sociability ratings; for example, in Jones and Gosling's (2005) review of the dog temperament literature that was not restricted to working dogs, the mean validity coefficient for sociability ratings was 0.34 (with 95% confidence intervals of 0.19 and 0.45), even when including the 0.00 coefficient reported above. The strongest validity coefficient was associated with submissiveness (mean $r = 0.56$). Again, this dimension was assessed in only one study that reported two coefficient estimates. The second strongest validity coefficient was for fearfulness (mean $r = 0.47$). Again, only one study reported validity within this domain, but the paper cited six coefficients.

DISCRIMINANT VALIDITY

As noted above, in addition to convergent validity, a core component of construct validity is discriminant validity. Reports on studies of discriminant validity in working dogs are even sparser than reports on convergent validity. In fact, we could find only such report; Serpell and Hsu (2001) found generally good evidence for the discriminant validity of the measures, although there were a few exceptions (e.g., an unpredicted association between attachment and stranger fear/stranger aggression).

SUMMARY OF VALIDITY FINDINGS

Past research generally supports the convergent and discriminant validity of temperament assessments in working dogs. However, the literature on working dogs remains somewhat small and we still have a lot to learn about the validity of temperament assessments. The few studies that have been done suggest that validity varies somewhat across the dimensions and across the traits measured within each dimension so drawing any firm conclusions about validity would be premature at this stage. In fact, the clearest lesson to emerge from the literature is that further research on validity is desperately needed to determine what contributes to the variability in validity coefficients and provide a robust estimate of the validity of these assessments.

SUMMARY AND CONCLUSIONS

Bringing together the research on temperament in working dogs allowed us to summarize what is known about canine temperament and to identify some trends and gaps in the field. Below we summarize our conclusions and, where appropriate, we highlight directions for future research.

1. An extensive literature search identified 13 empirical investigations of working dogs and their temperament. These articles covered an array of working dog types, including dogs working in police, military, and civilian contexts, engaged in patrol, detection (drugs and explosives), and guide work.

2. Of the assessment methods used, test batteries were the most common, a finding consistent with a review of the broader dog temperament literature (Jones and Gosling, 2005). Unlike the broader literature, expert ratings of breed prototypes were not used in any of the working dog studies reviewed.

3. Of the articles included in the review, 62% of them were restricted to only three breeds: German Shepherd, Labrador Retriever, and Golden Retriever. Within these studies, 54% assessed German Shepherds, 62% examined Labrador Retrievers, and 23% Golden Retrievers (percentages total more than 100 because some studies covered more than one breed). A few other breeds including Kelpies, Boxers, Belgian Tervurens, and a few cross breeds were also assessed.

4. The main focus of the studies reviewed was to examine the ability of dog temperament to predict success in specific working contexts.

5. Another result of the focus on working ability was the fact that most dogs assessed came from breeding programs designed to deliver dogs for particular tasks. These programs included the Swedish Dog Training Center (SDTC), Jackson Laboratories, the Australian Guide Dog Association, and the American Guide Dog Association.

6. The main focus of the reviewed studies was to assess the predictive ability of temperament and future work success, so most dogs assessed were very young. Sixty-two percent of studies first investigated dogs at 12 months of age or younger and virtually no dogs were assessed after the age of 24 months. Future research should investigate the effects that may have been overlooked as the result of this age bias in the literature (e.g., to what extent are traits stable into adulthood?).

7. The sexual status of dogs was rarely mentioned or recorded. In the few cases in which sexual status was mentioned and neutering occurred, the surgery generally took place right before sexual maturity, at 5 to 7 months of age. Future research is needed to examine the effects of spaying and neutering on dog temperament in general, and determine how age at castration affects later temperament.

8. Most of the traits investigated in the studies reviewed could be assigned to the seven categories identified by Jones and Gosling (2005): reactivity, fearfulness, activity, sociability, responsiveness to training, submissiveness, and aggression. Fearfulness and reactivity were investigated most frequently; submissiveness was least examined. Several other traits could not be assigned to any of the seven categories.

9. Overall, our review of reliability findings was promising. Consistent with previous reviews, studies to date suggest that temperament can be assessed reliably in working dogs.

10. Our review of validity also revealed generally promising findings, suggesting that it is possible to obtain valid estimates of temperament in working dogs. However, the findings were variable, suggesting that validity is not guaranteed. We also discovered that very little attention has been paid to issues of discriminant validity. These patterns of

findings underscore the need for future research to examine the reasons for the variability in validity findings and to provide a more complete picture of discriminant validity.

As the first major review of the temperament literature in studies of working dogs, this chapter provided an overview of the features of studies performed to date. In addition, it provides a roadmap that specifies useful directions for future research on this topic. Overall, the findings summarized support the viability and value of assessing temperament in working dogs.

REFERENCES

Campbell, D.T. and Fiske, D.W. (1959). Convergent and discriminant validation by the multi-trait–multimethod matrix. *Psychological Bulletin, 56*, 81–105.

Cronbach, L.J. and Meehl, P.E. (1955). Construct validity in psychological tests. *Psychological Bulletin, 52*, 281–302.

Devellis, R.F. (1991). *Scale Development.* Thousand Oaks, CA: Sage.

Diederich, C. and Giffroy, J.M. (2006). Behavioural testing in dogs: a review of methodology in search for standardization. *Applied Animal Behaviour Science, 97*, 51–72.

Fallani, G., Prato-Previde, E., and Valsecchi, P. (2006). Do disrupted early attachments affect the relationship between guide dogs and blind owners? *Applied Animal Behaviour Science, 100*, 241–257.

Fuchs, T., Gaillard, C., Gebhardt-Henrich, S., Ruefenacht, S., and Steiger, A. (2005). External factors and reproducibility of the behaviour test in German shepherd dogs in Switzerland. *Applied Animal Behaviour Science, 94*, 287–301.

Goddard, M.E. and Beilharz, R.G. (1982–1983). Genetics of traits which determine the suitability of dogs as guide dogs for the blind. *Applied Animal Ethology, 9*, 299–315.

Goddard, M.E. and Beilharz, R.G. (1984). A factor analysis of fearfulness in potential guide dogs. *Applied Animal Behaviour Science, 12*, 253–265.

Goddard, M.E. and Beilharz, R.G. (1986). Early prediction of adult behavior in potential guide dogs. *Applied Animal Behaviour Science, 15*, 247–260.

Gosling, S.D. (1998). Personality dimensions in spotted hyenas (*Crocuta crocuta*). *Journal of Comparative Psychology, 112*, 107–118.

Gosling, S.D. (2001). From mice to men: what can we learn about personality from animal research? *Psychology Bulletin, 127*, 45–86.

Gosling, S.D. (2008). Personality in non-human animals. *Psychology Compass, 2*, 985–1001.

Gosling, S.D. and John, O.P. (1999). Personality dimensions in non-human animals: a cross-species review. *Current Directions in Psychological Science, 8*, 69–75.

Gosling, S.D., Kwan, V.S.Y., and John, O.P. (2003). A dog's got personality: a cross-species comparative approach to evaluating personality judgments. *Journal of Personality and Social Psychology, 85*, 1161–1169.

Gosling, S.D., Lilienfeld, S.O., and Marino, L. (2003). Personality. In Maestripieri, D., Ed. *Primate Psychology: The Mind and Behavior of Human and Nonhuman Primates.* Cambridge: Harvard University Press, 254–288.

Gosling, S.D. and Vazire, S. (2002). Are we barking up the right tree? Evaluating a comparative approach to personality. *Journal of Research in Personality, 36*, 607–614.

Hart, L.A. (2000). Methods, standards, guidelines, and considerations in selecting animals for animal- assisted therapy. A: understanding animal behavior, species, and temperament as applied to interactions with specific populations. In Fine, A.H., Ed. *Handbook on Animal-Assisted Therapy: Theoretical Foundations and Guidelines for Practice.* San Diego: Academic Press, 81–97.

Hogan, R., Hogan, J., and Roberts, B.W. (1996). Personality measurement and employment decisions: questions and answers. *American Psychologist. 51,* 469–477.

Jones, A.C. and Gosling, S.D. (2005). Temperament and personality in dogs (*Canis familiaris*): a review and evaluation of past research. *Applied Animal Behaviour Science, 95,* 1–53.

Lefebvre, D., Diederich, C., Madeleine, D., and Giffroy, J.M. (2006). The quality of relation between handler and military dog influences efficiency and welfare of dogs. *Applied Animal Behaviour Science, 104,* 49–60.

Maejima, M., Inoue-Murayama, M., Tonosaki, K., Matsuura, N., Ktato, S., Saito, Y., Weiss, A., Murayama, Y., and Ito, S. (2006). Traits and genotypes may predict the successful training of drug detection dogs. *Applied Animal Behaviour Science, 107,* 287–298.

Morris, D. (2002). *Dogs: The Ultimate Dictionary of over 1000 Breeds.* North Pomfret, VT: Trafalgar Square Publishing.

Murphy, J.A. (1995). Assessment of the temperament of potential guide dogs. *Anthrozoos, 13,* 224–228.

Murphy, J.A. (1998). Describing categories of temperament in potential guide dogs for the blind. *Applied Animal Behaviour Science, 58,* 163–178.

Rooney, N.J., Gaines, S.A., Bradshaw, J.W.S., and Penman, S. (2007). Validation of a method for assessing the ability of trainee specialist search dogs. *Applied Animal Behaviour Science, 103,* 90–104.

Serpell, J.A. and Hsu, Y. (2001). Development and validation of a novel method for evaluating behavior and temperament in guide dogs. *Applied Animal Behaviour Science, 72,* 347–364.

Slabbert, J.M. and Odendaal, J.S.J. (1999). Early prediction of adult police dog efficiency: a longitudinal study. *Applied Animal Behaviour Science, 64,* 269–288.

Svartberg, K. (2002). Shyness–boldness predicts performance in working dogs. *Applied Animal Behaviour Science, 79,* 157–174.

Vazire, S., Gosling, S.D., Dickey, A.S., and Schaprio, S.J. (2007). Measuring personality in nonhuman animals. In Robins, R.W. et al., Eds. *Handbook of Research Methods in Personality Psychology.* New York: Guilford, 190–206.

Weiss, E. and Greenberg, G. (1997). Service dog selection tests: effectiveness for dogs from animal shelters. *Applied Animal Behaviour Science, 53,* 297–308.

Wilcox, B. and Walkowicz, C. (1995). *The Atlas of Dog Breeds of the World,* 5th ed. Neptune, NJ: TFH Publications.

Wilsson, E. and Sundgren, P.E. (1997). The use of a behavior test for the selection of dogs for service and breeding I. Method of testing and evaluating test results in the adult dog, demands on different kinds of service dogs, sex and breed differences. *Applied Animal Behaviour Science, 53,* 279–295.

5 Overview of Scent Detection Work
Issues and Opportunities

William S. Helton

CONTENTS

SCENT DETECTION TASKS

Dogs have been trained to perform a diverse set of detection tasks. They have been trained to detect estrus in dairy cows, cancer, contamination in aquaculture tank water, compact discs and DVDs, invasive species, accelerants, explosives, narcotics, insect infestations, microbial growth, wood rot, gas leaks, toxins, and scat of a wide range of species. Olfaction in dogs (see Chapter 8 by Goldblatt, Gazit, and Terkel) and other vertebrates involves the action of chemically sensitive sensory cells located in the olfactory epithelium in the nasal cavity.

The olfactory epithelium contains olfactory sensory neuron cells used in detection, along with supporting cells and stem cells used to replace damaged cells in the epithelium. A dog has an olfactory epithelium of approximately 170 cm^2 (varies by breed and individual dog) in comparison to the human's 10 cm^2 olfactory epithelium.

The dog's olfactory epithelium is also more densely innervated with more olfactory receptor neurons per unit of area. Olfaction research is underdeveloped compared to research on vision or hearing. Despite our gaps of knowledge in how olfaction works, we know dogs are or may act as very sensitive detectors based on some of their performance achievements. This chapter provides a general overview of the dog detection literature with a specific focus on current shortcomings in the literature from an ergonomic-applied perspective. In particular, the orientation is for agencies or organizations currently using dogs or interested in using dogs in detection work.

PERCEPTUAL THRESHOLDS

Psychophysics is a subdiscipline of perceptual science that investigates the relationship of the physical magnitudes of environmental stimuli and their perceived psychological magnitudes or precepts. Psychophysicists have developed a wide array of techniques useful in exploring how and when changes in real-world stimuli induce changes in perception. A primary pursuit of psychophysicists is the determination of detection and discrimination thresholds. From a classical psychophysical perspective, any sensory system involves a bare minimum of physical stimulation that will trigger a perception of a stimulus: the detection threshold. Below this limit, the physical stimulus is subliminal, or in other words, not detectable. Also, a minimum amount of physical change of a stimulus is required for detection or perception of change: the discrimination threshold.

Agencies pursuing the use of dogs as detectors are often interested in knowing the minimal amount of physical substance that can be detected by a dog or its olfactory detection threshold. Canine olfaction, as discussed in Chapter 8, is extremely complicated and includes unique receptor cells for different chemical compounds. Hence, thresholds vary for different chemicals and compounds. The literature includes significant disagreements about detection thresholds of dogs for the same chemical. These disagreements may arise from differences in the manner or tests used to determine thresholds, differences in dog training protocols, different breeds tested, and possible contamination of nonscent stimuli. Because dogs may be more sensitive than instrumental methods, it may be hard to confirm that nonscent stimuli are not contaminated with the target odor.

Recent work on scent detection thresholds in dogs demonstrates low thresholds. Johnston and colleagues (1999; William et al., 1997) determined a dog's limit of detection to be in the 10 parts per billion (ppb) range for benzoate, cyclohexanone, and nitroglycerin. In a recent report, the concentration of n-amyl acetate (nAA) was systematically lowered until chance detection performance was determined (Walker et al., 2006). The eventual detection threshold was 1 to 2 parts per trillion (ppt)—30 to 20,000 times lower than previously recorded thresholds. The dogs in this study were, however, given 6 months of training on the task, so the lowering of the threshold may have been due partially to perceptual learning.

Regardless of disagreements in the literature, recent research suggests that for the odorants tested dogs' detection thresholds are in tens of parts per billion or even in tens of parts per trillion. Data from the Canine Research Institute at Auburn University also indicates dogs' detection thresholds for target odors are still very low even when a co-presented masking odorant has much higher concentration (Johnston, 1999). Generally, dogs are extremely sensitive to odors, but the absolute threshold will vary for different odor compounds, thus making general statements about thresholds difficult to formulate.

PERFORMANCE

Many agencies and organizations would like to know how good a dog is at a detecting a certain object or substance. This is critical in determining whether dogs are the best detectors available, whether their level of performance makes their evidence admissible in legal cases, and for providing benchmarks for engineers to match with alternative detection technologies. I was contacted, for example, a few years ago by personnel at a government laboratory about dogs' explosive detection performance. Dogs are often considered the gold standard in the explosive detection community and the government scientists needed to know what an acceptable performance level was because they were working on alternative technological detection systems. Although I could point them to some sources, I was unable to easily answer their question. Unfortunately, despite a large number of studies examining canine detection performance on a diverse array of detection tasks, it is exceedingly difficult to make general comments about detection capacities for any given object type. Dogs are complex biological systems; they are not neatly constructed mechanical devices. Furthermore, every dog is unique and this makes generalizing inherently difficult.

In addition to individual differences among dogs, issues in the research literature impair our ability to assess dogs' capacities. Each research group typically addresses a specific set of scientific questions and their published papers report data relevant to those specific questions. The papers do not include all information relevant to outsiders, including agencies interested in general about detector dog performance. The same, of course, could be said about all areas of scientific research, so this is not a criticism unique to detector dog research. Problems currently in the literature for agencies interested in assessing dog performance include (1) a lack of uniformity in reporting performance, (2) a lack of uniformity in testing conditions, and (3) a lack of training information provided.

PERFORMANCE METRICS

Performance cannot be assessed with a single number. Speed and accuracy, for example, are clearly separate metrics of performance. Speed is how quickly a dog can detect a substance or make a response. Accuracy is how well a dog discriminates target substances (signals) from background nontarget substances (noise). These are not, obviously, the same parameters. A dog may quickly find substances, but may miss targets because of haste. A dog may also be very slow to find substances, but based on a methodical approach may miss few or no targets. This means a dog may be both quick and accurate or slow and inaccurate.

The relationship between speed and accuracy is often complicated by whether comparisons are made of a single dog on different trials or between dogs performing the same task. When human performance is examined, there is typically a negative correlation between speed and accuracy when a single person's performance is examined, but a positive correlation between speed and accuracy when multiple people are compared. A person can speed his or her performance, but accuracy typically suffers, or the individual can slow down and be more careful. When more than one person, however, is compared, this result is not typical. Highly skilled individuals are often both accurate and fast; poorly skilled individuals are often both slow and inaccurate. This has been demonstrated in agility dogs (Helton, 2007a) that, in comparison to less skilled dogs, are both quick and accurate, and both skill elements improve with practice (Helton, 2007b). Obviously, it would be helpful if both metrics were recorded and reported in the literature.

SPEED

When an agency or individual wants to know how effectively a dog detects a substance or object, the question usually refers to accuracy. Speed is, however, an important aspect of performance, especially for making decisions about the merits of dogs versus alternative sensors and technologies. In certain applications or scenarios, speed is as critical as accuracy. Avalanche search and rescue is a pertinent example (Slotta-Bachmayr, 2005; see Chapter 13 in this book). Time is a critical factor during an avalanche search time if the objective is to locate buried individuals who may still be alive. Any delay in finding these people is unacceptable because a buried person is only likely to have a minimal amount of time left to live. In this case, a dog that is extremely accurate but very slow would be an inappropriate choice.

Speed, of course, can entail a number of subcomponents: speed of orientation or initial detection, speed of movement to the scent source, and speed of appropriate response (letting the handler know of the detected substance). In the existing literature, if speed is reported, it is usually referred to as search time, which is a global summary of speed, starting from the initiation of the search, when the dog is given the command to search, and ending when the dog makes the appropriate overt response. The individual components of speed may be of interest and technology is improving our capability of measuring them.

An example of a new technology useful in assessing subcomponents of speed of detection is the use of global positioning systems (GPSs) to accurately track dogs in field experiments. Equipment manufacturers like Garmin produce and sell dog tracking GPSs to the public (Garmin's system is called Astro®). Cablk et al. (2008), for example, employed GPS tracking technology to record detection dogs' behavior while they searched for desert tortoises. To locate objects using olfaction, dogs follow odor plumes to their source. Odors do not disperse uniformly in the air so a dog has no linear gradient to follow. Instead, odor dispersal in air is influenced by turbulence, leading to discontinuous and complex patterns (Weissburg, 2000) that contain areas of varying concentrations of odor interspersed with areas of clean, no-odor air.

In addition to the inherent turbulent nature of fluids (air is a fluid), terrain features can influence how odor disperses. Odors may be trapped or pooled in certain areas as a result of interactions of air flow and terrain features (Cablk et al., 2008). We do not know exactly how dogs are able to track an odor in this spatially and temporally complex environment (Hepper and Wells, 2005). Some animals follow an odor plume to its source by integrating odor detection with rheotaxis. These animals orient to the direction of the current of the medium they inhabit, whether air or water, and move upstream or upwind. By integrating the detection of odor with the detection of flow, animals can quickly orient to an odor source. One problem for an animal is dealing with gaps of the odor signal in the plume, requiring the animal to move laterally (zig-zag) to reacquire the odor. By tracing the exact movements of a dog in relationship to a tortoise, terrain features, and the direction of air flow, Cablk and colleagues could ascertain the moment when the dog picked up the scent of the tortoise by noting an abrupt change in its movement toward the tortoise when downwind of the tortoise.

Their technique could be used to ascertain how quickly a dog first detects a scent in relationship to environmental variables such as humidity, temperature, wind movement, and terrain features. The use of this technology would also enable researchers to determine the time between initial scent detection and the dog's final overt response or signal to a handler. Part of this will involve the time required for a dog to trace an odor plume back to the scent source. Individual dogs may differ in the speed of this activity, allowing for better screening in time-sensitive environments. In addition, by parsing these subcomponents of speed from other skill components, better training techniques may be employed and trainers will be able to make precise diagnoses of individual dog skills.

Furthermore, with additional development, this technique may improve detection. It is possible that a dog may detect an odor by orienting to it in a plume but may not make an overt response because it loses the plume and cannot reacquire it. Because dogs cannot communicate their levels of certainty about the presence of an odor and can only convey a yes or no decision (odor target present or absent), we have no objective way of currently assessing an uncertain (maybe) response. A dog cannot tell a handler that an odor may be present; it cannot indicate uncertainty. Experienced handlers are probably able to gauge a dog's uncertain state from postural cues. Comparative psychologists since the early days of behaviorism noted that animals often waver back and forth between options at decision points (Tolman and Minium, 1942).

Rats, for example, when confronted with two pathways in a maze, often look back and forth between the two options or even partially enter one path and then follow the other and so on. This behavior is labeled by psychologists as vicarious trial and error (VTE). While many comparative psychologists were hesitant to label VTE as a marker of uncertainty because they feared being labeled anthropomorphists (see discussion in Chapter 1), recent work by comparative psychologists has examined animals' awareness of their uncertainty (Smith, Shields, and Washburn, 2003). Skilled handlers who are not so fearful of the anthropomorphic label attribute uncertain states to their dogs and probably use overt behavioral cues given by their dog partners. Tracking dogs' movements, however, provides a potentially objective technique. Tracking movements relative to airflow and terrain features may provide uncertainty information; for example, a dog may have oriented and appeared to follow a plume for a time but did not make an overt response. By looking for anomalies in a dog's search pattern and movements such as abrupt changes toward the direction of air flow (rheotaxis), we may be able to determine when a dog starts to follow an odor plume, even if the dog later loses the plume and cannot reacquire it. As discussed later in this chapter, trained dogs rarely send false alarms, so uncertain detections may in some applications be important to follow up with more careful searches.

Speed should not be underestimated as an important metric of performance. A number of other animal species possess olfactory detection senses accurate as those of dogs, for example pigs, but dogs are more agile and certainly faster. For certain tasks, trainers may also wish to focus on improving speed. Rewarding dogs when they work quickly, of course, requires knowing when a dog is working quickly or when it is not. It may not be necessary for individual researchers to record each aspect of a search related to speed. Video recording a searching dog would allow for this information to be ascertained later.

ACCURACY

Accuracy is another aspect of performance critical to assessing detector dogs. Typically when an individual wants to know how well a dog can detect a substance or object, the issue is accuracy, not speed. Unfortunately, accuracy cannot be assessed with a single number. Table 5.1 reveals the complexity of measuring accuracy. In any detection scenario, a target substance is physically present or absent and the detector (a dog in this case) can report that the target is present or absent.

Any detection task presents four possible outcomes: (1) the target is present and the detector reports its presence; this is known as a correct detection, a hit, or a true positive; (2) the target is not physically present and the detector reports it as not present; this is known as a correct rejection or a true negative; (3) the target is physically present and the detector reports it is not present; this is known as a miss or a false negative; (4) the target is not physically present and the detector reports it is present; this is known as a false alarm or a false positive. It is natural to focus on the hits or correct detections or the relative relationship of hits to misses, also known as the proportion of correct detections, hit rate, or sensitivity (in medical circles):

Proportion of Correct Detections = Sensitivity = Hits/(Hits + Misses)

While the hit rate or sensitivity is important, this value alone does not provide a complete picture of performance. False alarms are equally important in assessing

TABLE 5.1
Decision Matrix

		Dog Indicates...	
		Target	**No-Target**
	Target	Hit	Miss
Physically	No Target	False Alarm	Correct Rejection

performance. The relative relationship of false alarms to correct rejections is also critical, sometimes called the proportion of false alarms, false alarm rate, or when subtracted from 1, specificity:

Proportion of False Alarms = False Alarms/(False Alarms + Correct Rejections)

Specificity = Correct Rejections/(False Alarms + Correct Rejections)

Two other useful accuracy metrics, although not typically used in detector dog literature, are positive predictive power and negative predictive power:

Positive Predictive Power (PPP) = Hits/(Hits + False Alarms)

Negative Predictive Power (NPP) = Correct Rejections/
(Correct Rejections + Misses)

Both PPP and NPP are useful in assessing the diagnostic abilities of dogs. PPP assesses the proportion of the dog's target-present responses that are correct and NPP assesses the proportion of the dog's target-absent responses that are correct. These two measures are useful when an agency wants to know the likelihood of a target's presence if a dog indicates the target is present and the likelihood that a target is not present if a dog indicates no target is present (how predictive the dog is).

To properly assess detector dog performance, both the hit rate and the false alarm rate must be considered. A dog may achieve a high hit rate, missing relatively few targets, but if its false alarm rate is also high, then reporting that the dog is accurate is misleading. One dog may be biased to make many responses—lots of hits and lots of false alarms. Another dog may be biased toward few responses—more misses and fewer false alarms. Both dogs may exhibit approximately the same abilities to separate targets from distracters (or noise) but have completely different response biases. Perception and sensor researchers employ signal detection theory (SDT), a mathematical approach that attempts to separate perceptual sensitivity and response bias.

Typically SDT is used by researchers when they want to measure decisions made under conditions of uncertainty, e.g., when a signal is difficult to discern from background noise or distracters. Because uncertainty is involved, SDT mathematically separates perceptual sensitivity (how easy or hard it is to separate target signals from background noise or distracters) from response bias (how likely the detector is to report a target is present or not). The goal of SDT is to independently measure perceptual sensitivity and bias. A good detector has a very high perceptual sensitivity and in most cases is unbiased, although this quality depends on the decision context, for example, the consequences of the decision outcomes. If the failure to detect a target presents high consequences, a more lenient response bias may be preferable. The problem with a too lenient response bias is the high rate of false alarms—crying "wolf." Eventually the detector will be regarded as unreliable and its responses may be ignored.

A number of parametric and nonparametric measures of perceptual sensitivity and response bias are available but their use is somewhat misleading because all SDT metrics require the fulfillment of underlying mathematical assumptions. See

Macmillan and Creelman (2005) if you are interested learning more about signal detection theory. Common perceptual sensitivity measures are d' (parametric) and A' (nonparametric). Common bias measures are B, B", and c. A problem with parametric measures is they are incalculable if hit or false alarm rates are perfect (no misses, no false alarms) and require mathematical adjustments. For applied researchers or agencies, it is probably best not to consider SDT metrics as true invariant qualities of a detector, but instead use them as guides to overall detector performance. The use of SDT measures does allow comparisons of different detectors and may be useful in training.

For the benefit of external users, all four decision outcome rates (hits, misses, false alarms, and correct rejections), in addition to any of the derived performance metrics, should be reported. The derived sensitivity, specificity, PPP, NPP, and SDT measures of perceptual sensitivity and bias performance metrics are, moreover, not interchangeable. Recent research focusing on signal detection demonstrates that these different accuracy metrics all examine performance from different perspectives and a more complete picture can result if they are all used together (Szalma et al., 2006). Unfortunately, as will be described in the next section, many of these performance metrics require discrete trials. While hits, misses, and false alarms are countable, it can be difficult to ascertain in many test scenarios the number of correct rejections a dog makes. This means many accuracy metrics such as false alarm rate, specificity, NPP, and SDT are not calculable.

TESTING CONDITIONS

The literature discusses many variations in dog testing methods. The variations make comparing and integrating the results of several studies difficult. The situation is similar to administering different tests to students applying to universities in vastly different settings, some in quiet rooms and others outside, some tests requiring multiple choice responses and others requiring narrative answers. Determining a dog's ability depends on the context in which the test is given and the test employed. Two issues in the current literature are (1) the use or nonuse of discrete trials, and (2) the need to pay closer attention to distinguishing search failures from detection failures.

DISCRETE DECISION TRIALS

As noted earlier in the discussion of accuracy metrics, nondiscrete trials make certain accuracy metrics incalculable. For example, if a dog is given the task of detecting a few odor sources randomly placed in a complex environment to simulate reality, a researcher has no means of calculating or estimating the number of correct rejections. Hits, misses, and false alarms are calculable. The number of correct odor sources is, of course, known and the dog's responses, whether correct or incorrect, can be counted, but the number of incorrect odor sources is unknown. Thus the dog's false alarm rate is not known because of the lack of a base of correct rejections for comparison. How many opportunities for false alarms does the dog encounter? Performance can be assessed in such test scenarios because hits, misses, and false alarms can be counted, but a complete assessment of accuracy or inaccuracy cannot be made by such tests.

Some researchers employ discrete trial tests. These tests often sacrifice realism, for example, in broad area searches, but they accurately reflect some detection tasks such as odor diagnosis or scent lineups. In these tests, a dog is presented with a limited set of plausible odor sources, some of which have the target odor. Since all plausible odor sources, targets and distracters are countable, every accuracy metric discussed previously can be calculated, including correct rejections. The advantage of knowing how many correct rejections a dog makes may outweigh any sacrifice of realism if the goal is total assessment of detector ability, especially in comparison to other detection technologies.

REPORTS OF SEARCH-VERSUS-DETECTION ERRORS

Another issue in test variation is the care or lack thereof when considering the sources of errors. Many detection failures reported to dogs may have been misattributed and were errors on the parts of handlers. For example, if a test simulates a wide area search, as would be the case with narcotic, explosive, or invasive species detection, it must include some way of determining whether the dog failed to detect the scent source or was not given an opportunity to detect the scent source because of a constrained search pattern. This failure is most likely due to an impatient handler, not an unreliable dog.

In a set of tests conducted in Guam using Jack Russell terriers trained to detect brown tree snakes hiding in airplane cargoes, Engeman et al. (2002; see Chapter 10 in this book) carefully recorded the sources of detection errors. A third of the detection failures reported were attributed to a constrained search pattern (handler mistake), not inability of the dogs to detect the snakes. Researchers using field tests should be as careful as Engeman and his associates in assessing failures. Handlers often make mistakes misattributed to dogs. Unless care is taken in recording data, the abilities of a dog as a detector may be underestimated. Unfair criticisms of dogs are often made by technologists wishing to replace existing and reliable canine sensors with untried, expensive, and often inferior alternatives. Researchers should make every attempt to be impartial and determine the actual sources of the errors made during testing.

TRAINING

Underestimating the role of training in canine skill development and maintenance is a major problem (see Chapter 8 by Goldblatt, Gazit, and Terkel). As recent research focusing on expertise development demonstrates, training and practice can markedly improve skill to the point where experts can perform tasks others consider impossible (Ericsson, 2007). Researchers often do not report details of the amounts and types of training the dogs in their studies received. They often fail to report how much time has passed testing and training. Additionally, researchers rarely present data showing how performance metrics change with training. Forecasting beyond the data actually collected is difficult and risky in any setting. Nevertheless, detailed data

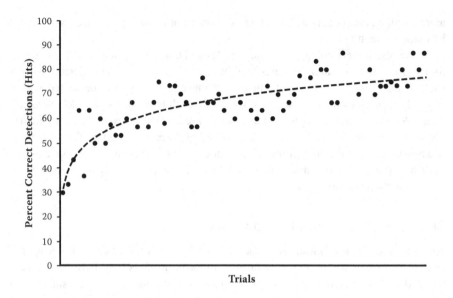

FIGURE 5.1 Hypothetical power function fit between skill metric (hits in this example) and training trials.

showing performance changes over practice sessions would provide some indication of what is likely to happen after further training.

Psychologists studying performance changes in motor and cognitive skills over time suggest a power law of learning (Anderson, 1982; Ericsson et al., 1993). When time spent practicing a skill is plotted against a metric of skill performance, a power function provides a reasonably good fit (Figure 5.1). Learning continues to improve over practice, but the benefits of time spent practicing decline over time. Investigations of perceptual skill also indicate a power law of improvement due to practice with reasonably good power or exponential fits between performance and practice time (Dosher and Lu, 2007). Research examining the functional form of canine olfactory learning is lacking, but a power-like function is probable given the overall literature on learning.

Figures indicating performance changes relative to training time could be used to project likely performance gains arising from additional training. If the best fit between performance and practice observed is a steep linear fit (straight line) instead of a power function, a dog may be far from its skill asymptote because a power or exponential fit is expected. If a leveling of performance is not seen when performance is plotted against a metric of training time, the inference is that performance may continue to improve with more practice. This knowledge would be useful for determining whether dogs tested will improve with additional practice and training. Many published reports of detector dog performance utilized dogs specifically trained for the study instead of working dogs subjected to a test. A reasonable question is whether the training of the test dogs was comparable in amount and quality to the training of the working dogs (Kauhanen et al., 2002). In studies of operational dogs, a crucial piece of information is the time elapsed between their training on the task and the type and extent of remedial practice training they received. Most skills degrade unless practice is maintained. Training must be considered in comparisons of dogs with other technologies because training is expensive and time consuming.

PERFORMANCE REPORTING

Despite problems with the dog detection literature, criticisms are not helpful to readers hoping to evaluate detector dog performance. Clearly, even considering limitations, detector dogs deserve their reputation as the gold standards of detection technology. Table 5.2 lists sensitivity and specificity data for a number of substances. Aside from the relatively low sensitivity, 75% for two dogs trained to detect microbially infected wood (Kauhanen et al., 2002), sensitivity overall was fairly high. Kauhanen et al., moreover, acknowledge limitations of their study such as the possibility of contamination of distracters with target odors. Overall dog specificity was very high, indicating few false alarms. When a trained detector dog alerts a handler, it is prudent to follow up the alert with a more thorough search. While 100% ratings for both sensitivity and specificity are desirable, perfection is an unlikely result for any detection technology in the real world.

For a comparative perspective, mammography screening for breast cancer typically has a sensitivity around 86.6% and a specificity around 96.8% (Banks et al., 2004). Dogs trained to detect cancer in breath samples show comparable ability— sensitivity around 88.0% and specificity around 97.7% (McCulloch et al., 2006). Furthermore, unlike mammography that may result in detrimental radiation exposure and is considered by many women to be uncomfortable because the breast tissue is compressed, breath sampling is a simple and comfortable procedure that has no

TABLE 5.2
Sensitivity and Specificity for Detection Tasks

Reference	Detection Task	Number of Dogs	Sensitivity	Specificity
Kiddy et al. (1978)	Estrus-related odors	6	81.60	81.80
Pickel et al. (2004)	Melanoma	2	100.00	100.00
Smith et al. (2003)	Kit versus red fox scat	4	100.00	96.20
Brooks, Oi, and Kohler (2003)	Termites	6	95.93	97.31
McCulloch et al. (2006)	Lung cancer	5	99.30	98.61
McCulloch et al. (2006)	Breast cancer	5	88.00	97.74
Kauhanen et al. (2002)	Microbial growth	2	75.00	90.10
Gazit, Goldblatt, and Terkel (2003)	C4 outdoor (in dark)	6	87.78	97.22
Gazit, Goldblatt, and Terkel (2003)	C4 outdoor (daylight)	6	93.83	100.00
Gazit, Goldblatt, and Terkel (2005)	TNT outdoor	7	93.74	100.00
Gazit, Goldblatt, and Terkel (2005)	C4 outdoor	7	92.86	100.00
Gazit, Goldblatt, and Terkel (2005)	PETN outdoor	7	91.29	100.00

side effects. Based on accuracy and lack of side effects, cancer detection dogs may represent a superior diagnostic approach.

Two points should be noted when comparing dogs to other detector technologies: (1) dog detection capabilities depend on training; performance improves markedly with proper training; and (2) dogs are extremely flexible. Unlike electromechanical sensors, dog performance of a detection task improves with repeated practice. As indicated by Gazit, Goldblatt, and Terkel (2005a), dog sensitivity to TNT over a short practice period increased from 80.3 to 93.74%. No one knows the limits of this link between training and performance, but sufficient training may allow dogs to achieve close to 100% sensitivity (they often achieve almost 100% specificity). Furthermore, dogs are far more flexible than electromechanical sensors. As the range of substances cited in Table 5.2 indicates, dogs can conceivably be trained to detect any substance that has an odor—from floating whale scat to DVDs, two objects dogs are currently used to detect.

OPPORTUNITIES: PERFORMANCE MODELS

Human factors and ergonomic researchers have developed human performance models. Some, with suitable alterations, may be useful in explaining dog performance. A case in point is the attention–situational awareness (A-SA) model developed by Christopher Wickens and colleagues (2003) to explain how people allocate their attention (where they look) during visual tasks such as flying an airplane or directing air traffic. While dogs are not typically employed for their visual abilities, this model could with modification be generalized to other senses including olfaction.

The A-SA (2003) combines a model of optimal attention allocation during visual tasks with an inference model based on the belief updating model of Hograth and Einhorn (1992). The model of optimal selective attention is based on the assumption that the acquisition or awareness of visual information is driven by four factors: (1) the salience (S) or conspicuity of the visual objects that may capture attention, (2) the effort (EF) required to shift attention from one object to another via visual saccade or head–body movement, (3) expectancies (EX) about the whereabouts of visual information in the scene, and (4) the value (V) or importance of the information for the task at hand at a given location in the visual scene. Essentially, human eye movements are determined both by a bottom–up process of attention capture, controlled primarily by object salience, and a top–down process generated by prior expectations and an internal representation of the likely locations of relevant information. Eye movements are inhibited by the effort required to move over longer distances.

$$P(A) = sS - efEF + (ex\ EX + vV)$$

The model indicates that the probability of attending to a visual object is related to the linear weights of the four components. The uppercase letters represent salience, effort, expectancy, and value; the lowercase letters describe weights assigned to those properties. This model of selective attention is combined with a belief updating model to determine current situational awareness (SA)—a fancy way of indicating what a person knows about a situation at a particular time. SA is updated each time an item or set of items is encountered. A change in SA is determined by a weighted mean of the value (V) of each item. Weighted V for each item is determined by the P(A) of the item and its value V. Thus, both the real value of the information to the task at hand (V) and the amount of attention allocated to it [P(A)] influence its contribution to current SA. A high value item may exert minimal impact on SA if, for example, the item has a low salience or requires a lot of effort to acquire. Current SA is used to calculate the probability of making erroneous decisions and predict where in a visual scene an operator directs his or her attention (dwell time). The A-SA has been tested in a taxiway simulation task to predict the dwell times of pilots on visual locations (Wickens et al., 2003); the model accounted for 90% of the variance in actual dwell times.

The A-SA model with some modifications may prove useful in predicting dog performance in odor detection tasks. With extensive search experience, a dog will develop an underlying sense of the likeliness that a target is present in a certain location (EX). For example, research by Gazit, Goldblatt, and Terkel (2005b) demonstrated that dogs form expectations regarding the likelihood that a previously searched path will or will not contain targets and that these expectations can interfere with target detection. Dogs also probably use environmental features to estimate whether a target is present. Odors, like visual stimuli, vary in salience (S); masking odors can cause interference and lower chemical concentrations can make an odor more difficult to distinguish. Odor detection by a dog also entails physical effort (EF): sniffing along with head and body movements. Following an odor plume, for example, may require a fair amount of physical effort. Effort may also be exacerbated by environmental variables such as thermal stress (heat).

The development of this model for canine olfaction will require further research, but it can be used even now as a potential guide or heuristic for understanding canine performance. The use of this model points out two often overlooked issues: (1) dogs are cognitive and will thus form expectations that assist them in the probability of earning rewards, e.g., they will use previous experiences to guide current performance, and (2) necessary physical effort will reduce the likelihood that a dog will become aware of a target. From a practical perspective, this means handlers and trainers must be very careful to make training searches as diverse as possible. For example, trainers should not repeatedly place targets in easily predicted places. If possible, the amount of physical effort required for a detection task should be reduced (an example is the remote explosive scent trace method discussed in Chapter 7). While these suggestions may seem like common sense measures, it is interesting to note how common sense may be easily overlooked in areas of critical importance (see Norman, 2002).

CONCLUSION

This chapter discusses certain limitations of the current scientific literature on detection dogs. End users, for example, groups interested in using detector dogs, need better presentations of performance results. On the positive side, the research conducted to date indicates that dogs are excellent detectors. In some domains, their detection capabilities match or exceed the capabilities of state-of-the-art technology and human experts and present no side effects. Despite the advances, a lot of territory should be further explored. Testing the limits of training in olfactory detection and the development of performance models are two projects awaiting future canine ergonomics researchers.

REFERENCES

Anderson, J.R. (1982). Acquisition of cognitive skill. *Psychological Review, 89,* 369–406.

Banks, E. et al. (2004). Influence of personal characteristics of individual women on sensitivity and specificity of mammography in the Million Women Study: cohort study. *British Medical Journal, 329,* 477–483.

Brooks, S.E., Oi, F.M., and Koehler, P.G. (2003). Ability of canine termite detectors to locate live termites and discriminate them from nontermite material. *Journal of Economic Entomology, 96,* 1259–1266.

Cablk, M.E., Sagebiel, J.C., Heaton, J.S., and Valentin, C. (2008). Olfaction-based detection distance: a quantitative analysis of how far away dogs recognize tortoise odor and follow it to source. *Sensors, 8,* 2208–2222.

Dosher, B.A. and Lu, Z. (2007). The functional form of performance improvements in perceptual learning: learning rates and transfer. *Psychological Science, 18,* 531–539.

Engeman, R.M., Vice, D.S., York, D., and Gruver, K.S. (2002). Sustained evaluation of the effectiveness of detector dogs for locating brown tree snakes in cargo outbound from Guam. *International Biodeterioration and Biodegradation, 49,* 101–106.

Ericsson, K.A. (2007). Deliberate practice and the modifiability of body and mind: toward a science of the structure and acquisition of expert and elite performance. *International Journal of Sport Psychology, 38,* 4–34.

Ericsson, K.A., Krampe, R., and Tesch-Romer, C. (1993). The role of deliberate practice in the acquisition of expert performance. *Psychological Review, 100,* 363–406.

Gazit, I., Goldblatt, A., and Terkel, J. (2005a). Formation of olfactory search image for explosives odours in sniffer dogs. *Ethology, 111,* 669–680.

Gazit, I., Goldblatt, A., and Terkel, J. (2005b). The role of context specificity in learning: the effects of training context on explosives detection in dogs. *Animal Cognition, 8,* 143–150.

Gazit, I. and Terkel, J. (2003). Domination of olfaction over vision in explosives detection by dogs. *Applied Animal Behaviour Science, 82,* 65–73.

Helton, W.S. (2007). Skill in expert dogs. *Journal of Experimental Psychology: Applied, 13,* 171–178.

Helton, W.S. (2007). Deliberate practice in dogs: a canine model of expertise. *Journal of General Psychology, 134,* 247–257.

Hepper, P.G. and Wells, D.L. (2005). How many footsteps do dogs need to determine the direction of an odour trail? *Chemical Senses, 30,* 291–298.

Hograth, R.L. and Einhorn, H.J. (1992). Order effects in belief updating: the belief adjustment model. *Cognitive Psychology, 24,* 1–55.

Johnston, J.M. (1999). *Canine Detection Capabilities: Operational Implications of Recent R&D Findings.* Auburn, AL: Auburn University.

Kauhanen, E., Harri, M., Nevalainen, A., and Nevalainen, T. (2002). Validity of detection of microbial growth in buildings by trained dogs. *Environment International, 28,* 153–157.

Kiddy, C.A., Mitchell, D.S., Bolt, D.J., and Hawk, H.W. (1978). Detection of estrus-related odors in cows by trained dogs. *Biology of Reproduction, 19,* 389–395.

Macmillan, N.A. and Creelman, C.D. (2005). *Detection Theory: A User's Guide.* Mahwah, NJ: Lawrence A. Erlbaum.

McCulloch, M. et al. (2006). Diagnostic accuracy of canine scent detection in early- and late-stage lung and breast cancers. *Integrative Cancer Therapies, 5,* 30–39.

Norman, D.A. (2002). *The Design of Everyday Things.* New York: Basic Books.

Pickel, D. et al. (2004). Evidence of canine olfactory detection of melanoma. *Applied Animal Behaviour Science, 89,* 107–116.

Slotta-Bachmayr, L. (2005). How burial time of avalanche victims is influenced by rescue method: an analysis of search reports from the Alps. *Natural Hazards, 34,* 345–352.

Smith, D.A. et al. (2003). Detection and accuracy rates of dogs trained to find scats of San Joaquin kit foxes (*Vulpes macrotis mutica*). *Animal Conservation, 6,* 339–346.

Smith, J.D., Shields, W.E., and Washburn, D.A. (2003). The comparative psychology of uncertainty monitoring and metacognition. *Behavioral and Brain Sciences, 26,* 317–339.

Szalma, J.L. et al. (2006). Training for vigilance: using predictive power to evaluate feedback effectiveness. *Human Factors, 48,* 682–692.

Tolman, E.C. and Minium, E. (1942). VTE in rats: overlearning and difficulty of discrimination. *Journal of Comparative Psychology, 34,* 301–306.

Walker, D.B. et al. (2006). Naturalistic quantification of canine olfactory sensitivity. *Applied Animal Behaviour Science, 97,* 241–254.

Weissburg, M.J. (2000). The fluid dynamical context of chemosensory behavior. *Biological Bulletin, 198,* 188–202.

Wickens, C.D. and Hollands, J. (2000). *Engineering Psychology and Human Performance,* 3rd ed. Upper Saddle River, NJ: Prentice Hall.

Wickens, C.D. et al. (2003). Attentional models of multitask pilot performance using advanced display technology. *Human Factors, 45,* 360–380.

Williams, M. et al. (1997). Determination of the canine odor detection signature for selected nitroglycerin based smokeless powder. *Proceedings of 13th Annual Security Technology Symposium and Exhibition,* Virginia Beach, VA.

6 Evaluating Learning Tasks Commonly Applied in Detection Dog Training

Lisa Lit

CONTENTS

INTRODUCTION

Dogs are currently used for scent detection in a wide variety of scenarios, for example, the detection of people (Harvey et al., 2006), cadavers (Oesterhelweg et al., 2008), drugs (Maejima et al., 2007), explosives (Gazit, Goldblatt, and Terkel, 2005; see also Chapter 8 of this book), snakes (Engeman et al., 2002; see also Chapter 10 of this book), and skin cancer (Pickel et al., 2004). Scent detection dogs can be trained to locate a single scent or search for multiple scents involving identification of unrelated scents (e.g., humans, guns, and drugs) or identify related scents (e.g., live humans and cadavers).

The requirement to identify multiple scents is generally referred to as cross-training. However, the term can carry meanings other than identification of multiple scents (Allegheny Mountain Rescue Group, 2008). For example, cross-training may refer to training a dog for multiple activities such as scent detection task and competition in the obedience ring. Additionally, cross-training may indicate training a dog across multiple scent detection disciplines, for example, to both detect scent (such as drug detection) and match scent (when tracking or trailing). The lack of a clear operational definition for cross-training arises in part from the many combinations of training variables such as number of trained scents, different commands, and specific alerts utilized in scent detection training.

A closer examination of scent detection training variables shows that dog training methodologies can be subdivided into these variables: (1) number of trained scents; (2) whether distinct commands (cues) are given for each scent; and (3) whether single or multiple alerts are required in response to individual scents. These combinations of variables are representative of well-recognized learning tasks that activate specific cognitive processes. Thus the variables utilized contribute to different cognitive factors underlying these alternative training paradigms. Importantly, performance expectations for these learning tasks vary widely as a function of such underlying variables.

The goal of this chapter is to consider in depth these training paradigms and the associated behavioral and cognitive factors potentially affecting scent detection dog performance. The components of scent detection training paradigms, corresponding learning tasks, and associated cognitive factors will be presented, and circumstances under which performance may be impacted will be considered. For readers interested in an overview of general cognitive and behavioral concepts, a brief discussion of these concepts is included at the end of the chapter to extend the scope of material presented.

The information in this chapter will allow handlers to develop a greater awareness and understanding of (1) the distinct learning tasks associated with alternate training variable combinations; (2) the similarities and differences of cognitive underpinnings for each of these tasks; and (3) the effects on performance across tasks. This understanding will assist handlers in making informed training choices supporting optimum performance upon deployment.

SCENT DETECTION TRAINING: DIFFERENT TASKS

The training for scent detection dogs, regardless of types of scents used, varies widely according to the number of commands, scents, and different alerts used for each dog. As a result, training paradigms entail distinct learning tasks based on combinations of these variables. Levels of performance and reliability vary across these learning tasks. This variability generates unique cognitive demands that arise from the interaction between task demands and situational factors. In some cases, cognitive demand increases as a function of situational factors. In other cases, tasks may be executed successfully only through use of a cognitive strategy and cognitive demand increases due to these task characteristics (demands). In all cases, handlers can expect that task performance, accuracy, and speed, will decline as levels of difficulty increase (Nippak et al., 2003; Woodbury, 1943). These tasks, organized according to combinations of commands, number of scents to be detected, and number of responses required, are outlined in a flow diagram (Figure 6.1). The six tasks shown are:

1. Go/no-go
2. Go/no-go using a learning set

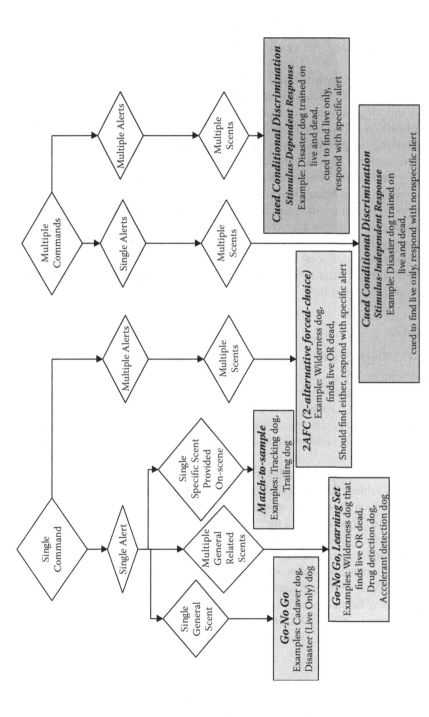

FIGURE 6.1 Task flow chart identifying learning tasks (shaded boxes) associated with combinations of commands, scents, and alerts commonly utilized in scent detection dog training.

3. Match to sample
4. Two-alternative forced choice
5. Cued conditional discrimination with stimulus-independent response
6. Cued conditional discrimination with stimulus-dependent response

Handlers can determine which task applies to their individual training situations by using Figure 6.1 to trace their requirements for commands, scents, and alerts. Although this diagram is by no means exhaustive, it serves to map six distinct training paradigms onto their respective learning tasks for dogs. This diagram may useful in assessing the probable cognitive demands placed on the animals and these demands in turn may be critical to predicting performance.

Training paradigms, associated learning tasks, and examples of field applications are described in detail below. A behavioral description of each task is provided, followed by the application of that task in scent detection training. Generally, *cue* corresponds to the command given by a handler to a dog. *Stimulus* is the target scent to be recognized by the dog. *Response* represents the trained alert given by the dog upon recognition of the target scent. Outcome, for purposes of this chapter, is assumed to be a reward, typically some form of positive reinforcement presented upon correct response such as play or food.

Go/No-Go

Go/No-Go: Single Scent

Task Description. The go/no-go task, considered a response inhibition task, requires performance of a desired response in the presence of a particular stimulus and inhibition of the desired response when alternate stimuli are present. This task essentially consists of a fixed contingency between a stimulus and the reinforcement for performing an operant behavior upon detection of that stimulus element (Sutherland and Rudy, 1989).

Detection Dog Application. The go/no-go task requires a dog to locate a single scent source (Table 6.1). This is the simplest example of single source scent detection. Upon location of the scent stimulus, the dog issues a trained alert in order to receive a reward. This represents the "go" portion of the task. If no scent is located, the dog withholds responding, thereby exhibiting the "no-go" segment. This type of task is executed by dogs trained to search for live human scent only. The go/no-go task utilizes a simple association with a fixed contingency: if the dog smells scent S, he or she responds with the trained operant response R to obtain reinforcement O. The activity may be under control of a verbal cue C, the discriminative stimulus in this case, so that $C \rightarrow S \rightarrow R \rightarrow O$. Other contextual cues may also be associated with this task, such as a specific collar worn by the dog during the search activity.

Go/No-Go: Learning Set

Task Description. This task is the same as a go/no-go task, with one difference. In the go/no-go task, a single stimulus is used. A learning set, on the other hand, consists of a group of stimuli, and the desired response is required in the presence of any

TABLE 6.1

Cues, Stimuli, and Responses in Six Learning Tasks Applied in Scent Detection Training[a]

Task	Cue	Stimulus	Response
Go/no-go	Single command	Single scent source	Single alert
Go/no-go learning set	Single command	Multiple scent sources	Single alert
Match to sample	Single command	Single specific scent source	Single alert
Two-alternative forced choice (2AFC)	Single command	Multiple scent sources	Trained alert specific to scent source
Cued conditional discrimination, stimulus-dependent response	Multiple commands specific to desired scent source and alert	Multiple scent sources	Trained alert specific to scent source
Cued conditional discrimination, stimulus-independent response	Multiple commands specific to desired scent source	Multiple scent sources	Single alert regardless of scent identified

[a] Outcome for all tasks was reward upon correct response.

one or more of the learning set components (Vauclair, 1996). In the absence of any of these stimuli, the desired response should be withheld. Learning the appropriate response does not involve creation of individual associations between the multiple stimuli and the response. Rather, a learning strategy is devised permitting consolidation of target stimuli (Vauclair, 1996).

Detection Dog Application. A group of scents is trained for recognition using a single command (Table 6.1). This group of scents requires a single, common alert upon location of any one or a combination of the target scents. When a dog is trained to recognize an assortment of scents, for example various drugs, and offers the same alert to the entire group of scents, this dog performs a go/no-go task using a learning set of scents, rather than a single scent. An alternate example is a wilderness search and rescue dog deployed to locate a lost person within a target area and trained to offer a single alert whether the person located is alive or deceased. The important components are a single command issued by the handler and a single trained alert provided by the dog upon recognition of any of the scents within the learning set. Research has demonstrated the ability of detection dogs to successfully recognize up to 10 scents in a learning set (Williams and Johnston, 2002).

There is essentially no difference between the go/no-go task with a single scent source and the go/no-go task with a learning set; the task sequence remains C→S→R→O but S represents multiple scent sources. Indeed, the single scent of a live human, for example, may be considered an agglomeration of numerous scents. Thus even a single scent source can serve as a learning set so that any go/no-go task faced by a detection dog involves a group of scents to which the dog will respond. Despite the multiple scents included in a learning set, the task involves simple association.

Cognitive Demands. The go/no-go task (single stimulus or learning set) initially appears to be a simple association task utilizing a single handler command, a single stimulus or set of stimuli, and a single alert by the dog in response to detection or discrimination of the target stimulus. However, the relative contributions of cognitive processes may increase based on training procedures and situational factors.

For example, the balance of go and no-go trials presented in training exercises can result in a prepotent (predominant) tendency to respond (Chikazoe et al., 2008) in that the number of searches conducted by dogs with scent compared with no scent present can generate a tendency to issue false alerts. If scent (go) trials are presented more frequently than no-scent (no-go) trials, a prepotent tendency to alert is established. No-scent (no-go) trials then require a component of response inhibition to overcome this prepotent tendency. Similarly, if no-scent (no-go) trials are presented more frequently, the prepotent tendency is no alert, and a component of processing infrequent stimuli is introduced when scent is present. These two cognitive processes have been shown to recruit separable neural regions (Chikazoe et al., 2008), providing evidence that multiple cognitive processes interact in the activation and inhibition of alerts. Paradoxically, if scent (go) and no-scent (no-go) trials are presented with equal frequency during training, the task essentially becomes response selection rather than response inhibition (see 2AFC discussion below). The specific effect of how the ratio of scent to no-scent trials affects performance in detection dogs has not been empirically examined. However, it has been investigated in humans performing visual go/no-go tasks. Performance in high go tasks is markedly different from performance in high no-go tasks (Helton, in press; see also Chapter 7 of this book). Whether this holds true for dogs in go/no-go scent detection tasks is an important issue to be addressed by canine ergonomists in the future.

Interestingly, in a recent study of explosive detection dogs, their tendencies to issue false alerts was not associated with trainers' ratings of their abilities (Rooney et al., 2007). It was suggested that the tendency to issue false alerts was a behavioral characteristic. However, it is probable that the balance of scent and no-scent trials experienced by the dogs would enhance any such characteristic, or indeed even be solely responsible for false alerts independent of individual characteristics.

In addition to the cognitive demands of response inhibition and processing of infrequent stimuli, context has been shown to affect performance of a go/no-go task (Gazit, Goldblatt, and Terkel, 2005). Trained explosive detection dogs repeatedly presented with no-go trials in a specific location demonstrated significantly reduced response in subsequent go trials in the same location (Gazit, Goldblatt, and Terkel, 2005; see also Chapter 8 of this book). Thus the go/no-go task, although the simplest of all scent detection tasks, involves varying cognitive demands depending at a minimum on frequency of scent presentation and context. These cognitive demands can affect the tendency of a dog to issue a false alert in the absence of scent or fail to alert when scent is present. Both failures could produce potentially devastating results.

Match to Sample

Task Description. The match to sample task involves initial presentation of one stimulus (sample stimulus) and subsequent presentation of other stimuli, one of which may be the same as the sample. A simultaneous match to sample occurs when the stimulus matching the sample is immediately available. If the matching stimulus is not immediately available, the task becomes a delayed match to sample task.

Detection Dog Application. Tracking dogs, trailing dogs, and scent identification lineup dogs perform match to sample tasks (Schoon, 1996). A dog is presented with a specific scent and required to find an individual, follow a track, or locate articles matching the specific scent (Table 6.1). Each deployment involves a single scent. Upon completion of the track or identification of a scent match, a dog may offer a single alert, like lying down upon locating the individual or article at the termination of the track, or may simply locate the desired target without a specific alert. In the event of a time lag between receipt of the initial scent cue and the dog's location of the track, item, or individual matching the cue, this task is more correctly designated a delayed match to sample.

The match to sample may be distinguished from other tasks discussed in this chapter by the requirement to match a specific provided scent compared to general scent recognition tasks (Davis, 1974; Syrotuck, 1972). Thus this task requires discrimination rather than detection or a combination of detection and discrimination. However, one command is utilized and a single alert is expected. The task sequence remains C→S→R→O and S represents the specific scent provided.

Cognitive Demands. The match to sample also appears to be a simple association task, with a single handler command, a single specific scent, and a single response upon detection of the scent. Certainly, the previous cognitive process requirements outlined for the go/no-go task apply to the match to sample task as well. In addition, because of the added requirement to match the provided scent, successful execution of the match to sample requires understanding of the concepts of "same" and "different" (Vauclair, 1996). When the initially provided scent is not readily available, the delayed match to sample task requires recruitment of working memory processes to retain a perceptual representation of the sample stimulus (Domjan, 1998).

It has not been clearly established which components of scent are utilized by dogs for matching (Taslitz, 1990). Bloodhounds tested for ability to discriminate between monozygotic (identical) twins, related persons, and unrelated persons demonstrated significantly less ability to discriminate between monozygotic twins than they demonstrated with related or unrelated individuals (Harvey et al., 2006). This suggests a genetic component of the scents utilized by the dogs in this study and also that the dogs indeed matched specific scents of individuals rather than merely tracking crushed vegetation or any human scent.

Two-Alternate Forced Choice (2AFC)

Task Description. The two-alternative forced choice (2AFC) is a response selection task involving presentation of two stimuli. Detection of either stimulus should generate a response specific to that stimulus.

Detection Dog Application. A dog performing a 2AFC task is required to detect either of two different scents and issue a specific alert depending on the scent located (Table 6.1). A wilderness search and rescue dog that issues distinct alerts depending on whether the target found is alive or deceased would execute a 2AFC task. The technique can be expanded to an *n*-alternative forced choice task by increasing the number of scents and scent-dependent alerts; *n* represents the number of scents (and corresponding specific alerts) subject to recognition by the dog. The handler issues a single command, and the dog issues an alert specific to the trained scent detected, regardless of which trained scent is identified. Thus, the task sequence would now appear as C→S1→R1→O and C→S2→R2→O where S1 represents a live find, R1 would be the alert for a live find, S2 represents a cadaver scent, and R2 would be the alert for a cadaver find. It is important to recognize that in a 2AFC exercise, the dog is to alert regardless of which trained scent is detected; however, the alert should be specific to the scent located. If the dog were required to give the same alert regardless of scent found, the exercise would become a go/no-go task with a learning set.

Cognitive Demands. The 2AFC is more difficult than the preceding tasks (go/no-go, match to sample) because it introduces an element of choice of response (response selection). The 2AF requires a dog to choose between stimulus-specific alerts, resulting in an increase in complexity that can affect scent discrimination accuracy (Abraham et al., 2004; Friedrich, 2006; Uchida and Mainen, 2003). Interestingly, this task can be solved with either a stimulus-response-driven elemental solution (separate simple association between each scent and alert) or a cognition-driven configural approach (recognition of the requirement to execute an explicit response choice). Some research suggests that once configural relationships are used to solve problems, they may then also be applied to problems that could be solved with simple associations (Alvarado and Rudy, 1992).

Dogs are typically trained in this paradigm with presentation of one target scent or the other one (successive presentation of individual odor stimuli) rather than simultaneous presentation of individual odor stimuli. In rats, successive presentation compared with simultaneous presentation negatively impacted response accuracy, presumably due to lack of an explicit stimulus-based comparison in response selection (Eichenbaum et al., 1988). It is possible that successive presentation generates elemental responding, ultimately less accurate than configural strategies applied with simultaneous stimuli presentation. An elemental solution would resemble two go/no-go problems, with the added need for relational processing, particularly if both trained scents were found simultaneously (Eichenbaum et al., 1988). In this case, the requirement to select a response in the presence of both scents may result in *response conflict*: simultaneous activation of incompatible responses (Figure 6.2; Haddon and Killcross, 2007). No current research clearly indicates which approach detection dogs use when presented with 2AFC tasks or whether varying task demands, breed of dog, or environmental factors affect this approach.

Cued Conditional Discrimination

Task Description. Responses to two (biconditional discrimination) or more (multiconditional discrimination) stimuli are trained. Stimulus-specific cues indicate whether response to a stimulus will be rewarded (Brown, Pagani, and Stanton,

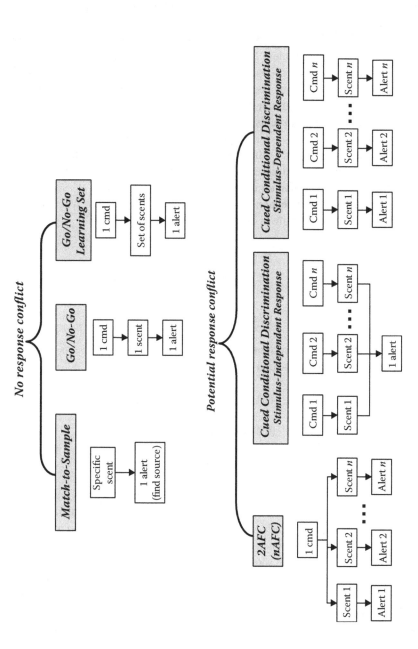

FIGURE 6.2 Learning tasks (shaded boxes) separated according to potential for response conflict arising from presence of a decision-making component.

2005). However, only responses to the cued stimulus are rewarded; responses to any uncued stimuli present should be suppressed. This task requires response selection depending on the stimulus identified and response inhibition in the presence of the uncued stimulus.

Detection Dog Application. A dog is trained to detect two (or more) different scents or groups of scents (Table 6.1). Unlike the 2AFC task described above, each scent and its corresponding response are under discriminative control of a distinct cue; that is, a different command is given to indicate the dog should respond to only the scent indicated by the command. Thus, if given verbal cue C1, a dog should search for scent 1 or group of scents 1 (S1) and ignore the presence of other trained scents. However, if given verbal cue C2, the dog should search for scent 2 or group of scents 2 (S2) and ignore other trained scents. If two scents have been trained, the technique is considered biconditional discrimination; if more than two scents have been trained, the task is multiconditional discrimination.

A **stimulus-independent response** indicates that a dog is to issue a single alert, regardless of scent target identified:

$$(C1 \rightarrow S1 \rightarrow R \rightarrow O)$$

$$(C2 \rightarrow S2 \rightarrow R \rightarrow O)$$

$$(Cn \rightarrow Sn \rightarrow R \rightarrow O)$$

Alternatively, a **stimulus-dependent response** occurs if a handler expects a specific alert depending on the handler-cued scent target:

$$(C1 \rightarrow S1 \rightarrow R1 \rightarrow O)$$

$$(C2 \rightarrow S2 \rightarrow R2 \rightarrow O)$$

$$(Cn \rightarrow Sn \rightarrow Rn \rightarrow O)$$

Regardless of whether the response is stimulus-dependent or stimulus-independent, and regardless of whether more than a single scent source is present, *the dog is required to withhold alerting to any scent source other than that indicated by the handler by the specific command issued.* This is an important difference compared with the 2AFC task in which the dog should alert regardless of which scent is identified. As with the tasks described above, contextual cues can also serve to enhance differences between cues. For example, one collar might be used when searching for scent S1, while a different collar might be used searching for scent S2.

The cued conditional discrimination task represents the most complex training task of the five presented in this chapter. The multiple behavioral and cognitive demands inherent in the nature of this task are discussed below. The added requirement of a stimulus-dependent response increases these demands compared to a stimulus-independent response.

Cognitive Demands. The cued discrimination task, in addition to requiring a dog to remember two or more alerts, requires attention to and maintenance of representations of and relationships of handler commands, scents, and alerts (Sutherland and Rudy, 1989). The complexity of this task, particularly compared to a go/no-go task, is reflected in the recruitment of numerous neural mechanisms compared with a simple association task.

The recruitment of these neural regions arises from attention and decision-making components present within the cued discrimination tasks. Importantly, the cued discrimination task requires a dog to make choices between previously rewarded responses. The dog must use contextual cues to decide between possible actions in order to receive a reward (Haddon and Killcross, 2007). *If multiple scents are present simultaneously, the dog must withhold one commonly reinforced alert to respond with the correct alert.* As with the 2AFC, this requirement to select an alert when presented with conflicting stimuli may generate response conflict (Figure 6.2; Haddon and Killcross, 2007). Performance of scent detection dogs trained to locate more than one scent with alerts based on handler's cue demonstrated negatively impacted accuracy when the animals faced response conflict (Lit and Crawford, 2006).

Odors predictive of reinforcement result in firing of a selective set of neurons (Setlow, Schoenbaum, and Gallagher, 2003). Such firing selectivity is reversed when odor outcome contingency is reversed, for example, when detection of a previously rewarded odor is no longer rewarded. Motor responses associated with odor cues are possibly encoded in the same neural region as well (Setlow, Schoenbaum, and Gallagher, 2003). Thus both firing selectivity and encoded motor response can contribute to difficulty in withholding a motor response to a previously rewarded odor cue. Indeed, an increased tendency to alert incorrectly in the absence of scent was seen in dogs trained to respond to cadaver and live scents, depending on handler cue (Lit and Crawford, 2006).

The potential for response conflict when multiple rewarded scents are present simultaneously is difficult to address through training. The more frequently that a combination of two odors is presented, the more the perceived similarity of the odors increases (Stevenson, Case, and Boakes, 2003). Subsequent presentation of the odors individually results in reduced odor distinctiveness and discriminative capability. Further, initial elemental training can interfere with later abilities to solve configural problems (Melchers et al., 2005), although these findings have not been explicitly investigated in dogs.

The cued conditional discrimination task represents a configural association requiring a cognitive strategy to solve correctly. The nature of the problem and the potential for response conflict together contribute to the difficulty of this scent discrimination paradigm. This type of problem affects performance in scent detection dogs in the presence of more than one scent (reduced accuracy) and in the absence of any trained scent (potentiated tendency for false alert; Lit and Crawford, 2006).

SUMMARY AND FUTURE AIMS

While it is important to further understand the specific biological mechanisms underlying canine olfaction (Overall and Arnold, 2007), it is equally critical to further examine effects of higher-order cognitive systems on performance of scent detection dogs. This chapter offers a first step toward this aim: the clarification of training paradigms utilized within the scent detection dog world and cognitive underpinnings of each paradigm.

As discussed in the introduction, "cross-training" is ill defined. It has been applied to dogs trained for go/no-go with a learning set task, 2AFC, and cued discrimination. The ambiguity surrounding this oft-used term emphasizes the importance of understanding the cognitive demands of the individual tasks, recognizing when the potential for response conflict exists, and acting to avoid the activation of response conflict, particularly in a deployment situation.

At certain times, dogs must recognize more than one scent, for example, a wilderness search and rescue focusing on a lost person who may be alive or deceased. The cognitive demands and associated impaired performance of the cued conditional discrimination and the possibility of reduced task accuracy with 2AF suggest that dogs required to detect multiple scents be trained instead with a go/no-go learning set. Multiple scents would then be under the discriminative control of a single command and require a uniform response. Certainly, a dog trained for 2AF or cued discrimination should not be deployed in a situation that may involve multiple trained scents present simultaneously or when consequences of resource allocation based on an incorrect alert would be potentially lethal.

Other, more general cognitive findings in dogs should be examined in the context of detection dog performance. For example, the breed-independent ability of dogs to recognize subtle, inadvertent social cues from humans is well documented (Gacsi et al., 2005; Hare and Tomasello, 1999; M. Pongracz et al., 2005; P. Pongracz et al., 2001; Topal et al., 2005). Whether this ability affects dogs trained to find scent differently from untrained dogs has not been investigated. This and other research could reveal a variety of cognitive-based factors affecting deployment decisions and success rates independently of training, breed, or handler trait characteristics. Execution of this research will require a commitment to ongoing communication and cooperation among researchers and handlers.

COGNITION AND BEHAVIOR: SOME BASICS

Cognition is a set of controlled processes encoding knowledge that are not necessarily associated with any particular behavior, including perception, attention, memory, decision making, and motivation. These controlled processes may be compared with automatic processes such as classical conditioning or instrumental/operant conditioning that link stimuli with responses (stimulus–response) and/or outcomes (response–outcome). Controlled and automatic processes interact in normal behavior (Hirsh, 1974), such that cognition modulates the strength of stimulus–response (S–R) and response–outcome (R–O) associations (Toates, 1997).

Therefore, any distinction between controlled cognitive processes and automatic S–R or R–O processes is relative, not absolute (Cohen, Dunbar, and McClelland, 1990). The relative contribution of cognition compared with S–R or R–O associations varies according to task demands. Well practiced behaviors repeated under identical conditions require minimal controlled cognitive input, while novel situations or situations involving decision making or requiring flexibility may require high-level cognitive processes to assess available options (Lewis, 1979). Indeed, it has been suggested that in more complex organisms, the provision of information contributing to decision flexibility may override the effects of conditioning trials (Kirsch et al., 2004). Further, some tasks require controlled cognitive processes for successful completion; they cannot be successfully resolved with S–R or R–O solutions; they cannot be totally automatized (Head et al., 1995).

Simple association tasks are elemental problems; that is, the relationship between a stimulus element and reinforcement for performing an operant behavior upon detection of that stimulus element is fixed (Sutherland and Rudy, 1989). Alternatively, configural association tasks are cognitive problems in which stimulus elements bear some relationship to each other, and it is necessary to understand associations of these elements to determine appropriate response and receive reinforcement (Pearce and Wilson, 1990).

Generally, cognitive tasks such as configural association problems produce similar patterns of neural activation (Duncan and Owen, 2000). These patterns differ from those involved in S–R tasks like simple associations. Importantly, neural requirements do not change for complex olfaction problems (Dusek and Eichenbaum, 1998; Larson and Sieprawska, 2002).

In dogs and other organisms, the use of cognitive strategies or S–R strategies depends on task demands, the level of experience with a task, and age (Chan et al., 2002; Nippak et al., 2003; Nippak et al., 2006). While dogs can learn to solve configural association problems, they learn simple discriminations more rapidly and reliably (Woodbury, 1943). Studies of dogs indicate that task accuracy declines with increasing levels of difficulty (Nippak et al., 2003), reflecting previous models showing increased cognitive demand associated with enhanced task difficulty (Toates, 1997; see also Chapter 7 by Helton). Although the preponderance of evidence specifically examining cognitive processes associated with olfactory function comes from human and rodent studies (Overall and Arnold, 2007), olfactory transduction and neural processing appear to be well conserved mechanisms operating across most species (Hildebrand and Shepherd, 1997). This suggests that research findings from other species are also pertinent to dogs.

ACKNOWLEDGMENTS

We acknowledge the researchers whose work has been reviewed in this chapter and colleagues for their inputs and manuscript review. The insightful reviews and helpful comments offered by Dr. Sylvia Bunge, Dr. Catherine Fassbender, and Debby Boehm are especially appreciated. The author's studies were supported by grants from ASI at California State University.

REFERENCES

Abraham, N.M., Spors, H., Carleton, A., Margrie, T.W., Kuner, T., and Schaefer, A.T. (2004). Maintaining accuracy at the expense of speed: stimulus similarity defines odor discrimination time in mice. *Neuron, 44*(5), 865–876.

Allegheny Mountain Rescue Group. (2008). AMRG website: A SAR Dog Lexicon. Retrieved April 30 from http://amrg.info/canine-sar/a-sar-dog-lexicon.html [electronic version].

Alvarado, M.C. and Rudy, J.W. (1992). Some properties of configural learning: an investigation of the transverse-patterning problem. *Journal of Experimental Psychology: Animal Behavior Processes,* 18(2), 145–153.

Brown, K.L., Pagani, J.H., and Stanton, M.E. (2005). Spatial conditional discrimination learning in developing rats. *Developmental Psychobiology,* 46(2), 97–110.

Chan, A.D. et al. (2002). Visuospatial impairments in aged canines (*Canis familiaris*): the role of cognitive behavioral flexibility. *Behavioral Neuroscience,* 116(3), 443–454.

Chikazoe, J. et al. (2008). Functional dissociation in right inferior frontal cortex during performance of go/no-go task. *Cerebral Cortex,* epub, April 28.

Cohen, J.D., Dunbar, K., and McClelland, J.L. (1990). On the control of automatic processes: a parallel distributed processing account of the Stroop effect. *Psychological Review,* 97(3), 332–361.

Davis, L.W. (1974). *Training Your Dog to Track.* New York: Howell.

Domjan, M. (1998). *The Principles of Learning and Memory,* 4th ed. Pacific Grove: Brooks Cole.

Duncan, J. and Owen, A.M. (2000). Common regions of the human frontal lobe recruited by diverse cognitive demands. *Trends in Neuroscience,* 23(10), 475–483.

Dusek, J.A. and Eichenbaum, H. (1998). The hippocampus and transverse patterning guided by olfactory cues. *Behavioral Neuroscience,* 112(4), 762–771.

Eichenbaum, H., Fagan, A., Mathews, P., and Cohen, N.J. (1988). Hippocampal system dysfunction and odor discrimination learning in rats: impairment or facilitation depending on representational demands. *Behavioral Neuroscience,* 102(3), 331.

Engeman, R.M., Vice, D.S., York, D., and Gruver, K.S. (2002). Sustained evaluation of the effectiveness of detector dogs for locating brown tree snakes in cargo outbound from Guam. *International Biodeterioration and Biodegradation,* 49(2–3), 101–106.

Friedrich, R.W. (2006). Mechanisms of odor discrimination: neurophysiological and behavioral approaches. *Trends in Neuroscience,* 29(1), 40–47.

Gacsi, M. et al. (2005). Species-specific differences and similarities in the behavior of hand-raised dog and wolf pups in social situations with humans. *Developmental Psychobiology,* 47(2), 111–122.

Gazit, I., Goldblatt, A., and Terkel, J. (2005). The role of context specificity in learning: the effects of training context on explosives detection in dogs. *Animal Cognition,* 8(3), 143–150.

Haddon, J.E. and Killcross, S. (2007). Contextual control of choice performance: behavioral, neurobiological, and neurochemical influences. *Annals of the New York Academy of Sciences,* 1104, 250–269.

Hare, B. and Tomasello, M. (1999). Domestic dogs (*Canis familiaris*) use human and conspecific social cues to locate hidden food. *Journal of Comparative Psychology,* 113(2), 173–177.

Harvey, L. M., Harvey, S. J., Hom, M., Perna, A., and Salib, J. (2006). The use of bloodhounds in determining the impact of genetics and the environment on the expression of human odor type. *Journal of Forensic Sciences,* 51(5), 1109–1114.

Head, E. et al. (1995). Spatial learning and memory as a function of age in the dog. *Behavioral Neuroscience,* 109(5), 851–858.

Helton, W.S. (2008). Impulsive responding and the sustained attention to response task. *Journal of Clinical and Experimental Neuropsychology,* in press.

Hildebrand, J.G. and Shepherd, G.M. (1997). Mechanisms of olfactory discrimination: converging evidence for common principles across phyla. *Annual Review of Neuroscience,* 20, 595–631.

Hirsh, R. (1974). The hippocampus and contextual retrieval of information from memory: a theory. *Behavioral Biology*, 12(4), 421–444.

Kirsch, I., Lynn, S.J., Vigorito, M., and Miller, R.R. (2004). The role of cognition in classical and operant conditioning. *Journal of Clinical Psychology*, 60(4), 369–392.

Larson, J. and Sieprawska, D. (2002). Automated study of simultaneous cue olfactory discrimination learning in adult mice. *Behavioral Neuroscience*, 116(4), 588–599.

Lewis, D.J. (1979). Psychobiology of active and inactive memory. *Psychological Bulletin*, 86(5), 1054–1083.

Lit, L. and Crawford, C.A. (2006). Effects of training paradigms on search dog performance. *Applied Animal Behaviour Science*, 98(3–4), 277–292.

Maejima, M. et al. (2007). Traits and genotypes may predict the successful training of drug detection dogs. *Applied Animal Behaviour Science*, 107(3-4), 287–298.

Melchers, K.G., Lachnit, H., Ungor, M., and Shanks, D.R. (2005). Prior experience can influence whether the whole is different from the sum of its parts. *Learning and Motivation*, 36(1), 20–41.

Nippak, P.M. et al. (2003). Response latency in *Canis familiaris*: mental ability or mental strategy? *Behavioral Neuroscience*, 117(5), 1066–1075.

Nippak, P.M., Ikeda-Douglas, C., and Milgram, N.W. (2006). Extensive spatial training does not negate age differences in response latency. *Brain Research*, 1070(1), 171–188.

Oesterhelweg, L. et al. (2008). Cadaver dogs: a study on detection of contaminated carpet squares. *Forensic Science International*, 174(1), 35–39.

Overall, K.L. and Arnold, S.E. (2007). Olfactory neuron biopsies in dogs: a feasibility pilot study. *Applied Animal Behaviour Science*, 105(4), 351–357.

Pearce, J.M. and Wilson, P.N. (1990). Configural associations in discrimination learning. Journal *of Experimental Psychology: Animal Behavior Processe*s, 16(3), 250–261.

Pickel, D., Manucy, G.P., Walker, D.B., Hall, S.B., and Walker, J.C. (2004). Evidence for canine olfactory detection of melanoma. *Applied Animal Behaviour Science*, 89(1–2), 107–116.

Pongracz, M., Miklosi, A., Vida, V., and Csanyi, V. (2005). The pet dog's ability for learning from a human demonstrator in a detour task is independent from the breed and age. *Applied Animal Behaviour Science*, 90(3–4), 309–323.

Pongracz, P., Miklosi, A., Kubinyi, E., Gurobi, K., Topal, J., and Csanyi, V. (2001). Social learning in dogs: the effect of a human demonstrator on the performance of dogs in a detour task. *Animal Behaviour*, 62, 1109–1117.

Rooney, N.J., Gaines, S.A., Bradshaw, J.W.S., and Penman, S. (2007). Validation of a method for assessing the ability of trainee specialist search dogs. *Applied Animal Behaviour Science*, 103(1–2), 90–104.

Schoon, G.A.A. (1996). Scent identification lineups by dogs (*Canis familiaris*): experimental design and forensic application. *Applied Animal Behaviour Science*, 49(3), 257–267.

Setlow, B., Schoenbaum, G., and Gallagher, M. (2003). Neural encoding in ventral striatum during olfactory discrimination learning. *Neuron*, 38(4), 625–636.

Stevenson, R.J., Case, T.I., and Boakes, R.A. (2003). Smelling what was there: acquired olfactory percepts are resistant to further modification. *Learning and Motivation*, 34(2), 185–202.

Sutherland, R.J. and Rudy, J.W. (1989). Configural association: role of the hippocampal formation in learning, memory, and amnesia. *Psychobiology*, 17(2), 129–144.

Syrotuck, W.G. (1972). *Scent and the Scenting Dog*. New York: Arner Publications.

Taslitz, A.E. (1990). Does the cold nose know? The unscientific myth of the dog scent iineup. *Hastings Law Journal*, 42(1), 15–134.

Toates, F. (1997). The interaction of cognitive and stimulus-response processes in the control of behaviour. *Neuroscience and Biobehavioral Reviews*, 22(1), 59–83.

Topal, J., Gacsi, M., Miklosi, A., Viranyi, Z., Kubinyi, E., and Csanyi, V. (2005). Attachment to humans: a comparative study on hand-reared wolves and differently socialized dog puppies. *Animal Behaviour,* 70, 1367–1375.

Uchida, N. and Mainen, Z.F. (2003). Speed and accuracy of olfactory discrimination in the rat. *Nature Neuroscience,* 6(11), 1224–1229.

Vauclair, J. (1996). *Animal Cognition: An Introduction to Modern Comparative Psychology.* Cambridge: Harvard University Press.

Williams, M. and Johnston, J.M. (2002). Training and maintaining the performance of dogs (*Canis familiaris*) on an increasing number of odor discriminations in a controlled setting. *Applied Animal Behaviour Science,* 78(1), 55–65.

Woodbury, C.B. (1943). The learning of stimulus patterns by dogs. *Journal of Comparative Psychology,* 35(1), 29–40.

7 Attention in Dogs
Sustained Attention in Mine Detection as Case Study

William S. Helton

CONTENTS

> A consistent and reliable sensor with the sensitivity of a well-trained dog would be of great value.
>
> **Trevelyan, 1997, p. 118**

INTRODUCTION

Detection dogs are highly accurate and flexible extensions of our senses. Despite improvements in machine intelligence and modern sensor technology, automated systems are still unable to match the operational effectiveness of trained canines in a number of activities such as explosive and narcotic detection (Fjellanger, Andersen, and McLean, 2002; Furton and Myers, 2001). There is, however, a persistent bias against using dogs in settings of consequence, like explosives detection, especially among engineers. Trevelyan (1997), who is an engineer, commented, "Dogs can smell explosive vapors, but the most sensitive detectors developed so far have orders of magnitude less sensitivity. Unfortunately, dogs are not consistent and treat their job as a game; they soon become bored."

Trevelyan provided no reference for his claim that dogs are boredom-prone, although the question of whether boredom is a problem for mine detection dogs has been raised more recently (GICHD, 2001). Dogs are not, however, alone in being criticized for inconsistent behavior; engineers have a tendency to disparage biological entities in general, including people. Typically, when industrial, medical, and transportation mishaps occur, the blame falls on human operators or human errors (Sanders and McCormick, 1993). Engineers and system designers often advocate complete automation: the total removal of an unreliable, presumably boredom-prone, biological entity from its role as a vital system component.

Fiction films and literature are populated with autonomous, nearly flawless, artificial agents. The popularity of these science fiction films and books undoubtedly leads some people to believe that foolproof technological fixes could lead to the exclusion of biological entities in modern systems if only someone were able to provide adequate funding. Unfortunately, the notion is highly mistaken, as our current engineering knowledge is rarely able to attain the required level of sophistication. Substantial funding has been allocated to research focusing on machine intelligence and pattern-recognition technologies, but they are still in their infancy. Animals, including people and dogs, are sophisticated pattern recognizers, able to recognize and react to an immense variety of items. The primary limitation on their use is in harnessing their skills in a specific context. Although that is primarily a problem of training and equipment, understanding the limits of sensory systems is also an essential feature of their use.

Animals receive vast amounts of sensory information from their surrounding environments. The total number of sensory receptors possessed by most animals, especially complex vertebrates, is truly staggering. Olfaction is only one sensory system among the many that animals possess, and each sensory system is composed of thousands to millions of receptor cells. Animals, therefore, constantly operate under conditions of sensory overload. To be useful in the control and direction of appropriate actions, the information from the sensory systems must be processed, integrated, and filtered. Information processing or computation, as any computer engineer knows, is costly. Computation requires resources in terms of both energy inputs and the information processing structures utilized. Animals work under rigorous economic constraints. On the one hand, they must process enough sensory information to make effective decisions but processing all the information would impose an unreasonable energetic and architectural burden. Additional processing power might help, but at some point the benefits of additional processing power are outweighed by resource costs. Based on these physical constraints, animals have evolved the ability to process sensory information conservatively. Psychologists and biologists studying this ability have named it *attention*, and have divided it into four categories:

- Selective attention
- Focused attention
- Divided attention
- Sustained attention

Because animals have a wide variety of sensory input channels, they require mechanisms for selecting which channel should be given processing priority: selective attention. After a channel has been selected, another mechanism must allow the animal to ignore the other channels of information: focused attention. At times, however, an animal must perform tasks simultaneously. For example, when foraging, it must also watch for predators. This constitutes divided attention. Finally, some rare or cryptic signals are critical (e.g., they indicate the presence of a valued food resource) and must be sought continuously. This is sustained attention (which can also be thought of as vigilance). Note that these four aspects of attention are functional descriptions; they are not proposed as fully independent processes in the neural architectures of animals.

Limitations of attention represent a major constraint in animal evolution and skilled behavior, but attention limitations in animals other than humans is currently understudied by behavioral scientists (Dukas, 2004). Dukas and Kamil (2000) found evidence of attention limitations in blue jays' search behavior. The birds were trained to perform an artificial prey detection task in which they indicated the presence of prey images presented on a touch screen. When the birds where given highly cryptic (low visual contrast with background) prey to search for within the center of the screen, their detection of peripheral targets decreased, as compared to when they were shown more easily detected prey in the center of the screen. Dukas (2004) makes a cogent argument that much of the search image literature about animals (see Goldblatt, Gazit, and Terkel's Chapter 8 for more information about search images) can be better understood as based on limitations of attention. Recent research on rats demonstrates a slowing of responses to signals presented in different sensory modalities, auditory and visual, in comparison to situations in which the signals are presented in only one modality (McGaughy, Turchi, and Sarter, 1994; Turchi and Sarter, 1997). These findings demonstrate the performance costs of divided attention. There is, in addition, a growing literature on sustained attention in nonhuman animals that will be discussed in more detail in the rest of this chapter.

SUSTAINED ATTENTION

Sustained attention and its relevance to the work of detection dogs is the primary focus of this chapter. Very little research has concentrated on sustained attention in working dogs. The primary source of information on sustained attention in animals comes from research on humans (Helton et al., 2005; Warm, 1993), with a growing body of research on rats (Bushnell, Benignus, and Case, 2003). Similarities in sustained attention performances of humans and rats have been identified. Based on the anatomical similarities of mammals in general, it is reasonable to predict that the results of this nondog work apply to large degree to dogs as well. At least a comparative perspective will provide a base on which to plan future research on dogs; we have no need to reinvent the wheel.

Head (1920 and 1923) first employed the *vigilance* term to refer to a state of maximum physiological and psychological readiness to react. Systematic research on vigilance did not begin, however, until Mackworth (1948 and 1950) was requested to work on a practical problem encountered by the Royal Air Force during World

War II. Pulse-position radio detection and ranging—radar—was a new technological innovation employed by airborne observers to locate surfaced Axis submarines. No doubt, airborne radar reconnaissance was a significant improvement over earlier techniques employed to detect U-boats, such as training sea gulls to flock around them (Lubow, 1977). However, the system had a serious operator problem. Regardless of their high levels of motivation and extensive training, the airborne observers began to miss the "blips" on their radar scopes indicating the presence of submarines after only 30 minutes on watch. The end result unless this situation were not corrected would have been increased losses of Allied vessels to U-boat action.

Mackworth tackled the vigilance problem experimentally by using a simulated radar display called a clock test. The display consisted of a black pointer that shifted discretely in a clockwise direction along the circumference of a white, blank-faced circle (similar to a clock face, hence the name). The regular movement of the pointer was 0.3 inches (7.5 mm), once each second. Occasionally, the pointer would double jump 0.6 inches (15 mm). The observer's task was to respond to the double jumps (critical signals) by pressing a key. In Mackworth's experiments, as in most of those that followed, observers were tested individually for prolonged and continuous periods (2 hours in Mackworth's case). The signals appeared unpredictably with a low probability of occurrence (<5%), and the observers had no control of the timing. Critical signals were clearly perceptible when observers were forewarned about them.

Using the clock test, Mackworth discovered that the abilities of observers to detect critical signals waned over time. Initial signal detection levels were high (85%) but declined by 10% after 30 minutes and continued to drop as vigils continued. The decline in performance efficiency over the period of watch is known as the vigilance decrement or decrement function. It has been replicated in many studies and is the most common effect noted in vigilance research (Helton et al., 2005; Matthews et al., 2000; Warm, 1993). The vigilance decrement has been found with experienced monitors such as those employed by Mackworth and with novices, in both operational settings and laboratories (Pigeau et al., 1995). It has also been found in studies using animal observers (rats: Bushnell, 1998; Bushnell, Benignus, and Case, 2003; McGaughy and Sarter, 1995; Robbins, 1998).

The long history of studies conducted since Mackworth's seminal investigations provide a convincing case that the central nervous system cannot sustain attention for indefinite periods. Bushnell, Benignus, and Case (2003) investigated humans and rats performing homologous vigilance tasks under varying conditions, using three levels of signal presentation (or event) rate and six levels of signal intensity. A nearly identical series of results were found for both human and rat observers. Thus, the vigilance decrement appears to generalize across mammalian species (see also Dukas and Clark, 1995). This limitation in sustained attention should play a role in determining work–rest cycles of animals in cases in which work is the active search for predators and prey and rest is a period of recovery.

Two of the presumed benefits of employing automated systems in operational settings are (1) relieving operators of the need to engage in highly demanding mental work, and (2) reducing task-induced stress. Unfortunately, the findings of many recent vigilance studies suggest that these benefits are not achieved (Warm and

Dember, 1998). Earlier researchers (Nachreiner and Hanecke, 1992) believed that vigilance duties, while boring, were fundamentally benign assignments with low demand characteristics. Recent research repudiates that view: such tasks are indeed boring (Scerbo, 1998), but they are by no means benign.

Vigilance tasks induce high levels of operator stress, measured in terms of both physiological and self-report indices. On a physiological level, elevated amounts of circulating catecholamines and corticosteroids (biochemical markers of stress; Parasuraman, 1984) have been noted during vigilance performance, along with stress-related changes in heart rate and galvanic skin response (Davies and Parasuraman, 1982). Other measures of vigilance-induced stress include increments in muscle tension, body movements, tremors, and the development of tension headaches during a vigil (Galinsky et al., 1993). The physiological findings are supported by many studies employing a variety of self-report measures in which observers described feeling less attentive and more strained, bored, irritated, sleepy, and fatigued after a vigil than before it (Galinsky et al., 1993; Helton et al., 1999, 2002, 2005, and 2008; Scerbo, 1998). These findings on stress in relation to sustained attention are troubling, given the potentially negative impact of task-related stress on productivity and worker wellbeing.

RESOURCE THEORY

Everyone has experienced the inability of fully concentrating on more than one task. When, for example, one's primary task becomes highly demanding, performance of secondary tasks typically suffers. Under normal conditions a well trained driver is usually able to drive (primary task) while simultaneously engaging in a conversation with a passenger (secondary task). When driving conditions become more difficult, for example during dense, unpredictable traffic or on slippery roads, the conversation must stop. The skill needed to perform the tasks is in limited supply; there is not enough of it to get both jobs done well. Psychologists functionally refer to this limited supply as attention or cognitive resources. Researchers have proposed that only limited resources are available for information processing (Kahneman, 1973; Matthews et al., 2000). Resource theories typically employ hydraulic or economic metaphors.

In the hydraulic metaphor, resources are considered reservoirs of mental energy dedicated to the performance of a task, analogous to a fuel that supplies energy (Hirst and Kalmar, 1987). The resources are held in a central tank and distributed to tasks using an allocation protocol controlled by an undefined central executive. The total amount of resources fluctuates, depending on the amounts used and the amounts replenished. The hydraulic metaphor is extremely useful in making inferences from resource theory because it is relatively easy to visualize (see Figure 7.1).

In the economic metaphor, resources are subject to the economics of supply and demand. This view is commonly employed in studies of divided attention in which performance of a secondary task causes performance impairment on a primary task. The secondary task presumably draws resources away from the primary task. Without the interference of the secondary task, primary task performance improves. The economic metaphor is useful in grasping that attentional resources are most likely not

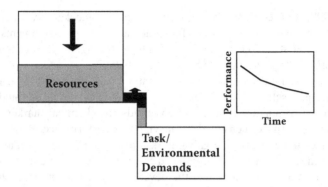

FIGURE 7.1   Hydraulic metaphor of resource theory applied to sustained attention.

singular physical entities. Like money that represents the value of resources of limited supply in our economy (physical assets, knowledge, labor), attentional resources represent items of limited supply in the control systems of animals (glucose, oxygen, neurotransmitters, neurons, neuronal groups). Neither metaphor fully elucidates the underlying complexities of biological systems but they do serve as relatively simple explanatory devices for studying attention.

The limited resource perspective is used frequently to explain the findings in vigilance tasks (Davies and Parasuraman, 1982; Dukas and Clark, 1995; Helton et al., 2002, 2004, and 2005; Helton and Warm, in press; Temple et al., 2000; Warm, 1993). During vigilance tasks, observers must make active, continuous signal–noise discriminations under conditions of uncertainty, without rest. The continuous nature of the vigilance activity does not allow replenishment of resources. Hence, using the hydraulic metaphor, the resource or energy pool depletes over time and the depletion is reflected as a decline in performance efficiency (see Figure 7.1). Demands upon the limited pool of resources increase as the difficulty or arduousness of a task increases.

The usefulness of resource theory in understanding the vigilance decrement and vigilance tasks in general is revealed when it is compared to the commonly held alternative explanation, the one proposed by Trevelyan (1997) related to the performance of dogs: boredom. Vigilance tasks are typically dull and repetitive. In the case of mine detection, an animal is given the task of constantly sniffing the environment. With each sniff the dog makes a perceptual discrimination: is the sought-after odor present or not? The task is, presumably, unexciting. The dog, seeking more stimulation, may distract itself with more interesting activities and stop performing its odor discrimination assignment. Boredom has been proposed as an explanation of the vigilance decrement in humans as well. This hypothesis fails miserably, however, when compared to resource theory in explaining the actual findings of vigilance research (Sanders and McCormick, 1993; Warm, 1993):

- **Signal Salience Effect** — When signals are made more difficult to detect by lowering the signals' contrast with their background environments, performance suffers (see Figure 7.2).

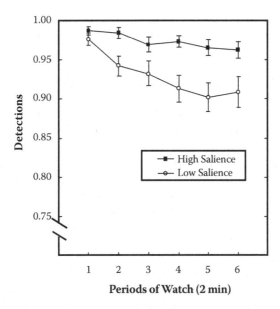

FIGURE 7.2 Signal salience effect in sustained attention. The proportion of correct detections declines in both high (easy) and low (difficult) salience vigilance tasks over the periods of watch. The decrement, however, is more prominent in the low salience condition, as typical in sustained attention studies that manipulate signal salience. (Adapted from Helton et al., 2002.)

- **Event Rate Effect** — The more quickly the signals are presented, the more performance suffers.
- **Spatial Uncertainty Effect** — The more uncertain the observer is of the spatial locations of signals, the more performance suffers.

The boredom hypothesis would predict exactly the opposite of these findings. Making a task more challenging by making a signal harder to discriminate (lowering signal salience or intensity), by making discriminations quicker (increasing event rate), or by adding a hide-and-seek element (spatial uncertainty) should make boredom less of an issue; performance in more challenging scenarios should be improved. From a resource theory perspective, however, any increase in the demands of the task will hurt performance; make the task more challenging and performance will suffer. This is exactly what happens with humans and rats. Although many details remain to be worked out, resource theory is currently the dominant explanatory model of sustained attention. Most likely it will provide useful explanatory power for understanding questions about attention in dogs as well.

STRESS

When placed in a situation of constant resource demand, animals react with stress responses (see Chapters 13 and 14 for more details on stress responses in dogs).

The term *stress* is often used loosely in the psychological literature. Traditionally, applied researchers interested in stress focused exclusively on environmental determinants of stress or on the resulting physiological reactions (Matthews et al., 2000). Engineering psychologists devoted considerable effort to manipulating situational variables labeled as stressors, e.g., cold, confinement, crowding, heat, noise, electric shock, sleep deprivation, time pressure, vibration (Hammond, 2000).

Physiological researchers assessed stress responses, defining stress as a set of body symptoms (Cannon, 1915; Selye, 1956). Although the research on stressors and their concomitant physiological responses was informative, certain troubling issues became apparent. Certain *a priori* stressors actually proved beneficial for task performance and were subjectively pleasant in certain situations (Poulton, 1976). Further, basic physiological indicators are not informative at the cognitive level (Matthews et al., 2000), and physiological and psychological responses often do not correlate highly among or between each other. In addition, animals respond to perceived threats or demands by employing a multitude of response strategies or coping mechanisms that attenuate stress symptoms (Wechsler, 1995).

In recognition of its complex nature, researchers have introduced a transactional approach to account for stress (Hammond, 2000; Lazarus and Folkman, 1984). From this perspective, psychological distress is the result of an individual's appraisals of an environment as taxing or exceeding the individual's resources and/or endangering its safety (Lazarus and Folkman, 1984). The appraisals of threatening or demanding environments are multidimensional, reflecting a wide array of emotional (affect), motivational (conation), and cognitive processes (Matthews, 2001). These appraisals do not have to be made at a conscious level.

As described by Matthews (2001), these assessments are separable into primary and secondary appraisals. Primary appraisals are assessments of overall relevance or significance of a confronted situation to an individual's goals (for example with dogs: eating, drinking, social interactions, and avoiding pain). The animal may make a primary appraisal that the task it is requested to do no longer fulfills its goals, in which case it would presumably stop performing. This is a motivational issue. Secondary appraisals are evaluations of coping capability. The animal appraises whether it is able to handle a task or not. This is not motivational; it is a resource supply-and-demand issue. In vigilance tasks, animals assess potential incongruities between environmental task demands and reservoirs of attention resources. If an animal recognizes an incongruity between task demand and available resources (a secondary appraisal) and comes up short, it will experience distress and become uncomfortable. With humans, unpredictable task demands further increase operator distress (Helton et al., 2004 and 2008). Presumably this is because the operator cannot make accurate predictions about resource demands, and this proves unnerving.

In a mine detection scenario, the transactional model may be useful in diagnosing performance problems. If a dog appears to stop performing, the cause may be a primary appraisal or a secondary appraisal. If the dog has made a primary appraisal that the task is no longer relevant to its goals, then the rectification is making the link between performance of the task and the dog's goals clearer to the dog. This is typically done with the methods of operant conditioning. On the other hand, a dog may appear to stop performing because of a secondary appraisal. The dog wants to perform the task, but feels ill equipped to get the job done; the assignment is too demanding or too unpredictable. This is not a motivational issue. The task needs to be made easier, more predictable, or the dog must given time to replenish its resources. How do you tell the difference? I suspect many handlers intuitively know this when a dog shows signs of distress. A distressed dog, unlike one that prances away from the task, most likely stops work because it perceives the task situation is too difficult and is trying to cope as best it can.

COPING

As previously noted, a key element of the transactional model is the focus that it places on the manner in which operators actively regulate their responses to stressful situations via the initiation of various styles of coping (Endler and Parker, 1990; Lazarus and Folkman, 1984; Matthews et al., 2000). Matthews and Campbell (1998) described three fundamental dimensions of coping in the human performance context.

- *Problem-focused coping* involves active attempts to formulate and execute a plan of action that will deal directly with a situation.
- *Emotion-focused coping* attempts to deal with a stressor by changing one's feelings or thoughts about it, for example by positive thinking or self criticism.
- *Avoidance coping* refers to attempts to deal with a stressful situation by specifically directing attention away from the problem, for example via distraction.

These coping styles can be interpreted as information processing strategies that influence the achievement of task goals (Kanfer and Ackerman, 1996; Matthews and Wells, 1996). Hancock and Warm (1989) proposed further that tasks alone often represent sources of stress, especially when they are appraised as taxing an operator's capacities. Hence, both performance efficiency and personal wellbeing may depend on the coping strategy adopted by operators in response to task demands (Matthews and Campbell, 1998). Problem-focused coping is adaptive in that it aids in the achievement of task goals in performance settings (Matthews and Wells, 1996). Dealing with emotional information hinders the achievement of task goals (Helton et al., 1999). Matthews and Campbell (1998) demonstrated that avoidance coping leads to poor performance of rapid information processing tasks and simulated driving tasks and is therefore potentially maladaptive.

One concern for working dog researchers is the role that task and environmental conditions play in eliciting adaptive or maladaptive coping. For example, in studies of humans, driving fatigue decreases the effectiveness of problem-focused coping (Desmond and Matthews, 1997), and failure experiences induced by making the task impossible during driving amplify emotion-focused coping (Matthews, Joyner, and Newman, 1998). Conversely, high information processing demands may actually increase the effectiveness of problem-focused coping (Desmond and Matthews, 1997). In a working dog scenario, these issues are also of concern. Fatigued dogs may be less able to deal with tasks. Their fatigue may be overcome by allowing rest or even by providing the dogs with performance enhancing drugs: psychostimulants. Many of us counter fatigue by self medicating with caffeinated beverages. In rats, for example, psychostimulants including nicotine, amphetamine, and caffeine reduced the vigilance decrement and improved sustained attention (Grottick and Higgins, 2002). This coping perspective may also help in diagnosing the underlying causes of performance failures. Performance problems attributed to a dog's boredom, for example, may actually be due to the dog's emotion-focused coping arising from feelings of task-induced distress.

A MODEL FOR WORKING DOGS

Decrements in performance over time combined with high levels of operator stress are unfortunate realities confronting biological systems engaged in vigilance or sustained attention tasks. Nevertheless, vigilance is and will remain a key element of such systems. Animals like dogs, rats, and humans will continue to be required to handle tasks that they perform more efficiently than machines, such as complex pattern recognition (Lund, 2001).

Sustaining attention during complex pattern recognition search tasks is relevant for many operations employing well trained detector dogs. Working canines are often assigned the task of continuously monitoring the environment for relatively rare critical signals. In naturalistic terms, this is analogous to searching for cryptic and rare prey. In the course of a land mine clearing task, an explosive detection dog must continually sample the air and judge whether the molecular signature of an explosive substance is present. This is akin to a human baggage inspector's task of continuously sampling visual scenes and judging whether a contraband item is present.

Resource theory and the transactional model of stress have helped us understand human vigilance performance. Individuals working with detector dogs should not be dissuaded by the cognitive natures of the two theories. The recent growth of research in animal cognition highlights the similarities of humans and animals in regard to complex information processing. Dogs are social animals. Their predatory success in natural environments is largely due to their ability to hunt and live cooperatively. As Dennett (1996) argues, dogs may be unique among animals in sharing human culture with humans; both species obey human cultural rules and ritual behaviors. Recent research indicates dogs are better at reading human signals, such as orientation of gaze and pointing, than great apes. Likewise, dogs communicate this directional information exceedingly well to humans (Agnetta, Hare, and Tomasello, 2000; Miklosi et al., 1998 and 2000). Dogs are clearly cognitive creatures.

Combining resource theory and the transactional model of stress leads to the performance model depicted in Figure 7.3. As shown in the figure, resources can be considered within a hydraulic metaphor. The resources are undisclosed pools of information processing energy or capacity that may be directed toward tasks or off-task activities. A task is determined by a trainer or employer, just as it would be with a human operator. After a dog makes a primary appraisal of the task and determines that it fits its goals, it will make secondary appraisals regarding the match between its available resources and the demands of the task. The outcome of the primary appraisal implicates motivation. The trainer must convince the dog that the task is

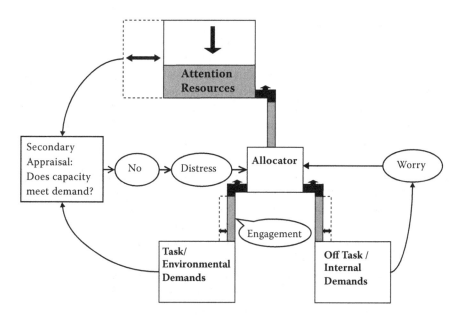

FIGURE 7.3   Model of operator performance in a sustained attention task.

important. This process of convincing the dog falls in the realm of traditional learning theory and operant conditioning. If the dog is motivated to perform the task, but feels it does not have enough resources to do the task, it will experience distress. The recognition of distress will feed back on the resource allocator. The more task demands placed on resources, the more likely the animal will feel distressed. The animal can alter the amount of resources applied to the task within limits. However, at a certain point, regardless of motivation, the resource pool will be depleted by the task demands. This depletion is the primary factor controlling the vigilance decrement. Although motivation plays an important role in resource allocation and thus can alter the progression of the decrement, it does not eliminate the reality of resource limitations.

Additional factors must be considered in the model. Dogs, for example, share many personality traits in common with humans (Gosling and John, 1999). Matzel and his colleagues (2003) have provided evidence for a general learning ability, similar to the human intelligence factor, in mice. Although it has yet to be demonstrated, presumably dogs share this factor as well. Canine basic cognitive abilities are similar enough to those of humans that they have become useful models for studying the effects of aging on cognition; dogs, like humans, vary in these abilities (Adams et al., 2000). In the model displayed in Figure 7.2, these individual differences in cognitive abilities and temperament are indicated by the exapandable areas (noted with dotted lines) in the resource reservoir and lines. Differences in basic cognitive abilities may alter the size of the resource reservoir, with more intelligent animals having more capacity, possibly due to their ability to process information at a faster rate. Other differences may arise from temperament and personality traits. These trait differences may be analogous to differences in the relative circumferences of the lines leading to task and off-task activities. Dogs with high work–play drives may in some sense have larger "fuel lines" to tasks. This would allow more resources to flow to a task with less effort. Dogs with distractible dispositions may have restricted lines to a task or a bigger line toward off-task activities. Extensive research has demonstrated the role of these traits in human sustained attention, and there is no reason to believe dogs differ significantly (Helton et al., 1999; Matthews et al., 2001).

A closer examination of sustained attention of dogs due to individual differences in genes, early life experiences, and training, may, as discussed in Chapter 1, provide a win–win benefit for those employing working dogs and those seeking an understanding of sustained attention in people. Dogs may prove useful in this regard as they are subject to genetic control and their early life experiences can be manipulated (Schmutz and Schmutz, 1998; Slabbert and Rasa, 1997). Recent research, furthermore, indicates dogs differ in attention abilities and some dogs suffer from a condition analogous to attention deficit hyperactivity disorder (Vas et al., in press). Dogs may prove useful in studying attention disorders.

Despite the limited research focusing on vigilance performance in dogs, animal behaviorists studying canine workers are aware of potential limitations in the dogs' capacity to sustain effort over time. They, however, often tend to interpret decrements in performance efficiency as motivational problems (e.g., boredom), not due to fundamental limitations of attentional resources (Garner et al., 2000). It is common for organizations employing canines to limit their searches to 30 minutes. In a study funded by the U.S. Federal Aviation Administration, Garner and colleagues (2000) investigated the ability of scent detection dogs to work for extended search periods (more than 2 hours). The dogs did not seem to show significant declines in performance over time. This, however, may be because the detection task was very easy; the authors noted that they opted for an easily detected scent.

Performance efficiency of humans has been demonstrated to vary directly with the psychophysical strength or ease of detection of a stimulus (Bushnell, Benignus and Case, 2003; Temple et al., 2000). Garner and colleagues interpreted their findings to imply that scent detection dogs are capable of sustaining their performance over long periods without costs, but in light of the human and rat vigilance research, this interpretation is most likely mistaken. Unless a signal is easy to detect—probably not the case under real working conditions—dogs, like humans and rats, should show performance decrements over time. Garner et al.'s findings indicate that the dogs' performance is not limited by their boredom. If that were the case, any easy task would presumably cause dogs to lose interest relatively quickly.

Most likely, although only future research will fully determine this, the reason most organizations limit searches to fewer than 30 min is because they require the dogs to search for hard-to-detect signals. The dogs, regardless of motivation, are unable to avoid the reality of depleted resources. They need to rest and replenish frequently. Garner et al.'s (2000) research, although only one study, convincingly rules out the boredom hypothesis. Apparently, well trained dogs rarely become bored, or even if they do, they do not let the boredom interfere with their jobs.

REDUCING ATTENTION DEMANDS: REST SETTING

New advances in the dog detection arena, like the remote explosive scent tracing (REST) concept, fit well with the attention model in Figure 7.3. In the REST scenario, air and dust samples are collected in filters from areas potentially containing land mines or unexploded ordinance. These cartridges are then brought back to a testing facility. REST dogs are trained to sniff the filters and signal when they detect an explosive scent.

REST dogs work in a relatively controlled environment. A REST analysis center can be designed to laboratory standards, minimizing distractions, allowing optimized presentation of odor (e.g., by heating the filter), and allowing the dog to concentrate on the task at hand. In contrast, field dogs must work under prevailing environmental conditions, potentially involving large thermal fluctuations, differing levels of noise, and many other distractions. The terrain may be uneven, requiring the dog to expend mental effort on navigating to avoid falling into holes or stepping on sharp objects. The potentially demanding nature of this navigation task should not be underestimated. A handler in the field may also become impatient and push a dog to work more quickly than is comfortable.

If the boredom model of canine performance decrement is correct, then the REST concept may be disastrous. In the relatively unstimulating environment of a REST procedure, a dog would be highly susceptible to boredom. The more stimulating environment of field work would make the animal less susceptible to boredom. The field gives a dog plenty of places to search (spatial uncertainty) and because the environment is potentially dangerous, a dog may be rushed by a trainer (high event rate). This should make the exercise very stimulating for the dog. On the other hand, these are exactly the conditions that place high resource demands on a dog. If the resource model is correct, then the REST procedure is a major advance in canine ergonomics. It reduces environmental complexity, spatial uncertainty, and possibly event rate; the dog is not being pushed by a trainer. A future experiment would entail comparing dog performance in a REST environment to performance in a field environment with comparable, equally salient, signals. My expectation is that this study would demonstrate that resource theory trumps boredom as an explanation of performance decrements in detection dogs; the REST procedure would improve performance relative to the field condition.

One problem with the current REST technique is safely collecting samples from the field. An environmental sampler must be able to extract samples from varying and difficult terrain and operate safely in mined areas. At Michigan Tech, I have been developing an aerial sampler: the aerial robotic remote scent tracer (ARREST; Helton et al., 2007). I chose a slow moving, lighter-than-air platform, a remote controlled blimp, because it was relatively cheap and easy to maintain and operate. Other research groups have worked on similar blimp samplers (Gosschnick and Harms, 2002). This system will not be discussed at length here, but is instructive to note that the collaboration of canine ergonomics, human ergonomics, and engineering can lead to affordable and safe technologies. It provides a case study in interdisciplinary scientific collaboration centered on working dogs. The future will undoubtedly see increases in augmentative technology used to assist the work of dogs (see Chapter 11 by Ferworn for more examples).

MONITORING ATTENTION RESOURCES: REAL-TIME BRAIN IMAGING

Working dogs and their human handlers will increasingly be placed in complex environments mediated by enhanced technology—conditions such as the REST scenario described in the previous section. Clearly, animals including dogs have information processing limitations. Augmented cognition is a recently developed field of research seeking to extend the information processing capacities of operators by using real-time assessments of cognitive states and employing these assessments as inputs in technological systems (Schmorrow and Kruse, 2004). An operator's cognitive state can be assessed via a variety of sensor technologies detecting behavioral, psychophysiological, and neurophysiological data acquired in real-time. This data can then be used to adapt or augment the technological interface to significantly improve operator performance. Currently the operator is a person, but there is no reason to limit this technological approach to people. Augmented cognition can be applied to dogs as well.

Augmented cognition techniques may enable handlers to predict or detect when dogs will make errors and mistakes. This would lead to a dramatic reduction in detection failures. This approach could be used to devise new predictive models of dog performance and lead to a better understanding of when dogs are likely to make errors. Training and operating procedures could thereby be adapted and better fit to dogs.

Additionally, if augmented cognition technology is affordable, robust, comfortable, reliable, and nonintrusive, it could be deployed in field settings. This would provide real-time feedback to handlers, the technological system, and possibly the dogs as well. The feedback could be used to assist the dogs and their handlers in real-time during actual field conditions.

Cognitive and attention resources can be conceptualized as undisclosed pools of information processing capacity that can be directed toward tasks or off-task activities (see Figures 7.1 and 7.3). A dog will make appraisals regarding the match between available cognitive resources and the demands of a current task. If the dog is motivated to perform the task but feels it lacks sufficient resources to do so, it will become distressed. The more task demands are placed on cognitive and attention resources, the more likely a dog will feel distressed. The dog can alter the amount of resources applied to the task within limits, but at some point, regardless of initial motivation, the cognitive demands of the task will overwhelm the dog. This cognitive and attention demand overload is detectable by changes in physiology and behavior. Sensors could be used to detect this overload.

Peripheral physiological markers can be used to gauge the amount of resources expended, for example, heart rate (see Chapter 13 by Schneider and Slotta-Bachmayr). In addition, several studies in humans have shown a close relation between mental activity and cerebral blood flow velocity as measured by transcranial Doppler sonography. Cerebral blood flow velocity (CBFv) is more rapid when observers engage in a wide variety of information processing activities compared with rest or baseline periods (Stroobant and Vingerhoets, 2000; Tripp and Warm, 2007). Recent signal detection studies (Hitchcock et al., 2003; Schnittger et al., 1997; Warm and Parasuraman, 2007) indicate that the temporal decline (vigilance decrement) of signal detection that typifies vigilance performance is accompanied by a parallel decline in CBFv. They also indicate that the absolute level of CBFv in vigilance tasks is positively related to the psychophysical and cognitive demands placed upon observers and that these effects are lateralized to the right cerebral hemisphere (Helton et al., 2007; Hitchcock et al., 2003; Warm and Parasuraman, 2007), a result consistent with PET and fMRI studies indicating right hemisphere functional control of vigilance performance in humans (Parasuraman, Warm, and See, 1998). CBFv can be measured in dogs as well. Near-infrared spectroscopy technology has also been used to study dogs (see Chapter 11 for more details).

An alternative to transcranial Doppler sonography and near-infrared spectroscopy is the measurement of cerebral temperature using near-infrared tympanic membrane temperature. Temperature measurement at the tympanic membrane is a reliable measure of core body temperature (Schuman et al., 1999) because of its close proximity to the blood supplied directly to the brain. Additionally, tympanic membrane temperature can be used to infer the flow of blood supplying the nearest cerebral hemisphere (Cherbuin and Brinkman, 2004 and 2007; Helton et al., in press; Meiners and Dabbs, 1977; Swift, 1991). Differences in temperatures taken from the right and left ears reflect underlying differences in hemispheric activation. Increased blood flow to the nearest hemisphere increases the radiation of heat and this is reflected in a noticeable temperature cooling.

Since many psychological components are cerebrally lateralized, these changes in hemispheric activation may be used like transcranial Doppler sonography and other brain imaging devices to warn a handler when the dog is susceptible to errors. They may also be useful as inputs in the control of automation and as decision aids. Near-infrared tympanic temperature presents a number of advantages over transcranial Doppler sonography and other potential brain imaging devices, such as near-infrared spectroscopy. Commercial veterinary units such as Advanced Monitors' Vet-Temp® and Pet-Temp® are very affordable in comparison to these other technologies (the cost magnitude difference is about 1,000×). Furthermore, the measurement technique is qualitatively simpler as the ear canal serves as a physical guide for placing the sensor. Both transcranial Doppler sonography and near-infrared spectroscopy require skilled technicians to ensure that the measurements are taken in the right position. Finally, the ear canal also serves as a natural holding device that would prevent sensor movement during operations. One limitation of all these devices, however, is the need to make them comfortable for dogs to wear for prolonged periods (see Chapter 11).

These physiological metrics, with further development, may provide objective measures of task difficulty and be useful for assessing training goals. Detection tasks should become easier for a dog with experience and practice. This increasing ease of detection should be noticeable physiologically and serve as a marker of skill development. The possibility of real-time monitoring of dog physiology should not be overlooked as a potential assistive device for training handlers to recognize behavioral cues regarding cognitive states. New techniques using objective posture and movement analysis may also prove useful. Augmented cognition applied to dogs may lead to development of better team communications between dog and handlers. This may be critical in cultural settings in which handlers may have had few close interactions with dogs and limited understanding of dog behavior and cognition.

CONCLUSION

We can respond to technicians, like Trevelyan (1997), who argue, "... dogs are not consistent and treat their job as a game; they soon become bored," that well-trained dogs do not get bored easily or are professional enough to not let boredom get in the way of their work. Dogs have limitations, but with further research these limitations can be made predictable. Additionally, we can develop settings that facilitate their capacities and reduce their limitations, such as work in a REST setting. There is a consistent and reliable sensor with the sensitivity of a well trained dog: the well trained dog. Despite their limitations, dogs are capable of performing operations that dwarf the capacities of known artificial systems. Instead of trying to eliminate the dog from its role in detection work and replace it with engineering fantasies, a better approach may be to facilitate the existing abilities of dogs. Trevelyan (1997) admits that most designs for mine-detecting robots are hopelessly flawed and cost-inefficient. Perhaps, dogs need to be recognized as the experts they have proven to be (Helton, 2004 and 2005). Systematic research should be directed at making their jobs easier and safer.

REFERENCES

Adams, B., Chan, A., Callahan, H., and Milgram, N.W. (2000). The canine as a model of human cognitive aging: recent developments. *Progress in Neuro-Psychopharmacology and Biological Psychiatry, 24,* 675–692.

Agnetta, B., Hare, B., and Tomasello, M. (2000). Cues to food location that domestic dogs (*Canis familiaris*) of different ages do and do not use. *Animal Cognition, 3,* 107–112.

Bushnell, P.J., (1998). Behavioral approaches to the assessment of attention in animals. *Psychopharmacology, 138,* 231–259.

Bushnell, P.J., Benignus, V.A., and Case, M.W. (2003). Signal detection behavior in humans and rats: a comparison with matched tasks. *Behavioral Processes, 64,* 121–129.

Cannon, W.B. (1915). *Bodily Changes in Pain, Hunger, Fear, and Rage.* New York: Appleton.

Cherbuin, N. and Brinkman, C. (2004). Cognition is cool: can hemispheric activation be assessed by tympanic membrane thermometry? *Brain and Cognition, 54,* 228–231.

Cherbuin, N. and Brinkman, C. (2007). Sensitivity of functional tympanic membrane thermometry (fTMT) as an index of hemispheric activation in cognition. *Laterality, 12,* 239–261.

Davies, D.R. and Parasuraman, R. (1982). *The Psychology of Vigilance.* London: Academic Press.

Dennett, D.C. (1996). *Kinds of Minds.* New York: Basic Books.

Desmond, P.A. and Matthews, G. (1997). Implications of task-induced fatigue effects for in-vehicle countermeasures to driver fatigue. *Accident Analysis and Prevention, 29,* 513–523.

Dukas, R. (2004). Causes and consequences of limited attention. *Brain, Behavior, and Evolution, 63,* 197–210.

Dukas, R. and Clark, C.W. (1995). Sustained vigilance and animal performance. *Animal Behaviour, 49,* 1259–1267.

Dukas, R. and Kamil, A.C. (2001). The cost of limited attention in blue jays. *Behavioral Ecology, 11,* 502–506.

Endler, N. and Parker, J. (1990) Multi-dimensional assessment of coping: a critical review. *Journal of Personality and Social Psychology, 58,* 844–854.

Fjellanger, R., Andersen, E.K., and McLean, I. (2000). A training program for filter-search mine detection dogs. *International Journal of Comparative Psychology, 15,* 277–286.

Furton, K.G. and Myers, L.J. (2001). The scientific foundation and efficacy of use of canines as chemical detectors of explosives. *Talanta, 54,* 487–500.

Galinsky, T.L., Rosa, R.R., Warm, J.S., and Dember, W.N. (1993). Psychophysical determinants of stress in sustained attention. *Human Factors, 35,* 603–614.

Garner, K.J., Busbee, L., Cornwell, P., Edmonds, J., Mullins, K., Rader, K., Johnston, J.M., and Williams, J.M. (2000). Duty cycle of the detection dog: a baseline study. Final Report. Federal Aviation Administration, Washington, D.C.

GICHD (2001). *Designer Dogs: Improving the Quality of Mine Detection Dogs.* Geneva: GICHD.

Gosling, S.D, and John, O.J. (1999). Personality dimensions in nonhuman animals: a cross-species review. *Current Directions in Psychological Science, 8,* 69–75.

Gosschnick, J. and Harms, M. (2002). Land mine detection with an electronic nose mounted on an airship. *Proceedings of NATO Advanced Research Workshop on Detection of Explosives and Landmines, Physics and Chemistry, 66,* 83–91.

Grottick, A.J. and Higgins, G.A. (2002). Assessing a vigilance decrement in aged rats: effects of pre-feeding, task manipulation, and psychostimulants. *Psychopharmacology, 164,* 33–41.

Hammond, K.R. (2000). *Judgements under Stress.* New York: Oxford University Press.

Hancock, P.A. and Warm, J.S. (1989). A dynamic model of stress and sustained attention. *Human Factors, 31,* 519–537.

Head, H. (1920). *Studies in Neurology.* Oxford: Oxford University Press.

Head, H. (1923). The conception of nervous and mental energy II. Vigilance: a physiological state of the nervous system. *British Journal of Psychology, 14,* 126–147.

Helton, W.S. (2004). The development of expertise: animal models? *Journal of General Psychology, 131,* 86–96.

Helton, W.S. (2005). Animal expertise: conscious or not? *Animal Cognition, 8,* 67–74.

Helton, W.S., Begoske, S., Pastel, R., and Tan, J. (2007). A case study in canine–human factors: a remote scent sampler for landmine detection. *Proceedings of the Human Factors and Ergonomics Society, 51,* 582–586.

Helton, W.S., Dember, W.N., Warm, J.S., and Matthews, G. (1999). Optimism, pessimism, and false failure feedback: effects on vigilance performance. *Current Psychology, 18,* 311–325.

Helton, W.S., Hollander, T.D., Tripp, L.D., Parsons, K., Warm, J.S., Matthews, G., and Dember, W.N. (2007). Cerebral hemodynamics and vigilance performance. *Journal of Clinical and Experimental Neuropsychology, 29,* 545–552.

Helton, W.S., Hollander, T.D., Warm, J.S., Matthews, G., Dember, W.N., Wallart, M., Beauchamp, G., Parasuraman, R., and Hancock, P.A. (2005). Signal regularity and the mindlessness model of vigilance. *British Journal of Psychology, 96,* 249–261.

Helton, W.S., Kern, R.P., and Walker, D.R. (2008). Tympanic membrane temperature, exposure to emotional stimuli and the sustained attention to response task. *Journal of Clinical and Experimental Neuropsychology*, in press.

Helton, W.S., Shaw, T.H., Warm, J.S., Matthews, G., Dember, W.N., and Hancock, P.A. (2004). Workload transitions: effects on vigilance performance and stress. In Vincenzi, D.A. et al., Eds. *Human Performance, Situation Awareness and Automation: Current Research and Trends.* Mahwah, NJ: Lawrence A. Erlbaum, pp. 258–262.

Helton, W.S., Shaw, T., Warm, J.S., Matthews, G., and Hancock, P.A. (2008). Effects of warned and unwarned demand transitions on vigilance performance and stress. *Anxiety, Stress and Coping, 21,* 173–184.

Helton, W.S. and Warm, J.S. (2008). Signal salience and the mindlessness theory of vigilance. *Acta Psychologica,* in press.

Helton, W.S., Warm, J.S., Mathews, G., Corcoran, K., and Dember, W.N. (2002). Further tests of the abbreviated vigil: effects of signal salience and noise on performance and stress. *Proceedings of the Human Factors and Ergonomics Society, 46,* 1546–1550.

Hirst, W. and Kalmar, D. (1987). Characterizing attentional resources. *Journal of Experimental Psychology: General, 116,* 68–81.

Hitchcock, E.M., Warm, J.S., Matthews, G., Dember, W.N., Shear, P.K., Tripp, L.D., Mayleben, D.W., and Parasuraman, R. (2003). Automation cueing modulates cerebral blood flow and vigilance in a simulated air traffic control task. *Theoretical Issues in Ergonomic Science, 4,* 89–112.

Kahneman, D. (1973). *Attention and Effort.* Englewood, NJ: Prentice Hall.

Kanfer, R. and Ackerman, P.L., (1996). A self-regulatory skills perspective to reducing cognitive interference. In Sarson, I.G. et al., Eds. *Cognitive Interference: Theories, Methods, and Findings.* Mahway, NJ: Lawrence A. Erlbaum, pp. 153–174.

Lazarus, R.S. and Folkman, S. (1984). *Stress, Appraisal, and Coping.* New York: Springer.

Lubow, R.E. (1977). *The War Animals.* New York: Doubleday.

Lund, N. (2001). *Attention and Pattern Recognition.* East Sussex: Routledge.

Mackworth, N.H. (1948). The breakdown of vigilance during prolonged visual search. *Quarterly Journal of Experimental Psychology, 1,* 6–21.

Mackworth, N.H. (1950). Researches on the measurement of human performance. Medical Research Council Special Report 2680. London: H.M.S.O. (Reprinted in *Selected Papers in the Design and Use of Control Systems,* Sinaiko, H.W., Ed. 1961, New York: Dover).

Matthews, G. (2001). Levels of transaction: a cognitive science framework for operator stress In Hancock, P.A. and Desmond, P.A., Eds. *Stress, Workload and Fatigue.* Mahwah, NJ: Lawrence A. Erlbaum, pp. 5–33.

Matthews, G. and Campbell, S.E. (1998). Task-induced stress and individual differences in coping. *Proceedings of the Human Factors and Ergonomics Society, 42,* 821–825.

Matthews, G., Davies, D.R., Westerman, S.J., and Stammers, R.B. (2000). *Human Performance: Cognition, Stress and Individual Differences.* East Sussex: Psychology Press.

Matthews, G., Joyner, L.A., and Newman, R. (1998). Stress, age, and hazard perception in simulated driving. Paper presented at International Conference on Applied Psychology, San Francisco.

Matthews, G. and Wells, A. (1996). Attentional processes, coping strategies and clinical intervention. In Zeidner, M. and Endler, N.S., Eds. *Handbook of Coping: Theory, Research, Applications.* New York: John Wileyt & Sons, pp. 573–601.

Matzel, L.D., Han, Y.R., Grossman, H., Karnik, M.S., Patel, D., Scott, N., Specht, S.M., and Gandhi, C.C. (2003). Individual differences in the expression of a "general" learning ability in mice. *Journal of Neuroscience, 23,* 6423–6433.

McGaughy, J. and Sarter, M. (1995). Behavioral vigilance in rats: task validation and effects of age, amphetamine, and benzodiazepine receptor ligands. *Psychopharmacology, 11,* 340–357.

McGaughy, J., Turchi, J., and Sarter, M. (1994). Cross-modal divided attention in rats: effects of chlordiazepoxide and scopolamine. *Psychopharmacology, 115,* 213–220.

Meiners, M. and Dabbs, J. (1977). Ear temperature and brain blood flow: laterality effects. *Bulletin of the Psychonomic Society, 10,* 194–196.

Miklosi, A., Polgardi, R., Topal, J., and Csanyi, V. (1998). Use of experimenter-given cues in dogs. *Animal Cognition, 1,* 113–121.

Miklosi, A., Polgardi, R., Topal, J., and Csanyi, V. (2000). Intentional behavior in dog–human communication: an experimental analysis of "showing" behavior in the dog. *Animal Cognition, 3,* 159–166.

Nachreiner, F. and Haneke, K. (1992). Vigilance. In Smith, A.P. and Jones, D.M., Eds. *Handbook of Human Performance, Vol. 3.* London: Academic Press, pp. 261–288.

Parasuraman, R. (1984). The psychobiology of sustained attention. In Warm, J.S., Ed. *Sustained Attention in Human Performance.* Chichester: John Wiley & Sons, pp. 61–101.

Parasuraman, R., Warm, J.S., and See, J.E. (1998). Brain systems of vigilance: In Parasuraman, R., Ed. *The Attentive Brain.* Cambridge: MIT Press, pp. 221–256.

Pigeau, R.A., Angus, R.G., O'Neill, P., and Mack, I. (1995). Vigilance latencies to aircraft detection among NORAD surveillance operators. *Human Factors, 37,* 622–634.

Poulton, E.C. (1976). Arousing environmental stresses can improve performance, whatever people say. *Aviation, Space, and Environmental Medicine, 47,* 1193–1204.

Robbins, T.W. (1998). Arousal and attention: psychopharmacological and neurophysiological studies in experimental animals. In Parasuraman, R., Ed. *The Attentive Brain.* Cambridge: MIT Press, pp. 189–220.

Sanders, M.S. and McCormick, E.J. (1993). *Human Factors in Engineering and Eesign, 7th ed.* New York: McGraw-Hill.

Scerbo, M. (1998). What's so boring about vigilance? In Hoffman, R.B. et al., Eds. *Viewing Psychology as a Whole: The Integrative Science of William N. Dember.* Washington: American Psychological Association, pp. 145–166.

Schmorrow, D.D. and Kruse, A.A. (2004). Augmented cognition. In Bainbridge, W.S., Ed. *Berkshire Encyclopedia of Human–Computer Interaction.* Great Barrington, MA: Berkshire Publishing, pp. 54–59.

Schmutz, S.M. and Schmutz, J.K. (1998). Heritability estimates of behaviors associated with hunting in dogs. *Journal of Heredity, 89,* 233–237.

Schnittger, C., Johannes, S., Arnavaz, A. and Munte, T.F. (1997). Relation of cerebral blood flow velocity and level of vigilance in humans. *NeuroReport, 8,* 1637–1639.

Schuman, M., Suhr, D., Gesseln, B., Jantzen, J., and Samii, M. (1999). Local brain surface temperature compared to temperatures measured at standard extracranial monitoring sites during posterior fossa surgery. *Journal of Neurosurgical Anesthesiology, 11,* 90–95.

Selye, H. (1976). *The Stress of Life, Revised Edition.* New York: McGraw-Hill.

Slabbert, J.M. and Rasa, O.A.E. (1997). Observational learning of an acquired maternal behaviour pattern by working dog pups: alternative training method. *Applied Animal Behaviour Science, 53,* 309–316.

Stroobant, N. and Vingerhoets, G. (2000). Transcranial doppler ultrasonography monitoring of cerebral hemodynamics during performance of cognitive tasks: a review. *Neuropsychology Review, 10,* 213–231.

Swift, A., (1991). Tympanic thermometry: an index of hemispheric activity. *Perceptual and Motor Skills, 73,* 275–293.

Temple, J.G., Warm, J.S., Dember, W.N., Jones, K.S., LaGrange, C.M., and Matthews, G. (2000). The effects of signal salience and caffeine on performance, workload and stress in an abbreviated vigilance task. *Human Factors, 42,* 183–194.

Trevelyan, J. (1997). Robots and landmines. *Industrial Robot, 24,* 114-125.

Tripp, L.D. and Warm, J.S. (2007). Transcranial Doppler sonography. In Parasuraman, R. and Rizzo, M., Eds. *Neuroergonomics: The Brain at Work.* New York: Oxford University Press, pp. 82–94.

Turchi, J. and Sarter, M. (1997). Cortical acetylcholine and processing capacity: effects of cortical cholinergic deafferentation on cross-modal divided attention in rats. *Cognitive Brain Research, 6,* 147–158.

Vas, J., Topal, J., Pech, E., and Miklosi, A. (2008). Measuring attention deficit and activity in dogs: a new application and validation of a human ADHD questionnaire. *Applied Animal Behaviour Science*, in press.

Warm, J.S. (1993). Vigilance and target detection. In Huey, B.M. and Wickens, C.D., Eds. *Workload Transition: Implications for Individual and Team Performance*. Washington: National Academy Press, pp. 139–170.

Warm, J.S. and Dember, W.N. (1998). Tests of a vigilance taxonomy. In Hoffman, R.B. et al., Eds. *Viewing Psychology as a Whole: The Integrative Science of William N. Dember*. Washington: American Psychological Association, pp. 87–112.

Warm, J.S. and Parasuraman, R. (2007). Cerebral hemodynamics and vigilance. In Parasuraman, R. and Rizzo, M., Eds. *Neuroergonomics: The Brain at Work*. New York: Oxford University Press, pp. 146–158.

Wechsler, B. (1995). Coping and coping strategies: a behavioral view. *Applied Animal Behaviour Science, 43,* 123–134.

8 Olfaction and Explosives Detector Dogs

Allen Goldblatt, Irit Gazit, and Joseph Terkel

CONTENTS

INTRODUCTION

Dogs are widely considered to have an exceptional sense of smell that has evolved for the detection of odors relevant for their survival and reproduction. A dog's ability to detect and discriminate odors varies as a function of the importance of the odor. It is not a given that a dog's ability to detect explosives is as acute as its ability to find and discriminate individual human odors or dog odors. It is important to eliminate the mythology associated with dogs' noses and begin to understand the science behind the canine ability to detect explosive odors.

The explosives detector dog relies on olfaction to find the subtle cues of explosives. An understanding of olfaction is essential for the training and maintenance of these dogs. The most important development in the field of olfaction and working dogs is an understanding of the role of perceptual learning in olfaction. Specifically, the effects of experience on the increased sensitivity toward the training odor and the narrowing of the generalization gradient with continued exposure of the target odor may require changes in the manner in which explosives detector dogs (EDDs) are trained and maintained.

The understanding that olfaction is primarily a synthetic sense is also important for the optimal training of EDDs. Although no specific research focuses on dogs, research on rats, mice, and humans has shown that when a subject is first exposed to a complex mixture of previously unexperienced odors, the subject will learn the characteristics of the mixture and not the individual components constituting the mixture. Therefore, until thorough research on learning mixtures for dogs is available, it is recommended that animals be trained on individual explosives odors and not on mixtures of various explosives. After an animal learns to detect a specific explosive's odor, no problem ensues when the specific odor is mixed with other odors. Because of perceptual learning, the more experience an EDD has with the explosives odor, the easier it will be for the animal to detect it among other odors.

The olfaction sense is considerably less developed in humans. Therefore it is very difficult for people to appreciate the olfactory world of a dog. This can result in the inadvertent contamination of training odors because a handler may not be aware of the contamination. It is therefore essential that extreme care be used in handling trainer aids throughout the entire training and maintenance program. This includes the use of controls and periodic blind assessments of performance. Ideally, these assessments should be done by neutral observers and should involve trainer aids from a different source than that used during training.

The canine sense of smell is legendary and, as with many legends, it is sometimes hard to find the sources and separate scientific facts from urban legends. For example, the *Wikipedia* on-line encyclopedia has this to say about olfaction:

In vertebrates smells are sensed by olfactory sensory neurons in the olfactory epithelium. The proportion of olfactory epithelium compared to respiratory epithelium (not innervated) gives an indication of the animal's olfactory sensitivity. Humans have about 10 cm^2 of olfactory epithelium, whereas some dogs have 150 cm^2. A dog's olfactory epithelium is also considerably more densely innervated, with 100 times more receptors per square centimetre.... Dogs in general have a nose approximately a hundred thousand to a million times more sensitive than a human's. Scent hounds as a group can smell one to ten million times more acutely than a human, and the Bloodhound, which has the keenest sense of smell of any dog, has a nose ten to a hundred million times more sensitive than a human's. It was bred for the specific purpose of tracking human beings, and can detect a scent trail a few days old. The second most sensitive nose is possessed by the Basset Hound, which was bred to track and hunt rabbits and other small animals.

Unfortunately, it is difficult to determine whether the above statements are backed by scientific evidence. For many people, canine olfactory abilities have attained the same mythological status as a dolphin's intelligence and inherent goodness.

Scientific data, however, have documented some of the incredible performances of the canine nose. The dog has been shown to be able to discriminate between a touched and untouched glass slide three weeks after it was touched and placed on a countertop (King et al., 1964). Law enforcement agencies have investigated the abilities of Bloodhounds and have found that they can trail a specific person even when that person is in a crowd of others who all suddenly diverge onto separate paths. In a controlled experiment, the FBI gave a Bloodhound a letter written by a woman who had moved to a new house in a different state 6 months prior to writing the letter. Using the scent from the letter, the dog was able to select the house where she had previously lived even though she had not approached the house for 6 months (Jones, 2006). Stokham, Slavin, and Kift (2004) found that the odor of a specific individual who planted an explosive device was still recognizable on the device even after it exploded, and a dog was able to use the odor remaining on the device to trail the perpetrator.

Dogs have been shown to detect cancer earlier than traditional methods of diagnosis (McCulloch et al., 2006; Willis et al., 2004). They also have the ability to detect the odors of specific species of animals such as turtles, snakes, termites, screw-worms, and ferrets, and to discriminate between a target species and other species (Brooks et al., 2003; Cablk and Heaton, 2006; Engeman et al., 1998; Reindl-Thompson et al., 2006; Welch, 1990).

Dogs have the incredible ability to determine the direction that a person has walked because the fact that the odor in the direction walked is always fresher than the odor in older sections of the trail. Dogs are even capable of determining the direction of travel by detecting the concentration gradient when the difference in concentration of the scent at the start of the trail was only two seconds older than the scent at the end of the trail (Hepper and Wells, 2005). Finally, dogs have the ability to perform olfactory match-to-sample tasks. They can smell a specific novel object and match the odor of the object to the odor of a person who previously held the object (Schoon, 1994 and 1998).

Even though these abilities are extremely impressive, they do not necessarily generalize to other olfactory tasks. The olfactory abilities of all animals are evolved skills, honed by millions of years of natural selection for the ability to utilize olfactory cues that enhance probability of reproduction and survival. Thus the ability to detect and discriminate trace odors of humans and other animals does not necessarily mean that dogs have the same skills for the detection of explosives odors. To understand EDDs, it is necessary to understand olfaction, its evolution, genetic basis, and anatomy and physiology. It is further necessary to understand olfactory learning and olfactory training as applied to dogs. In the following sections we will review those features of the genetics, physiology, and psychology of olfaction relevant to understanding the EDD.

OLFACTION, EVOLUTION, AND NATURAL SELECTION

Olfaction is probably the oldest of the senses in that it is found in some form or other in all living cells (Buck, 2000). Basically, olfaction is the response of a living organism to chemicals in the environment. It is essential for the procurement of nutrients and avoidance of hazardous environmental conditions. As an example, the primitive soil nematode shows the ability to learn to avoid odors associated with pathogenic bacteria (Zhang, Lu, and Bargmann, 2005).

All cells have the capability and the necessity of sensing the external chemical environment. During evolution, this ability evolved into concentrations of cells specialized to sense the external environment. This chemical sensory system became the olfactory system. It is considered conservative and some of its features are common in round worms, invertebrates, and vertebrates (Eisthen, 1997). Evidence of convergent evolutionary processes shows that the various olfactory systems are more analogous than homologous (Dryer, 2000; Strausfeld and Hildebrand, 1999). In other words, there seem to be a limited number of ways to process odors and widely varying species independently evolved the same mechanisms (Hildebrand and Shepherd, 1997).

Because of natural selection, every species has developed a sensitivity to odors that facilitate survival and successful reproduction. In every generation, those individuals that could better utilize olfactory information were selected for. The present result of this ongoing selection is an olfactory system that is extremely good at detecting and discriminating odors important for survival. The ability is not general; thus, the thresholds for detection of the same odor differ among species and within any given species different odors have different thresholds for detection. This threshold is a function of both the physical properties of the odor and the importance of the odor for the species.

The discussion of olfaction in *Wikipedia* stated that the olfactory system of the dog was physically much larger than that of humans and contained many more receptors. This was used as the explanation for the unique olfactory abilities of the dog. We feel that this explanation is misleading. The rat and mouse olfactory systems are comparable to the dog olfactory system in both odor detection thresholds and the ability to discriminate odors (Laska, Seibt, and Weber, 2000; Salazar, Laska, and Luna, 2003), yet the noses of mice and rats are far smaller than those of humans.

Traditionally, it was convenient to divide mammals into microsmatic and macrosmatic groups, based on the size of the olfactory system. Primates (and especially humans) were considered classic examples of microsmatic species and it was assumed that odors were not important and could not be detected at concentrations detectable to macrosmatic animals such as dogs. Many recent studies have shown that so-called microsmatic animals can detect some odors much better than so-called macrosmatic animals (Hubener and Laska, 1998; Laska and Hudson, 1993a and b; Laska and Freyer, 1997; Laska, Seibt, and Weber, 2000; Laska, Hofmann, and Simon, 2003; Laska et al., 2004). These results led several authors to suggest that the terms macrosmatic and microsmatic are misleading (Laska et al., 2000; Smith et al., 2004). It is important to note that the odors for which monkeys show a lower threshold than dogs and rats are often odors associated with feeding on fruits—common foods for microsmatic monkeys. This supports the role of natural selection in determining the thresholds of specific odors (Dominy et al., 2001).

A more direct proof of the role of natural selection in determining odor thresholds was found in rats. Laska and co-workers (2005) found that rats were able to discriminate concentrations between 0.04 and 0.10 parts per trillion (ppt) of 2,4,5-trimethylthiazoline, an odor associated with fox feces. The fox is a natural predator of rats. Other species not threatened by foxes exhibited much higher thresholds. According to the authors this is by far the lowest olfactory detection threshold for an odorant reported in rats. These data suggest that dogs should be very good at detecting biologically important odors but it does not imply a carry-over to the detection of odors of biologically irrelevant compounds such as explosives.

GENETICS OF OLFACTION

Although evidence indicates that some olfactory receptor cells are fairly specific for relevant biological odors, this belief seems to be the exception rather than the rule (Hildebrand and Shepherd, 1997; Lin et al., 2005; Wilson and Mainen, 2006). Based on the tremendous number of possibly relevant odors, it would be very expensive to determine specific genetic receptors for each odor. This means that some evolutionary mechanism enables animals to easily and rapidly learn those odors that are important for survival. To determine how animals are capable of learning new odors, it is necessary to take a close look both at how noses process odors and the genetic differences among species in regard to olfaction.

More genes are involved in the sense of smell than in any other system (Fuchs et al., 2001; Mombaerts, 1999). Each of these genes codes for a specific receptor that is associated with one member of a superfamily of G-protein-coupled olfactory receptor proteins (Firestein, 2004; Friedrich, 2004; Leon and Johnson, 2003; Young et al., 2002). In other words, each gene codes for one protein expressed by one type of olfactory receptor. Each olfactory receptor expresses only one specific receptor protein.

Current wisdom holds that each olfactory receptor protein binds with a specific physio-chemical characteristic of an odorant. Some of these characteristics can be seen in Figure 8.1 (Johnson and Leon, 2007). Different odorants are characterized by varying carbon chain lengths and functional groups that determine the extent of binding to different olfactory proteins and thus activate different olfactory receptor neurons. It has proven very difficult to determine the specific characteristics of an odorant that binds to a specific olfactory receptor and it may be that the bonds are much less specific than is currently thought (Wilson and Mainen, 2006).

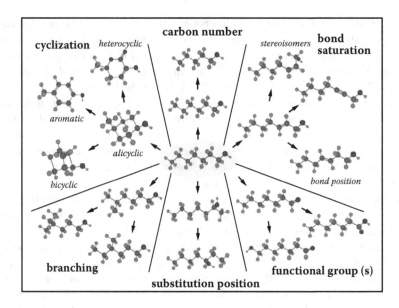

FIGURE 8.1 Chemical dimensions that may influence binding of odorants. (From Johnson and Leon, 2007. Reprinted with permission of John Wiley & Sons.)

Since each olfactory gene codes for a specific psycho-physical characteristic of an odorant, it is possible that the greater the number of olfactory genes, the greater the diversity of odors that can be detected and the finer the ability to discriminate odors. Unfortunately, this hypothesis has not been tested. What is known is that species exhibit large variations in numbers of functional olfactory genes. For example, Ache and Young (2005) report that chickens have 78 functional genes, humans have 388, chimpanzees have 450, and frogs have 410. They report that rats, mice, and dogs all have around 1000 functional olfactory genes, suggesting that they are about equal in the ability to discriminate odors. Although the number of genes probably does not determine the threshold for detection of specific odors, it seems reasonable that it does influence the ability of an animal to make fine discriminations between very similar odors.

FUNCTIONING OF THE NOSE

Before discussing how the nose works, it is first necessary to distinguish odorants from odors. Odorants are molecules objectively definable in terms of their physio-chemical characteristics. Each odorant binds, more or less selectively, to one or more of the olfactory receptor proteins. An odor is the perception resulting from the sum total of the activities of odorants.

ODOR INTAKE

For a dog to respond to an odorant, the odorant must enter the nostrils, travel up to the olfactory mucosa, be transported through the mucosa to the olfactory receptor, and

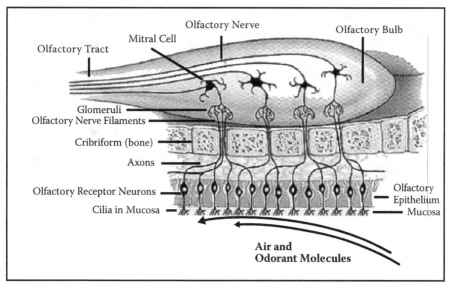

FIGURE 8.2   The human nose. Odorants are absorbed by the mucosa, and migrate through the mucosa to the cilia of the olfactory receptor neurons. The axons of the receptor neurons terminate in a glomerulus. Mitral and tufted cells in the glomeruli send axons via the olfactory tract to the olfactory cortex. (From Leffingwell, 2002.)

bind with one or more olfactory receptors that may activate neurons in the olfactory bulb and send a signal to the cortex. Figure 8.2 illustrates this path. However, getting the odorant into the nose is not simple and several factors must be understood. First, the odorant must enter the atmosphere, either as a vapor or attached to a particle of dust. It must then enter the nostrils where, depending on whether the dog is sniffing or not, it will be carried to the olfactory mucosa overlying the olfactory epithelium.

ODOR PLUME

Assuming the odor is in the atmosphere, it will be carried by wind currents in the form of plumes from the point of origin. This is not simple diffusion in which intensity decreases as a function of the distance from the point of origin (Vickers, 2000; Vickers et al., 2001). These plumes are more like twisting rivers of odors with areas of high concentration of odors bordered by areas containing no odor. Figure 8.3 illustrates an example of an odor plume. The eddies and changes in direction of fairly narrow plumes means that an animal attempting to follow a plume to the odor source will continually find and lose the plume. The animal must compensate for this by moving laterally so it can criss-cross the plume and reacquire the odor. Both animals (Porter et al., 2005) and humans (Porter et al., 2007) have shown this ability to navigate an olfactory plume.

SNIFFING

Once the animal enters an odor plume, the next stage is getting the odor into its nose and onto the olfactory mucosa. To achieve this, dogs have developed a complex

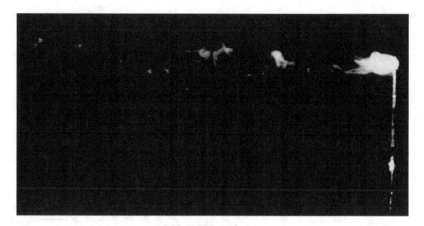

FIGURE 8.3   Example of odor plume in air. (From Ache and Young, 2005. With permission of Mark Willis.)

behavior called sniffing—an activity that has been the subject of laboratory and field research and is well understood. Dogs can sniff up to 20 times per second while hunting (Steen et al., 1996). Field studies have revealed that if a dog pants more because of heavy exercise, it is less likely to sniff and consequently less likely to detect explosives (Gazit and Terkel, 2003). When a dog is searching for an olfactory cue, the amount of sniffing increases (Steen and Wilsson, 1990) and the more difficult the task, the greater the sniff rate (Steen and Wilsson, 1990; Thesen et al., 1993). Laboratory studies of the physiology and anatomy of sniffing have shown that the difference in concentration of the odor between the two nostrils is also an important cue for following a trail (Porter et al., 2007).

The sniff has several functions; first, the outward puff of air disturbs the ground and raises a cloud of dust, including attached odorants. Second, the sniff humidifies and heats the odorant, facilitating its absorption by the mucosa. Third, the sniff diverts the airflow from a relatively straightforward path to the lungs to a convoluted path over the olfactory epithelium. These pathways have very recently been elucidated in an interesting article (Craven et al., 2007) that revealed the intricacy of a dog's nose and the different pathways used in sniffing and regular breathing.

ODORS AND OLFACTORY EPITHELIUM

A large amount of research has focused on odorant interactions with the olfactory receptors of the olfactory epithelium (Buck, 1996 and 2000; Chess et al., 1992; Firestein, Breer, and Greer, 1996; Firestein, 2001; Leon and Johnson, 2003; Mombaerts, 1999a and b; Zhao and Firestein, 1999). It is not our intention to provide yet another description of the circuitry involved in olfaction. This can be more easily obtained by reading one or more of the above references; therefore we will be very brief and discuss what we consider to be the essentials.

Figure 8.2 illustrates the wiring of the olfactory system. All the olfactory neurons that express a specific olfactory protein, i.e., selectively respond to a specific characteristic of an odorant, send their axons to one and only one glomerulus in the olfactory bulb. Thus, each glomerulus contains the axons from several thousand olfactory sensory neurons, which, regardless of the location in the soma of the epithelium, all express the same odorant receptor. For some unknown reason, all olfactory sensory neurons expressing the same odorant receptor typically target two symmetric mirror areas in each olfactory bulb (Salcedo et al., 2005). The glomeruli are the functional units of olfaction and the glomerular arrangement in the olfactory bulb provides an anatomical foundation for the encoding of odorant quality and intensity (Salcedo et al., 2005). Anatomically, glomeruli are some of the most distinctive structures in the brain, analogous to barrels or columns in the cerebral cortex. In the mammalian olfactory bulb, they are compactly packed almost two dimensionally near the olfactory bulb surface. The mean numbers and diameters of glomeruli are about 1800 and 85 mm in diameter in mice and 2400 and 190 mm in diameter in rats (Kosaka et al., 1998). The number of glomeruli is important in that it correlates with the number of odorant characteristics that the olfactory system is capable of selectively binding.

Electrophysiological recordings and imaging studies of the olfactory bulb show that all odorants activate several glomeruli. Some odors activate more glomeruli than others. Also, for some odors, the number of glomeruli activated is a function of the intensity of the odorants. This can be seen in Figure 8.4 which shows the output of axons from the frog's olfactory system and in Figure 8.5 which shows the pattern of activation of glomeruli to some representative odors.

The outputs of the glomeruli are the axons of mitral and tufted cells. These axons leave each glomerulus and go directly to the periform cortex where odorant information is processed and converted to the sensation of odor. The olfactory bulb also projects to other brain areas, some of which are concerned with emotional responses (Buck, 1996). As stated, the axons from the olfactory bulb terminate on neurons in the periform cortex. This area of the brain is very rich in axons and dendrites and is an area of high interconnectivity. Abundant evidence indicates that the neurons of the periform cortex are plastic in regard to the odorants that will activate them (Wilson, 2003; Zou, Li, and Buck, 2005; Zou and Buck, 2006). Zou and Buck (2006) found that binary odorant combinations stimulate cortical neurons that are not stimulated by their individual component odorants. They suggest that some cortical neurons require combinations of receptor inputs for activation and that merging the receptor codes of two odorants provides novel combinations of receptor inputs that stimulate neurons beyond those activated by single odorants. Wilson (2003) found that neurons were plastic in regard to their neural activity to the components of a mixture and the mixture itself. Initially, cells responded to a component of a binary mixture as if it were the same as the binary mixture, but after the exposure the cells responded to the component of the mixture as if it were different from the mixture. This plasticity in the periform cortex seems to be the biological basis for the ability to learn novel odors.

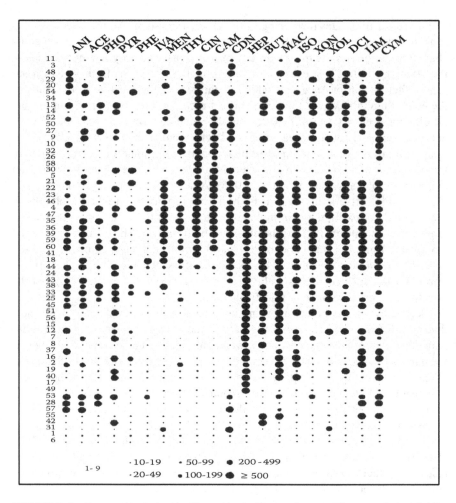

FIGURE 8.4 Responses of single olfactory neurons to odorants. The recorded activities of individual frog olfactory neurons in response to different odorants (top) are shown. Black spots indicate excitatory responses; spot size is proportional to response intensity. Note that most neurons respond to multiple odorants. (From Sicard and Holley, 1984; in Buck, 1996.)

PERCEPTUAL LEARNING AND OLFACTION

Wilson and Stevenson (2003a and 2003b and 2006) provide abundant evidence that experience is essential for odor discrimination, recognition, and maybe even detection. They maintain that memory and perceptual learning serve as the bases for olfactory perception. They provide data showing that exposure to an odor or odor mixture changes the subsequent electrophysiological and behavior response to the mixture and its components. These findings have many implications (Wilson and Stevenson, 2006) in regard to the role of perceptual learning and olfaction as applied to explosive detecting dogs (EDDs). These implications will be discussed in the next section.

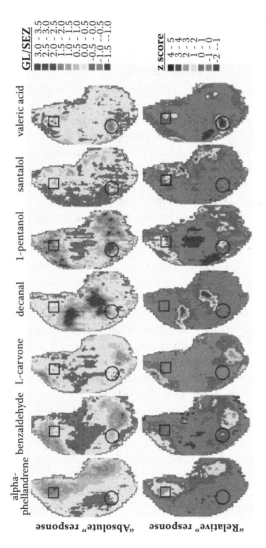

FIGURE 8.5 Although individual odorants can stimulate much of the glomerular layer, relative levels of activity are very distinctive for different odorants. Color-coded contour charts are used to illustrate uptake evoked by seven different odorants. In the top row, responses are scaled as a ratio of uptake in each glomerular layer location to uptake measured in a consistent part of the subependymal zone, which is not expected to display odorant-dependent activity. Yellow or warmer colors indicate uptake that was above what was measured when rats were exposed to air vehicle only. In the bottom row the same set of patterns is shown in a z-score color scale. Yellow or warmer colors indicate uptake that is more than one standard deviation above the average uptake calculated across the glomerular layer. These patterns are very distinctive for different odorants. Outlined regions highlight consistent locations on different charts. While responses to different odorants in a given location do not seem as distinctive in the top row, the same locations are very differently active when viewed as relative (z-score) responses in the bottom row. (From Johnson and Leon, 2007. Reprinted with permission of John Wiley & Sons.)

SYNTHETIC PERCEPTION OF ODOR MIXTURES

Two types of experiments demonstrate that mixtures of odors are perceived as synthetic wholes and not analytically. The first type of experiment exposes a subject to a mixture of odorants and then investigates the responses to the individual components. It is normally found that the subjects respond to the individual components as if they are new odors. The second type of experiment first exposes the subject to an individual odorant or odor and then investigates the response of the subject to that odor when it is included in mixtures of varying complexity. Again, subjects normally fail to recognize even very familiar odorants when they are part of a complex mixture. Some of these experiments will be discussed below.

Wilson and Stevenson (2006) describe experiments showing that the perception of complex odors is synthetic or configurational, not analytic. In other words, a mixture of odors is treated as a whole and it is very difficult to tease out its individual components. When rats are first exposed to an odor composed of four odorants they do not recognize the individual odorants (Staubli et al., 1987). Similarly, Linster and Smith (1999) trained rats on an odorant A, then tested to see whether the animals would generalize to odorant A when it was in a binary mixture. They found that the rats showed little generalization to the mixture containing the odorant A to which they had been trained. Recent studies on humans found that when four odorants were mixed it was very difficult to identify the individual components, even when the odorants were very familiar and the subjects received preliminary training to identify each of the four odorants used. Interestingly, when the odorants were mixed, previous descriptors of the individual odorants were no longer relevant (Jinks and Laing, 2001). Livermore and Laing (1998) extended these results from studies of individual odorants to complex odors such as chocolate and cheese. They found no significant difference in the ability to discriminate a complex odor from a mixture of complex odors and an odorant from a mixture of odorants.

A substantial body of research investigated the ability to discriminate or detect one odorant within a binary mixture. Humans have substantial difficulty in recognizing one well-known odorant when it is in a simple binary mixture (Laing and Francis, 1989; Laing and Glemarec, 1992; Livermore and Laing, 1998). In experiments on rodents, several groups of researchers found that, following initial training on simple binary mixtures, rodents often did not respond to the component odorants. The results indicate that binary mixtures may be treated analytically or configurationally, depending on the degree of similarity and relative salience of the two odorants (Cleland et al., 2007; Kay, Lowry, and Jacobs, 2003; Kay, Crk, and Thorngate, 2005; Kay et al., 2006; McNamara, Magidson, and Linster, 2007; Wiltrout, Dogra, and Linster, 2003). However, Kay and her colleagues (2005) recently suggested that the shift between synthetic and analytic processing is gradual and very much influenced by the relative salience of the two components of the mixture. They suggest that the primary mechanism determining the recognition of a component odorant in a binary mixture is salience or overshadowing. They provide evidence that depending on the relative concentrations of the two odorants, one will usually overshadow the other. This is supported by Sokolic et al. (2007), who found that when both butanal and heptanal were present in a binary mixture at the same concentration, the rats responded to the mixture as if only butanal was present. These results support the premise that in odor mixtures, a single salient component overshadows the other odorants. This results in the subject's response to only the dominant odorant in a binary mixture.

Studies of more complex mixtures indicate that overshadowing is important but that selective attention can influence a specific odorant to which an animal responds. For example, Takiguchi et al. (2008) investigated the ability of mice to distinguish the odors of different wines. They found evidence for both overshadowing and selective attention. Mice were trained to discriminate between a specific red wine (A-red) and distilled water. They were then tested for their ability to discriminate between A-red and other liquors including white wine, rosé wine, sake, plum liqueur, and another red wine (B-red). Whereas all of the mice could discriminate between the A-red and other types of liquors, some mice could not discriminate between the two types of red wines. Through the use of various manipulations and experiments Takiguchi concluded that individual mice directed attention to different subsets of volatile components emanating from the rewarded red wine. Mice whose attention was directed to components that were similar or identical in the two red wines failed to make the discrimination, but those mice whose attention was directed to odorants that were not common to the two wines were able to make the discrimination. Takiguchi et al. (2008) further showed that in mixtures of three odors, if one odorant was dominant or overshadowed the other components, the mice directed their attention to that dominant odor and, in tests of their ability to discriminate the mixture from its components, failed to discriminate between subsets when both contained the dominant component.

The overall result of these experiments on olfactory mixtures is that in complex mixtures of four or more odorants, a subject responds to the odor as a whole and not as a collection of different odorants. In humans, Laing and his colleagues found that well trained subjects exposed to familiar odors could not evaluate even the number of odorants within a mixture (Laing and Francis, 1989; Laing and Glemarec, 1992; Livermore and Laing, 1998). Several researchers found emergent properties in mixtures of odorants and noted that a small change in the ratio of the odorants could change the perception of the mixture (LeBerre et al., 2003; McNamara, Magidson, and Linster, 2007). Finally, as mentioned earlier, Jinks and Laing (2001) asked subjects to determine the dominant characteristics of four well-known odorants. When these odorants were mixed, the subjects were again asked to determine the dominant characteristics of the mixture. They found that the dominant characteristic descriptors of unitary odors were altered when the same odor was in a mixture. In other words, the characteristic descriptors of the individual odorants were no longer available to aid detection and identification from within the mixture.

RELEVANCE OF SYNTHETIC PROCESSING OF MIXTURES TO EDDS

To explain the relevance of the synthetic processing of odor mixture to the training and maintenance of an EDD, it is necessary to diverge for a moment and discuss the compositions of explosives. Explosives are mixtures of different chemicals, some of which are more salient than others.

Explosives Odors and Vapor Pressures

Explosives are chemical compounds that explode. That is about the only commonality that exists among explosives. Many different families of explosives exist and each family contains many different compositions and dozens of different types. Table 8.1 shows some of the families and representative members. The large variety of explosives means that no single odor is representative. Theoretically EDDs should require training on all available explosives. The ATF (Alcohol, Tobacco and Firearms Agency) suggested a training shortcut. The agency maintains that each class of explosives has a unique chemical characteristic. If the dog learns the odor of the unique characteristic, it will be able to detect all members of that class. Unfortunately, the ATF theory has never been tested on animals.

TABLE 8.1
Main Chemical Classes of Explosives and Examples of Classes

Explosive Classes	Mandatory Explosives$_2$	Mandatory Commercial Types of Mandatory Explosive Materials
NITRO ALKANES (C-NO$_2$)	None	None
NITRO AROMATICS (Ar-NO$_2$)	2,4,6-trinitrotoluene (TNT)	TNT, flake TNT, demilled military flake nitropel (TNT)
NITRATE ESTER (C-O-NO$_2$)	Pentaerythitol tetranitrate (PETN)	PETN (unadulterated)$_4$ Detasheet A (85% PETN + binder) red commercial form PETN detonating cord (100% PETN) cross-section usually white in color *(PIMALINE, PRIMACORD, PRIMASHEAR, OPTICORD, GEOSEIS, LOW FLEX, FIRELINE CORD)* Primasheet 1000 (PETN + plasticizers) SEMTEX A (PETN + plasticizers)
NITRAMINES (C-N-NO$_2$)	Trinitro-triazacylohexane (cyclonite or RDX)	RDX (unadulterated) RDX Det Cord Composition C4 (RDX + plasticizers) Datasheet (Flex -X) (RDX + plasticizers) Demex 200 (RDX + plasticizer) PE-4 (RDX + plasticizer) Primasheet 2000 (RDX + plasticizers)
ACID SALTS (NH$_{4+}$, NO$_3$.)	Ammonium nitrate (AN) Potassium nitrate (PN)1	Black powder (PN + charcoal + sulfur) PN (pure) procured from chemical supply, Kine-Pak and Kine-Stick solid component (AN) 34-0-0 agricultural fertilizers (pure AN)
PEROXIDES$_6$ (C-O-O-C)	None	None

Note: These explosives are considered mandatory for the training of explosive detection dogs by the SWGDOG committee. The table is from the SWGDOG website, http://www.swgdog.org.

For a dog to detect the active element of an explosive, the element must be volatile. Table 8.2 illustrates published vapor pressures of a variety of explosives. It can easily be seen that the vapor pressures of many commercial and military explosives are very low. Dogs are probably not able to detect these compounds (Furton and Myers, 2001; Lorenzo et al., 2003). Furton and Myers state:

> It is notable that there are seven orders of magnitude difference in the vapor pressure of EGDN and RDX. The very low vapor pressures of many explosives, including PETN, RDX and HMX, make the detection of the parent molecule unlikely, particularly at room temperature. Logically, one would expect an explosives vapor detector (EVD), including dogs, to utilize the more abundant chemicals in the headspace of target explosives and, therefore, the isolation, identification and quantization of these chemicals are very important. Unfortunately, few studies of this nature have, to date, been reported.

WHAT CONSTITUTES AN EXPLOSIVE?

The explosive element is not the only compound in an explosive. As shown in Table 8.2, compounds other than the parent explosive (that may be much less abundant

TABLE 8.2

Common Explosives and Properties

Explosive Class		Explosive	Vapour Pressure at 25°C (Torr)
Acid salt		Ammonium nitrate	5.0×10^{-6}
Aliphatic nitro		Nitromethane	2.8×10^{1}
	DMNB	2,3-Dimethyl-dinitrobutane	2.1×10^{-3}
Aromatic nitro	o-MNT	2-Nitrotoluene	1.5×10^{-1}
	p-MNT	4-Nitrotoluene	4.1×10^{-2}
	DNT	2,4-Dinitrotoluene	2.1×10^{-4a}
	TNT	2,4,6-Trinitrotoluene	3.0×10^{-6}
	Picric acid	2,4,6-Trinitrophenol	5.8×10^{-9}
Nitrate ester	EGDN	Ethylene glycol dinitrate	2.8×10^{-2}
	NG	Trinitroglycerin	2.4×10^{-5}
	PETN	Pentaerythritol tetranitrate	3.8×10^{-10}
	NC	Nitrocellulose	N/A
Nitramin	Tetryl	Tetranitro-N-methylamine	5.7×10^{-9}
	RDX	Trinitro-triazacyclohexane	1.4×10^{-9}
	HMX	Tetranitro-tetrazacyclooctane	1.6×10^{-13a}
	CL20	Hexanitro-hexaazaisowurzitane	N/A
Peroxide	TATP	Triacetone triperoxide	3.7×10^{-1a}
	HMTD	Hexamethylene triperoxide diamine	N/A

N/A = not available.
A = extrapolated value.
(Modified from Harper et al., 2005.)

TABLE 8.3

Differences in Percentages of TNT Components Found in Solid and Vapor States

Compound	Solid Phase Composition (%)	Vapor Phase Composition (%)
2,4,6 TNT	99.80	58
2,3,5 TNT	0.08	Trace
2,3,4 TNT	0.02	3
2,4 DNT	0.08	35
2,5 DNT	<0.01	4
3,5 DNT	<0.01	Trace
3,4 DNT	Trace	Trace
2,6 DNT	Trace	Trace
Other Impurities	None detected	Not analyzed

(Modified from Phelan and Webb, 2003.)

than the explosive) may be much more volatile. Table 8.3 illustrates the percentages of components of military TNT in both the solid and vapor states. While TNT is by far the most prevalent material in the solid state, other chemicals, particularly 2,4 DNT account for much of the headspace. Thus, while supposedly learning to detect explosives, an EDD may actually be learning to detect the volatile, nonexplosive component instead of the parent explosive.

Several studies have measured the headspace of TNT of various origins (Phelan and Webb, 2002). In some samples, certain isomers of TNT or DNT were not detectable. This work also showed that the average concentration of 2,4-DNT was about 10 to 30 times greater in vapor concentration than that of 2,4,6-TNT.

The differences in the headspace between different samples of TNT can be quite large (Table 8.4). Some differences in the headspace of TNT of different origins can be seen in a gas chromatograph (Figure 8.6). As another example, Russian-manufactured

TABLE 8.4

Differences in Percentage of TNT and DNT as Function of Origin

Source of Explosive	Headspace Vapor Concentration (ng/L) at 22°C		
	1,2 DNB	2,4, DNT	2,4,6 TNT
U.S. Military 1996 (TNT)	0.35	0.55	0.070
Yugoslavian PMA-1A (TNT)	4.6	1.4	0.078
Yugoslavian PMA-2 (TNT)	9.7	0.28	0.077

(From Phelan and Webb, 2003.)

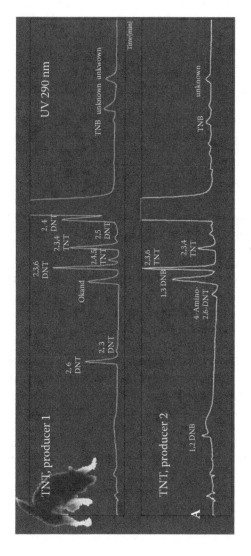

FIGURE 8.6 Gas chromatograph of two different types of TNT from two laboratories. (A is from FOI, the Swedish Defense Research Agency. Data were presented by Akerblom and Karlsson at a workshop held in Morogoro, Tanzania in 2007.

TABLE 8.5
Composition of U.S. Military C4

Ingredient	Percent Weight
Cyclonite or hexogen (RDX)	89.9–92.0
Polyisobutylene	2.1–2.5
Dioctyl adipate or dioctyl sebacate	5.3–5.9
Petroleum oil	1.4–1.6

(Data from Wasserzug et al., 2003.)

TNT contains no DNT (personal communication). Environmental factors can also influence the component odors found in headspace. For example, Lorenzo et al. (2003) reported that some examples of weathered TNT do not contain DNT. The real differences to be found in headspaces of different samples of TNT can readily be seen. These differences arise both from manufacturing processes and weathering under different climatic conditions (Phelan and Webb, 2002). For a thorough review of the environmental factors influencing the decomposition and migration of explosives through air and water, see Phelan and Webb (2002).

Another explosive investigated in terms of its components and headspace is C4. The explosive material in C4 is RDX, which has a very low vapor pressure. C4 types differ widely, based on the manufacturing process. An analysis of U.S. military C4 showed that four chemicals are present in the solid form. The chemicals and their percentages are shown in Table 8.5 (Wasserzug et al., 2002). Figure 8.6 shows gas chromatographs of the volatiles of C4. It can be readily seen that RDX is not present in the headspace (Figure 8.7).

What is true for TNT and C4 applies also to all other commercial explosives. Analysis of commercial explosives shows that they always contain chemicals in addition to the mother explosive and the other chemicals usually account for most of the headspace (Harper, Almirall, and Furton, 2005). Interestingly, some improvised explosives, i.e., those made by individuals in their kitchens, including TATP, have much higher vapor pressures than commercial explosives. However, even the headspace of TATP varies as a function of the specific manufacturing process (personal communication). It can be concluded that the headspace of commercial odors consists predominantly of chemicals other than the explosive. The implications of this for the training of EDDs will be considered in the next section.

EDD DETECTION

The "holy grail" of the EDD is the determination of the actual odor the animals use to detect explosives. A lot of research and speculation have been devoted to this question and we are still far from a definitive answer. Two approaches have been used in attempts to determine to what odor EDDs respond. The first approach is to train a dog with various components found in the headspace of an explosive and then test to

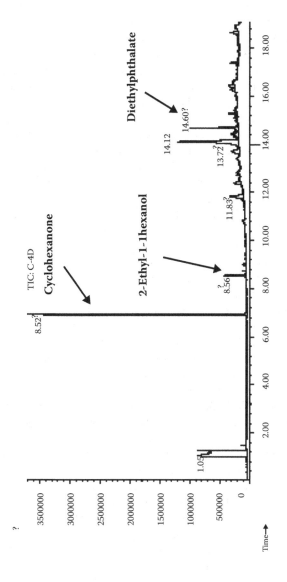

FIGURE 8.7 Headspace of C4. RDX is not present in the headspace. (From Wasserzug et al., 2002.)

determine whether the dog responds to the real explosive. This approach primarily has been used in the development of nonexplosive trainer aids (Harper, Almirall, and Furton, 2005; Lorenzo et al., 2003; Reaugh and Kury, 2002; Wasserzug et al., 2002).

The second approach is to train a dog to detect the real explosive and then present it with the various components and mixtures. If the dog responds as if the component is the real explosive, then that component could be the odor to which the dog is actually responding. A variation of this technique used by Johnson and colleagues (1998) allows the dog three possibilities: (1) response for no odor, (2) response for the odor of the explosive, or (3) response for an odor different from the explosive. They then present the dog with an explosives odor and various components and mixtures. When the dog responds in the same way to a component or mixture of components as it does to the explosive, the odor can be considered characteristic of the explosive.

In general, the only consistency in reporting results of training a dog on a component or components of explosives is inconsistency. For example, Wasserzug et al. (2002) trained dogs to respond to mixtures designed to imitate TNT or Semtex H. The dogs trained on the imitation of TNT did not respond well to the real explosive; the dogs trained on the genuine TNT responded to the simulation. The results were opposite for Semtex H. Dogs trained to respond to the odor of the trainer aid responded to genuine Semtex and dogs trained to respond to genuine Semtex did not respond well to the simulant.

When dogs are trained on a genuine explosive and tested on the components, the results are also confusing. Lorenzo (2003) found that when well-trained dogs were tested on a material designed to imitate the odor of TNT or C4, the dogs responded correctly. However, when the same laboratory performed a similar experiment a few years later the results were the opposite; none of the dogs trained on the real explosive responded to the trainer aid (Harper, Almirall, and Furton, 2005).

This question has also been addressed in more controlled laboratory settings. Williams et al. (1998) trained dogs on TNT, NG, and C4 and then determined how the dogs responded to components and mixtures of components found in the headspace. They identified the four most abundant components in the headspace of C4 as cyclohexanone (76%), 2-ethyl-1-hexanol (5%), toluene (7%), and cyclopentane (7%). They trained four dogs to respond to C4 and then presented the dogs with the above chemicals and various mixtures of the chemicals. In all cases the dogs were asked to respond to the odor by indicating (1) no odor present, (2) C4 odor, or (3) a different odor. Each dog responded differently. One responded to both cyclohexanone and 2-ethyl-1 hexanol as if they were C4; one dog responded to only cyclohexanone; one responded only to 2-ethyl-1-hexanol; and one did not respond to either odor as if it were C4. In addition, the authors varied the concentrations of various mixtures and found an interesting result. One dog responded to a concentration of 135 ppb of cyclohexanone as if it were C4 but also responded to lower (78 ppb) and higher (230 ppb) concentrations of cyclohexanone as if they were different odors (Williams et al., 1998).

Essentially the same results were found for TNT and commercial dynamite, in that each dog responded to the components differently (Williams et al., 1998). These results are understandable in the context of the previous section on perceptual learning. Explosives such as TNT and C4 constitute bouquets of odorants, some of which are much more salient than others. After an animal learns the bouquet of C4, it will treat the individual components as different from C4. This also explains the results of Wasserzug et al. (2002). Although they closely mimicked the olfactory headspace of C4, it was not enough to fool the dogs into responding as if the trainer aid were the real explosive.

On a practical level, in regard to the learning of olfactory stimuli, the correct training approach would be training an animal on a volatile component of the explosive that is always present. If the animal undergoes enough training (and *enough* must be empirically determined), it should detect that odor in the real explosives odor bouquet and respond positively (Wilson and Stevenson, 2006).

The only problem with this approach, as noted above, is that different explosives bouquets may consist of different odorants. Although Harper et al. (2005) recommend using DNT as a trainer aid for TNT, animals trained with such an aid will not detect TNT that does not contain DNT. (TNT manufactured in Russia does not contain DNT.) The optimal solution would be to first determine the relevant odorant for most explosives. Ken Furton and his colleagues provide a good start in this direction (Harper, Almirall, and Furton, 2005; Lorenzo et al., 2003). However, if this technique is used, it is still necessary to train with explosives that do not contain the specific odorant. It is therefore very important to perform an intelligence assessment of the types of explosives that are relevant to a specific theater of operations and train on those explosives.

The message is clear in relation to EDD training. Explosives are mixtures of many different odorants. If a dog is trained on the mixture, it will learn to respond to that mixture and not necessarily to any specific component of the explosive odor. If the dog is then required to choose a single component within the mix of explosives odors, the dog will usually choose the most salient (overshadowing) odorant, but this is not always true. Theoretically, a dog can learn to respond to any volatile component and it is impossible to predict which component it will learn. If mixtures are to be used, a dog should first be trained on the individual explosive odor.

IMPLICATIONS FOR EDDs: INITIAL TRAINING

One of the common procedures used to train dogs with no previous olfactory discrimination training to recognize different types of drugs and explosives is to mix four odors together, with the expectation that the dog will learn the individual explosives (personal communications with trainers). It is difficult to find scientific support for this procedure and it is contraindicated by the literature. For example, in Kay's experiments, even when a rat recognized a component following training on the compound, it usually responded only to the most volatile of the two odors (Kay et al., 2006). Linster and colleagues investigated this directly by varying the relative saliences of the two odors. They found that when one odor was substantially more volatile than the second, only the salient odor of the mixture was remembered (McNamara, Magidson, and Linster, 2007). Laing also found this to be true with humans (Laing et al., 1984). For those reasons, the practice of mixing many odors in an effort to speed up the learning should be discouraged. However, it must be noted that no single experiment on dogs has systematically investigated the effects of initial training on a bouquet of explosives on the detection of the individual explosives constituting the mix.

EFFECTS OF EXPERIENCE ON DISCRIMINATION AND DETECTION

Another derivative of the perceptual learning of odors is that the ability to discriminate between odors improves with continued experience with the odors. This phenomenon of improved discrimination following exposure to a stimulus has been found in several sense modalities (Gibson, 1953) and confirmed in several experiments using olfactory stimuli. In humans, Laing and his colleagues found that the more familiar an odorant is, the easier it is to recognize when it is included in a mixture (Laing and Francis, 1989; Laing and Glemarec, 1992; Livermore and Laing, 1998). In rats, Fletcher and Wilson (2002 and 2003) found that the discrimination of a simple component odor from a mixture containing that component improves with experience. Wilson and Stevenson (2006) generalized from the perceptual learning literature and suggested that the discrimination of odorant from background (e.g., masking odors) should also improve with continued experience with a target odor.

In an interesting experiment on the effects of olfactory experience, Mandairon et al. (2006) investigated the ability of rats to discriminate between two closely related odors. They presented rats with a mixture of two similarly smelling odors and then looked for differential habituation of the individual components. When the rats no longer investigated the mixture (habituation), they also did not investigate the individual components, showing that the components smelled the same as the mixture (a not unexpected result considering that the researchers selected closely smelling odors). They then exposed the rats to the individual components for 2 hours per day for 20 days. After the 20 days, they retested the rats' responses to the mixture and the individual components. They found that following the exposure, the rats did discriminate between the mixture and the individual components.

Few experiments have focused on masking odors. Of some relevance may be studies that investigated the ability of a subject to detect a known odor when present in a mixture. As an example of masking, Laing and Francis (1989) state:

> As demonstrated in studies with binary odor mixtures, a difficulty that needs to be overcome when investigating the perceptual capacity of humans is that of odor masking, i.e., where one odor reduces the perceived intensity of another when both are present in a mixture, since this effect can confound the results. For example, some odors can reduce the intensity of others to a level where the suppressed component cannot be perceived, yet when the suppressed component is presented alone it is quite perceptible and identifiable.

An example of this phenomenon is the masking of spoiled fish by fennel and cloves. Takahashi et al. (2004) found that fennel and clove odors suppressed the normal responses of mitral cells to spoiled fish odor. Conversely, Jinks and Laing (1999) found some ability to recognize an odor when it was mixed with as many as eight dissimilar odors.

Masking probably involves at least two mechanisms. First, is the competition for binding sites among different odorants. If two odorants bind to the same glomeruli and one has a greater affinity, that odor may block or alter the perception of the other (Wilson and Mainen, 2006). The second mechanism is based on the synthetic nature of olfaction. If odor mixtures are coded as unitary odors, then the ability to detect components within the mixture is very limited. This limitation can be overcome by first providing extensive experience with the specific component of interest. If, as discussed above, previous experience with an odor improves the ability of a subject to discriminate that odor from others, previous experience should also improve the ability to detect an odor from a rich olfactory background. In fact, at least in dogs, it has been found to be very difficult to prevent the detection of drugs and explosives by using maskers (personal communications with handlers and trainers).

To our knowledge, only one published study covers the detection by dogs of odors in the presence of maskers (Waggoner et al., 1998). After dogs were trained to detect an odor, the odor was paired with varying concentrations of two types of maskers. The first was a simple odorant and the second was a bouquet of odorants (because of security considerations, the authors were not allowed to state the chemical compositions of the maskers. Regardless of the specific identity, it was found that very high concentrations of the blocker were needed to reduce the detection of the target chemical (see Figure 8.8). The results also show that complex B was the more effective masker.

EFFECTS OF EXPOSURE ON THRESHOLD OF DETECTION

Many studies on perceptual learning have shown that with experience, the threshold of detection decreases and the subject is able to detect at thresholds much lower than at the initial testing (Ghose, 2004; Gibson, 1953; Goldstone, 1998). This has also been found with olfaction. In humans, several experiments have shown that exposure to anderostenone can reduce the threshold of detection and also enable nonsmellers to smell the odor (Moller, Pause, and Ferstl, 1999; Stevens and O'Connell, 1995; Wang, Chen, and Jacob, 2003). Dalton et al. (2002) found that exposure to other odors also reduces the threshold for detection in humans and especially in females. Rats exposed to amyl acetate or anderostenone showed significant decreases in detection threshold after exposure (Wang, Wysocki, and Gold, 1993; Yee and Wysocki, 2001).

Olfactory bulb imaging studies on the effects of exposure in mice revealed that extensive training on a specific odorant causes an increase in the number of glomeruli activated by the odorant (Salcedo et al., 2005). This increased activity in the olfactory bulb may be responsible for the observed decrease in the threshold of detection. Although very little research focused on the reduction of threshold as a function of experience in dogs, Gazit et al. (2005) found that the probability of detecting 30 g of TNT placed along a path increased with experience from 80 to 94%.

IMPLICATIONS FOR EDDS

The improved ability to discriminate between odors and the lowered threshold following continued exposure present both beneficial and detrimental implications for

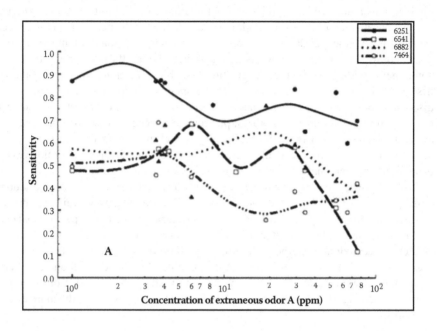

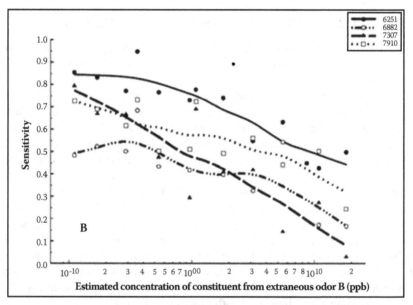

FIGURE 8.8   Effects of a simple masking odor (A) and a complex masking odor (B) on the detectablity of the target odor. (From Waggoner et al., 1998.)

EDDs and their training. The major benefit is that an EDD is able to detect smaller quantities of explosives and at greater distances. In addition, the animal can better detect the explosive odor in the presence of strong but irrelevant olfactory stimuli (maskers). Unfortunately, the detrimental effects are also important.

The first reason is practical; if a dog is well trained and can detect extremely low concentrations of explosives, in many cases, residual odors that remain after the explosive is removed will provide enough of a signal that the dog will indicate that explosives are present when they are no longer in the area. This provides a dilemma for a trainer: should he reinforce or not? The second reason is that because of the improved ability to discriminate odors, extensive practice with one specific explosive odor may in effect lock the dog onto the one odor used in training and it will not respond to other sources of the same odor. This may be the reason that EDDs trained with specific in-house explosives (trainer aids) often cannot detect explosives or trainer aids from other sources.

In addition, no matter how similar explosives simulants are to the genuine odor, extensive experience with the simulant in the absence of training on the genuine odor may prevent an animal from detecting explosives during operational searches. It is very important to overcome the sharpened generalization gradient caused by overtraining with only one exemplar of an explosive. It is therefore suggested that all trainers frequently change their trainer aids and make sure each dog is trained on as many exemplars of each explosive as possible.

EFFECTS OF INTENSITY ON ODOR DISCRIMINATION

An important concept of perception is perceptual constancy. An odor should smell the same regardless of its intensity. We know that the activation of olfaction glomeruli is in part a function of the intensity of an odor. This can easily be seen in Figure 8.9. For many odors, regardless of the pattern of activation caused by intensity differences, the perception of the odor remains unchanged. A subject can recognize the odor regardless of its intensity and changing patterns of activation (Cleland et al., 2007). This is of crucial importance for EDDs which, during searches, first encounter very low concentrations of a target odor; and may encounter much greater intensities as they approach their targets.

However, this is not true for all odors. For many odors, as the concentration increases, the perceived quality of the odor changes (Wilson and Stevenson, 2006). This has been intensively studied in humans and is well documented. An experiment by Gross-Isseroff and Lancet (1988) found that humans rated odor as different when the concentration differed. Laing et al. (2003) asked subjects to attach descriptors to the same odors at different concentrations. They found that the descriptors used to identify an odor changed markedly as a function of concentration.

Similar findings have been reported in animals. Most striking are the results of Coureaud et al. (2004) who found that young rabbits responded to a mammary pheromone only within a very narrow concentration range. When intensities were outside that range, the kits failed to respond.

These effects of intensity on the quality of odor may have implications for EDDs. It is known that dogs trained only on small amounts of explosives sometimes have

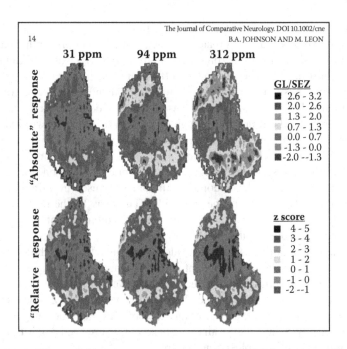

FIGURE 8.9 Effects of intensity on recruitment of glomeruli in the rat olfactory bulb. Absolute measures of glomerular metabolic activity show increases in both intensity and area with increasing odorant concentration (top row), whereas responses relative to other responses in the glomerular layer are constant across odorant concentration (bottom row). Shown are contour charts averaged across three animals exposed to each of three concentrations of the 1-pentanol odorant (Johnson and Leon, 2000a). In the top row, 2DG uptake is expressed as a ratio of glomerular layer (GL) uptake to uptake occurring in a portion of the subependymal zone (SEZ), a region containing immature neurons that do not respond to odorants. Green and warmer colors indicate uptake exceeding that detected in the same locations in animals exposed to air vehicle. In the bottom row, glomerular layer uptake at each location is expressed as a z-score relative to the means and standard deviations of all measurements across the glomerular layer. Green and warmer colors indicate uptake exceeding the mean uptake across the glomerular layer in the same bulbs. Each color bin corresponds to one standard deviation. (From Johnson and Leon, 2007. Reprinted with permission of John Wiley & Sons.)

trouble indicating very large amounts. Conventional wisdom maintains that an animal could recognize the higher intensity odor but could not find the location of the explosive, thus behaving as if it should do something but did not know where to make an indication. Gazit found that a dog trained on 30 g of an odor responded to 5 kg with great interest but would not sit (positive response). However, when it was encouraged to sit by placing a 30 g source next to the 5 kg, it immediately began to sit on the 5 kg samples (unpublished report). An alternative explanation could be that explosives at high intensities smell differently from explosives at low intensities. This suggestion receives some support in the research of Williams et al. (1998) who, as mentioned previously, looked at the responses of dogs to components of C4 at varying concentrations. They found that at least one dog trained to detect C4 responded to cyclohexanone at a concentration of 135 ppb as if it smelled like C4 but at concentrations above or below that value, the same odor was not reported to smell like C4 (Williams et al., 1998). No matter what the mechanism of the concentration effect, it is suggested that EDDs be trained and maintained on a range of intensities of explosive odors they may encounter during operations.

OTHER OLFACTORY VARIABLES INFLUENCING EXPLOSIVES DETECTION

NUMBER AND DURATION OF ODORS STORED IN MEMORY

Williams and Johnston (2002) found that dogs are capable of learning to respond to at least 10 odors and maintain those odors in their working memories. They found no decreases in the performances of the dogs as the number of odors increased. In addition, they found that the dogs could maintain these odors in working memory as long as 4 months between tests.

TRAINER AIDS

A trainer aid is a packaged odor used in training and maintaining training levels. Training aids are essential for teaching animals to respond to relevant odors during operations. Thus, a trainer aid should have the same odor expected to be encountered during operational searches. The use of a correct trainer aid is critical to successful training of EDDs. Unfortunately, trainer aids have a tendency to become contaminated due to faulty handling, storage, weathering, and aging. Increasing numbers of studies indicate the problems of contamination with nontarget odors during olfactory training.

Kay et al. (2005) state that likely influences on mixture perception are contaminants that may have higher vapor pressures than the intended components and thus be more salient. They suggest that even in purchased chemicals of high purity, contaminants may overshadow the target odor. This is supported by unpublished research concerning explosive odor detection by rats. In a series of experiments, Verhagen and Cox (personal communication) found that the probability that a giant African pouched rat would respond positively to a low concentration sample of TNT was strongly influenced by the person who prepared the sample. As shown in Figure 8.10, each sample preparer generated a different level of hits. Since no procedural differences were detected during sample preparation, a possible explanation for these differences among preparers is that the individuals had different intensities of body odor that may have contaminated the samples of TNT odor.

After use for some time, trainer aids become contaminated and must be renewed. We have seen dogs supposedly trained to indicate on an explosive odor respond positively to the wallet of the person who usually handled the explosives. Trainers have also reported that their dogs respond very well to their own trainer aids, but when they receive trainer aids from another organization for some reason the dogs fail to indicate. In addition to the problems of perceptual learning of a specific trained odor, there is always the possibility of contamination of the trainer aid by human scent.

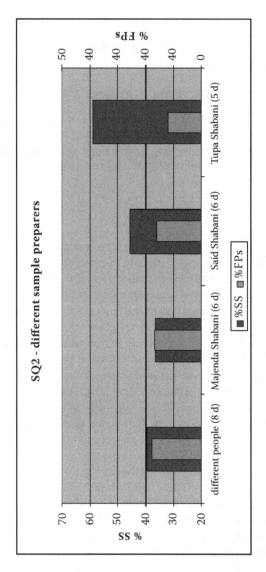

FIGURE 8.10 Probability of a successful detection (SS) and false positive (FP) of a low concentration of TNT as a function of the person preparing the sample. Large individual differences can readily be seen. Intense observation of the preparation procedure failed to detect any procedural differences. (Data from a presentation by Verhagen and Cox delivered at a workshop in Morogoro, Tanzania in 2007.)

Since human odors are much more salient and important to dogs than explosive odors, it is possible that human odors overshadow the odors of the explosives. Thus a dog that is always (inadvertently) trained on a combination of human and explosive odors may not learn to respond to the odor of the explosive and may instead respond to the human odor on the sample.

Although this can be checked by using trainer aids without explosive odors, the presence of human odor causes a more serious problem. If both the trainer aid with explosive odor and the trainer aid without explosive odor are always used, the dog should learn to respond positively only to the trainer aid with explosive. However, the dog could easily learn to search only those areas that contain human scent; in this case, the dog will fail to detect explosives that are not contaminated by human scent. One indication of this possibility comes from reports of trainers that dogs will detect freshly planted explosives but not explosives present in the same location for several days before the search (personal communications).

Another common problem with trainer aids is that if they are stored with other explosives, especially dynamite, they will absorb the odors of other explosives (Furton and Myers, 2001). Thus if the explosives are stored together in a bunker, a dog that is supposedly trained to respond to different explosives such as RDX and dynamite will actually respond to the RDX sample because the sample has absorbed the odor of the dynamite; the dog will not respond to the odor of RDX alone.

A final problem with trainer aids is the lack of a sufficient variety for the same explosive from different vendors and points of origin. The data discussed previously in regard to TNT clearly show that the variability among manufacturers is large enough that a TNT detecting dog may not respond to a sample that differs in origin or age from the trainer aid with which it trained. The same is true for the many types of C4 found on the terrorist market.

NONOLFACTORY TRAINING VARIABLES INFLUENCING DETECTION AND DISCRIMINATION

Much of this section is based on conversations with trainers and talks from conferences. It is difficult to find scientific references that focus on training.

REINFORCEMENT (PLAY VERSUS FOOD)

Two types of reinforcements have been used for training EDDs: play and food. The most prevalent reinforcer is play/retrieval. However, several organizations such as the ATF use food reinforcement. In many cases, the decision arises from tradition since no scientific reason supports preference for one type of reinforcement over another. Comparing the different types of reinforcement is useless because dogs are selected on the basis of motivational considerations. For example, if a dog is selected because it has a very high play drive, it would probably not be as food motivated as a dog selected based on its very high food drive.

According to the ATF (personal conversation), food motivation is used for several reasons. First, the agency trains dogs for use by foreign personnel who are largely unskilled and may be unfamiliar with dogs. In these cases food is much easier to deliver and the probability of not correctly reinforcing the dog is less than if play is used. Additionally, food reinforcement enables many more trials per day. If the trials are handled correctly, a large number of trials presents a definite advantage. It is easier to find food-motivated dogs than play-motivated dogs. On the other hand, trainers who use play/retrieval reinforcement maintain that the drive to find odor is much greater in dogs motivated to find and retrieve. Unfortunately, no good research has studied the relative merits of food versus play (this issue is also raised in Chapter 9 by Smith and Hurt).

DENSITY OF REINFORCEMENT

During EDD training, the question of how many trainer aids should be placed in the training location always arises. During initial training, an animal should encounter as many positive targets as possible in order to establish initial discrimination. However, after the dog knows the odor, the density of targets should be similar to that found in operational situations. Gazit investigated this and found that it was important to present a varying number of positive targets during training. If the number of targets is fixed, search efficiency is diminished (Gazit, unpublished report). Gardner et al. (2001), while investigating the duty cycles of searching dogs, also found that target density was a critical variable. It had to be high enough to keep the dog motivated to search but low enough to allow the dog to search large areas with no target stimuli and maintain its motivation without becoming frustrated.

SCHEDULES OF REINFORCEMENT

Normally, when an animal is differentially reinforced for responding correctly to a stimulus, reinforcement should be delivered with every correct detection (Bailey, personal communication). This maintains the reliability of the discrimination. However, when a dog makes a correct response during an operation, the handler does not know whether the dog has responded to an explosive order or responds because of frustration or simply wants reinforcement. This creates a real dilemma as to whether positive responses in operations should be reinforced. No scientific answer to this question exists. It is likely that the answer depends on the frequency with which the animal gives false positive responses during training. If the frequency is sufficiently low, then we would probably reinforce the animal since a missed target is much more expensive than a false positive.

However, if the normal practice is to not reinforce a positive response during operations (assuming the dog is not responding to a planted trainer aid), it is important to train the animal using a schedule of reinforcement. This enables a dog to learn that it can make a positive response without receiving reinforcement. After a dog learns this concept, lack of reinforcement during an operation will not frustrate the animal or cause extinction of the response. The schedule of reinforcement used is usually around 80%, which means that a dog receives reinforcement for 80% of correct indications during training. This technique is commonly used in probe trial experiments. For example, in the experiment by Williams et al. (1998) on the odor signatures of explosives, the dog was trained with reinforcement for 75% of the correct detections of C4. When the various components of C4 were presented to the animal, it was not reinforced if it responded positively. (Otherwise, the first positive reinforcement on the component would have trained the dog to always respond to that component!)

CONTEXT EFFECT

It is well known that learning is influenced by the context in which it occurs. If an animal is trained in one context to respond to a stimulus and then tested to the same stimulus in a different context, a decrease of the stimulus control often occurs in the new context—almost as if the animal must be partially retrained. The context effect is even stronger in regard to extinction. If an animal is trained in context A and the behavior is extinguished in context B, when the animal is replaced in context A or in a new context C, the behavior reappears. This phenomenon, called renewal, is very robust. It is also used to explain spontaneous recovery in which a previously extinguished response reappears some time after the extinction. See relevant references on the context effect (Bouton and Ricker, 1994; Bouton, Nelson, and Rosas, 1999; Bouton, 2002; Brooks et al., 1995).

The context effect can also affect EDDs. Gazit et al. (2005) trained dogs to find explosives on a stretch of road A. On alternate days, the dogs were told to search for explosives on road B, but no explosives were ever placed on road B. After 10 sessions on road A and road B, explosives were placed on road B at a frequency of one target every third day. The results showed that although the dogs searched very efficiently on road A, they detected fewer than 50% of the explosives on road B. The investigators discounted the effects of the low target density by introducing a new road, C, on which explosives were placed once every third day. The dogs were very efficient at detecting explosives on road C. Although the dogs then received new training on road B (explosives were placed every day and the dogs trained to detect them), the motivation to search road B never recovered.

This result has important implications for training. For example, if it is not possible to place explosives as trainer aids in a shopping mall or airport lounge and dogs are continually required to search that area, they will rapidly learn that no targets are present in the area. The result is reduced ability to detect a genuine target in that area despite continual reinforcement of the dog for correct detections in a secure but different area. This issue is also discussed in Chapter 7 by Helton.

SEARCH IMAGE DEVELOPMENT

The search image is a concept originally formulated by Tinbergen (1960) who studied foraging birds (cited by Croze, 1970). He saw that after a bird started catching an abundant but cryptic insect, it continued to find and prefer that insect even when its abundance decreased and the density of other cryptic insects increased. Tinbergen explained the results by developing the concept of search image, i.e., the bird developed a mental image of the insect that facilitated its location and capture. The search image has been extensively studied with vision in birds but not with olfaction in dogs. Gazit et al. (2005) investigated the possibility of an olfactory search image in EDDs. They reasoned that an olfactory search image may influence the probability of detecting known but less frequently appearing explosives.

EDDs were given extensive training on TNT, PETN, and RDX until all explosives were learned and detected in field studies. When a dog received several days of training on RDX only, it was less able to detect TNT. There are two possible reasons. First, the dog may have developed a search image for RDX that interfered with its ability to detect TNT. Alternatively, since the TNT was harder to detect than the RDX, the dog may have learned to search more rapidly when the target was only RDX. When TNT was placed in the field, its search behavior was such that it missed the TNT (which required a more systematic search). Based on the potential importance of search image to EDDs, it is important to ensure that the dogs are trained and maintained on all relevant explosives.

CUEING AND PROPER CONTROLS

Many things can go wrong in training. One of the most important control procedures is that the trainer/handler be blind as to the locations of targets, both positive and negative. If the trainer knows the correct answer, he or she provides involuntary cues to the animal that enable it to respond correctly. This is called the "Clever Hans" effect, named for a horse that was trained to perform arithmetic. After many years of mystifying scientists, it was discovered that the horse was very alert to subtle nonverbal cues emitted by its trainer or its audiences. The horse used the cues to provide the correct answers. One dog trainer stated that a dog always performs better when the trainer knows the results. The different responses to simulants found by Harper et al. (2005) and Lorenzo et al. (2002) may have arisen from a trainer who knew where the targets were located (personal communication). It is imperative in training that the trainer not know where the trainer aids are planted.

Another problem that often arises is a failure to maintain strict standards after a dog has learned the odors. During initial training, all trainers understand the need for controls and uncontaminated training aids. Unfortunately however, after a dog completes the course and is certified, maintenance is often neglected and/or performed haphazardly, without proper attention to correct procedures. This results in a dog for which, although explosive odor originally had stimulus control over the positive indication, during maintenance, the control may shift to human odors or the odors of contaminants.

A final problem relates to training aids. While trainers and handlers know that the aids should be replaced on a fixed and predetermined basis, they may not have the time to prepare the paperwork or the aids may not be available. In our opinion, the use of contaminated aids is more detrimental than not running a dog at all.

DISEASE

Myers has done extensive work on the effects of diseases on olfaction (Ezeh et al., 1992; Myers et al., 1985 and 1988; Myers et al., 1988). This work is summarized in an article by Furton and Myers (2001). In general, Myers' research shows that many diseases can have permanent (e.g., canine distemper) or temporary (e.g., flu, colds) deleterious effects on a dog's ability to detect olfactory stimuli. This can cause a serious problem because the dog behaves as if it were searching but its nose is not working. The only way to ensure that a dog is really using its nose is to calibrate the dog on a known target before performing operational searches.

POTENTIAL LOCI OF TRAINER MISTAKES

The training and maintenance of an EDD is somewhat like negotiating a minefield. The potential for mistakes exists at every stage, from initial training (imprinting), through search training, advanced training for certification, maintenance, and operations. A misstep at any stage can result in an EDD that does not detect explosives in operational settings. Any attempt of a trainer to rush through a stage may cause a dog to learn something other than what it was supposed to learn. One of the most important training lessons is that it is essential to proceed slowly and carefully; a mistake in training is very difficult if not impossible to correct. A learned behavior (correct or incorrect) is permanently embedded in the brain. It is possible to train over the mistake, but it is impossible to eliminate it (Bouton, 2002).

INITIAL TRAINING

The potential for mistakes starts with the initial training of the explosives odor. It is essential that the odor be uncontaminated and representative of the explosive the dog will have to detect. If the original training odor is contaminated, the situation is known as GIGO (garbage in, garbage out). If the original training is not handled correctly, nothing can be done later to correct it.

At this initial stage, it is important to have already decided whether the training will focus on the bouquet of the explosive or on a dominant component. If the training is targeted to the bouquet then, as discussed previously, the odorants the dog learns will depend on selective attention, i.e., its behavior is not very predictable. If the decision is made to train the dog on a dominant component, then training must also cover exemplars of the explosive that do not contain that component.

It is also important at this stage to run blind trials, not during every session, but at each milestone, to make sure the dog responds to the correct stimuli. It is also important to make sure that the animal will also respond to training aids from other sources and to newly manufactured training aids. The more and varied the training aids, the more likely the dog will learn to respond to the target odor and not to some peculiarity of the training aid used.

SEARCH TRAINING

After a dog learns an odor, it is trained to search for the odor. At this stage it is important to ensure that the dog responds only to the odor of the explosive, and not to the odor of the person placing the training aid. A dog must be trained with different amounts of explosives and their placement should gradually reflect the operational requirements. Periodic blind trials are also important at this stage; the context should vary according to operational demands.

MAINTENANCE

Assuming a dog searches for and responds to the odors of explosives, a trainer must make sure to maintain this behavior. Complacency and sloppiness are the biggest enemies of a good EDD. Stimulus drift refers to the gradual shift of control of a behavior from one cue to another. Stimulus drift is undesirable for any trained behavior and is intolerable for an EDD. The major causes of stimulus shift are sloppy training and maintenance. Great care must be placed on maintaining good training aids and requiring good search behavior. Care must also be taken to continually change the type and amount of explosive a dog seeks.

A problem that sometimes arises at this stage is caused by not rotating training aids of the same type of explosive. Even if a trainer is very careful and does not contaminate an aid, a dog can learn to respond to that particular trainer aid and not to respond to other trainer aids of the same type of explosive. This too is due to perceptual learning. The more practice an animal has on very specific stimuli, the less the generalization it will show. The way to remedy this problem is to utilize as wide a collection of trainer aids as possible.

OPERATIONS

Even if maintenance is handled correctly, problems can arise during operations. It is important that a dog not distinguish between operations and maintenance training. The density of positive targets during an operation should be maintained at the same level used in maintenance. The probability of reinforcement for a positive detection should also be the same as used in maintenance. It is important that a handler not push a dog harder or longer than the extent and duration it experienced during maintenance. Finally, operational searches are not replacements for maintenance training. An operational dog that is not maintained properly will not detect explosives as well as it could.

SUMMARY AND CONCLUSIONS

EDDs rely on olfaction to find the subtle cues of explosives. An understanding of olfaction is essential to train and maintain these dogs. The most important development in the field of olfaction in relation to working dogs is an understanding of the mechanisms of olfaction and the role of perceptual learning in olfaction. Specifically, the effects of experience on both the increased sensitivity toward the training odor and the narrowing of the generalization gradient with continued exposure of the target odor require that training aids be changed often and that they include many exemplars.

The understanding that olfaction is primarily a synthetic sense is also important for the optimal training of EDDs. Specifically, it is recommended that animals train on individual explosives odors and not on mixtures of explosives. After an animal has learned to detect a given explosive's odor, then the odor may be mixed with other odors. The more experience an EDD has with an explosive's odor, the easier it will be for the animal to detect it from other odors.

The olfactory sense is considerably less developed in humans. For that reason it is very difficult to appreciate the olfactory world of the dog. This can result in the inadvertent contamination of training odors without the handler knowing about the contamination. It is therefore essential that extreme care be used with trainer aids throughout the entire training and maintenance programs. This includes the use of controls and periodic blind assessments of performance. Ideally, these assessments should be performed by neutral observers and should involve trainer aids from a source different from the source used during training.

REFERENCES

Ache, B.W. and Young, J.M. (2005). Olfaction: diverse species, conserved principles. *Neuron* 48, 417–430.

Bouton, M.E. (2002). Context, ambiguity, and unlearning: sources of relapse after behavioral extinction. *Biological Psychiatry, 51,* 976–986.

Bouton, M.E., Nelson, J.B., and Rosas, J.M. (1999). Stimulus generalization, context change, and forgetting. *Psychological Bulletin,* 125, 171–186.

Bouton, M.E. and Ricker, S.T. (1994). Renewal of extinguished responding in a second context. *Animal Learning and Behavior, 22,* 317–324.

Brooks, D.C., Hale, B., Nelson, J.B., and Bouton, M.E. (1995).Reinstatement after counterconditioning. *Animal Learning and Behavior, 23,* 383–390.

Brooks, S.E., Oi, F.M., and Koehler, P.G. (2003). Ability of canine termite detectors to locate live termites and discriminate them from non-termite material. *Journal of Economic Entomology, 96,* 1259–1266.

Buck, L.B. (1996). Information coding in the vertebrate olfactory system. *Annual Review of Neuroscience, 19,* 517–544.

Buck, L.B. (2000). The molecular architecture of odor and pheromone sensing in mammals. *Cell, 100,* 611–618.

Cablk, M.E. and Heaton, J.S. (2006). Accuracy and reliability of dogs in surveying for desert tortoise (*Gopherus agassizii*). *Ecological Applications, 16,* 1926–1935.

Chess, A., Buck, L., Dowling, M.M., Axel, R., and Ngai, J. (1992). Molecular biology of smell: expression of the multigene family encoding putative odorant receptors. *Cold Spring Harbor Symposium Quantitative Biology, 57,* 505–516.

Cleland, T.A., Johnson, B.A., Leon, M., and Linster, C. (2007). Relational representation in the olfactory system. *Proceedings of the National Academy of Sciences of the United States of America, 104,* 1953–1958.

Coureaud, G., Langlois, D., Sicard, G., and Schaal, B. (2004). Newborn rabbit responsiveness to the mammary pheromone is concentration-dependent. *Chemical Senses, 29,* 341–350.

Craven, B.A. et al. (2007). Reconstruction and morphometric analysis of the nasal airway of the dog (*Canis familiaris*) and implications regarding olfactory airflow. *Anatomical Record: Advances in Integrative Anatomy and Evolutionary Biology, 290,* 1325–1340.

Croze, H. (1970). Searching image in carrion crows. *Zeitschrift fur Tierpsychologie, 5,* 1–86.

Dalton, P., Doolittle, N., and Breslin, P.A. (2002). Gender-specific induction of enhanced sensitivity to odors. *Nature Neuroscience, 5,* 199–200.

Dominy, N.J., Lucas, P.W., Osorio, D., and Yamashita, N. (2001). The sensory ecology of primate food perception. *Evolutionary Anthropology, 10,* 171–186.

Dryer, L. (2000). Evolution of odorant receptors. *Bioessays, 22*, 803–810.

Eisthen, H.L. (1997). Evolution of vertebrate olfactory systems. *Brain, Behavior, and Evolution, 50*, 222–233.

Engeman, R.M., Rodriquez, D.V., Linnell, M.A., and Pitzler, M.E. (1998). A review of the case histories of the brown tree snakes (*Boiga irregularis*) located by detector dogs on Guam. *International Biodeterioration and Biodegradation, 42*, 161–165.

Ezeh, P.I., Myers, L.J., Hanrahan, L.A., Kemppainen, R.J., and Cummins, K.A. (1992). Effects of steroids on the olfactory function of the dog. *Physiology and Behavior. 51*, 1183–1187.

Firestein, S.A (2004). Code in the nose. *Sci. STKE.* 2004, e15.

Firestein, S. (2001). How the olfactory system makes sense of scents. *Nature, 413*, 211–218.

Firestein, S., Breer, H., and Greer, C.A. (1996). Olfaction: what's new in the nose? *Journal of Neurobiology. 30*, 1–2.

Fletcher, M.L. and Wilson, D.A. (2002). Experience modifies olfactory acuity: acetylcholine-dependent learning decreases behavioral generalization between similar odorants. *Journal of Neuroscience, 22*, RC201.

Fletcher, M.L. and Wilson, D.A. (2003). Olfactory bulb mitral-tufted cell plasticity: odorant-specific tuning reflects previous odorant exposure. *Journal of Neuroscience, 23*, 6946–6955.

Friedrich, R.W. (2004). Odorant receptors make scents. *Nature, 430*, 511–512.

Fuchs, T., Glusman, G., Horn-Saban, S., Lancet, D., and Pilpel, Y. (2001). The human olfactory subgenome: from sequence to structure and evolution. *Human Genetics, 108*, 1–13.

Furton, K.G. and Myers, L.J. (2001). The scientific foundation and efficacy of the use of canines as chemical detectors for explosives. *Talanta, 54*, 487–500.

Garner, K.J. et al. (2001). Duty cycle of the detector dog. FAA Grant 97-G-020.

Gazit, I., Goldblatt, A., and Terkel, J. (2005). Formation of an olfactory search image for explosives odours in sniffer dogs. *Ethology, 111*, 669–680.

Gazit, I., Goldblatt, A., and Terkel, J. (2005). The role of context specificity in learning: the effects of training context on explosives detection in dogs. *Animal Cognition, 8*, 143–150.

Gazit, I. and Terkel, J. (2003). Explosives detection by sniffer dogs following strenuous physical activity. *Applied Animal Behaviour Science, 81*, 149–161.

Ghose, G.M. (2004). Learning in mammalian sensory cortex. *Current Opinion in Neurobiology, 14*, 513–518.

Gibson, E.J. (1953). Improvement in perceptual judgements as a function of controlled practice or training. *Psychological Bulletin, 50*, 401–431.

Goldstone, R.L. (1998). Perceptual learning. *Annual Review of Psychology, 49*, 585–612.

Gross-Isseroff, R. and Lancet, D. (1988). Concentration-dependent changes of perceived odor quality. *Chemical Senses, 13*, 191–204.

Harper, R.J., Almirall, J.R., and Furton, K.G. (2005). Identification of dominant odor chemicals emanating from explosives for use in developing optimal training aid combinations and mimics for canine detection. *Talanta, 67*, 313–327.

Hepper, P.G. and Wells, D.L. (2005). How many footsteps do dogs need to determine the direction of an odour trail? *Chemical Senses, 30*, 291–298.

Hildebrand, J.G. and Shepherd, G.M. (1997). Mechanisms of olfactory discrimination: converging evidence for common principles across phyla. *Annual Review of Neuroscience, 20*, 595–631.

Hubener, F. and Laska, M. (1998). Assessing olfactory performance in an Old World primate, *Macaca nemestrina*. *Physiology and Behavior, 64*, 521–527.

Jinks, A. and Laing, D.G. (1999). A limit in the processing of components in odour mixtures. *Perception, 28*, 395–404.

Jinks, A. and Laing, D.G. (2001). The analysis of odor mixtures by humans: evidence for a configurational process. *Physiology and Behavior, 72,* 51–63.

Johnson, B.A. and Leon, M. (2007). Chemotopic odorant coding in a mammalian olfactory system. *Journal of Comparative Neurology, 503,* 1–34.

Jones, P. (2006). Scents and Sense Ability. *Forensic Magazine,* April–May. 1–4.

Kay, L.M., Crk, T., and Thorngate, J. (2005). A redefinition of odor mixture quality. *Behavioral Neuroscience, 119,* 726–733.

Kay, L.M., Krysiak, M., Barlas, L., and Edgerton, G.B. (2006). Grading odor similarities in a go/no-go task. *Physiology and Behavior, 88,* 339–346.

Kay, L.M., Lowry, C.A., and Jacobs, H.A. (2003). Receptor contributions to configural and elemental odor mixture perception. *Behavioral Neuroscience, 117,* 1108–1114.

King, J.E., Becker, R.F., and Markee, J.E. (1964). Studies on olfactory discrimination in dogs: 3. ability to detect human odour trace. *Animal Behaviour, 12,* 311–315.

Kosaka, K., Toida, K., Aika, Y., and Kosaka, T. (1998). How simple is the organization of the olfactory glomerulus? The heterogeneity of so-called periglomerular cells. *Neuroscience Research, 30,* 101–110.

Laing, D.G. and Francis, G.W. (1989). The capacity of humans to identify odors in mixtures. *Physiology and Behavior, 46,* 809–814.

Laing, D.G. and Glemarec, A. (1992). Selective attention and the perceptual analysis of odor mixtures. *Physiology and Behavior, 52,* 1047–1053.

Laing, D.G., Legha, P.K., Jinks, A.L., and Hutchinson, I. (2003). Relationship between molecular structure, concentration and odor qualities of oxygenated aliphatic molecules. *Chemical Senses, 28,* 57–69.

Laing, D.G., Panhuber, H., Willcox, M.E., and Pittman, E.A. (1984). Quality and intensity of binary odor mixtures. *Physiology and Behavior, 33,* 309–319.

Laska, M. and Freyer, D. (1997). Olfactory discrimination ability for aliphatic esters in squirrel monkeys and humans. *Chemical Senses, 22,* 457–465.

Laska, M. and Hudson, R. (1993). Assessing olfactory performance in a new world primate, *Saimiri sciureus. Physiology and Behavior, 53,* 89–95.

Laska, M. et al. (2005). Detecting danger—or just another odorant? Olfactory sensitivity for the fox odor component 2,4,5-trimethylthiazoline in four species of mammals. *Physiology and Behavior, 84,* 211–215.

Laska, M., Hofmann, M., and Simon, Y. (2003). Olfactory sensitivity for aliphatic aldehydes in squirrel monkeys and pigtail macaques. *Journal of Comparative Physiology, 189,* 263–271.

Laska, M., Seibt, A., and Weber, A. (2000). Microsmatic primates revisited: olfactory sensitivity in the squirrel monkey. *Chemical Senses, 25,* 47–53.

Laska, M., Wieser, A., Rivas Bautista, R.M., and Hernandez Salazar, L.T. (2004). Olfactory sensitivity for carboxylic acids in spider monkeys and pigtail macaques. *Chemical Senses, 29,* 101–109.

Le Berre, E., Béno, N., Ishii, A., Chabanet, C., Etiévant, P., and Thomas-Danguin, T., (2008). Just noticeable differences in component concentrations modify the odor quality of a blending mixture. *Chemical Senses,* doi:10.1093/chemse/bjn006.

Leffingwell, J.C. (2002). Olfaction: update 5. *Leffingwell Reports, 2,* 1–34.

Leon, M. and Johnson, B.A. (2003). Olfactory coding in the mammalian olfactory bulb. *Brain Research Reviews, 42,* 23–32.

Lin, D.Y., Zhang, S.Z., Block, E., and Katz, L.C. (2005). Encoding social signals in the mouse main olfactory bulb. *Nature, 434,* 470–477.

Linster, C. and Smith, B.H. (1999). Generalization between binary odor mixtures and their components in the rat. *Physiology and Behavior, 66,* 701–707.

Livermore, A. and Laing, D.G. (1998). The influence of chemical complexity on the perception of multicomponent odor mixtures. *Perception and Psychophysics, 60,* 650–661.

Livermore, A. and Laing, D.G. (1998). The influence of odor type on the discrimination and identification of odorants in multicomponent odor mixtures. *Physiology and Behavior, 65,* 311–320.

Lorenzo, N. et al. (2003). Laboratory and field experiments used to identify *Canis lupus* var. *familiaris* active odor signature chemicals from drugs, explosives, and humans. *Analytical and Bioanalytical Chemistry, 376,* 1212–1224.

Mandairon, N., Stack, C., and Linster, C. (2006). Olfactory enrichment improves the recognition of individual components in mixtures. *Physiology and Behavior, 89,* 379–384.

McCulloch, M. et al. (2006). Diagnostic accuracy of canine scent detection in early- and late-stage lung and breast cancers. *Integrative Cancer Therapies, 5,* 30–39.

McNamara, A., Magidson, P.D., and Linster, C. (2007). Binary mixture perception is affected by concentration of odor. *Behavioral Neuroscience, 121,* 1132–1136.

Moller, R., Pause, B.M. and Ferstl, R. (1999). Inducibility of olfactory sensitivity by odor exposure of persons with specific anosmia. *Zeitschrift fur Experimentelle Psychologie, 46,* 53–59.

Mombaerts, P. (1999). Molecular biology of odorant receptors in vertebrates. *Annual Review of Neuroscience, 22,* 487–509.

Mombaerts, P. (1999). Seven-transmembrane proteins as odorant and chemosensory receptors. *Science, 286,* 707–711.

Myers, L.J., Hanrahan, L.A., Nusbaum, K.E., and Swango, L.J. (1985). Evaluation of canine olfactory function in health and disease by innate behavioral and electrophysiological techniques. *Chemical Senses, 10,* 411.

Myers, L.J., Hanrahan, L.A., Swango, L.J., and Nusbaum, K.E. (1988). Anosmia associated with canine-distemper. *American Journal of Veterinary Research, 49,* 1295–1297.

Myers, L.J., Nusbaum, K.E., Swango, L.J., Hanrahan, L.N., and Sartin, E. (1988). Dysfunction of sense of smell caused by canine parainfluenza virus infection in dogs. *American Journal of Veterinary Research, 49,* 188–190.

Phelan, J.M. and Webb, S.W. (2002). Chemical sensing for buried land mines: fundamental processes influencing trace chemical detection. SAMD 2002-0909.

Phelan, J.M. and Webb, S.W. (2003). Mine detection dogs: training, operations and odour detection. McLean, I.G., Ed. GICHD, Geneva, Switzerland.

Porter, J., Anand, T., Johnson, B., Khan, R.M., and Sobel, N. (2005). Brain mechanisms for extracting spatial information from smell. *Neuron, 47,* 581–592.

Porter, J. et al. (2007). Mechanisms of scent-tracking in humans. *Nature Neuroscience, 10,* 27–29.

Reaugh, J.E. and Kury, J.W. (2002). A measurement technique to determine the sensitivity of trained dogs to explosive vapor. UCRL-ID-148003.

Reindl-Thompson, S.A., Shivik, J.A., Whitelaw, A., Hurt, A., and Higgins, K.F. (2006). Efficacy of scent dogs in detecting black-footed ferrets at a reintroduction site in South Dakota. *Wildlife Society Bulletin, 34,* 1435–1439.

Rinaldi, A. (2007). The scent of life. *Embo Reports, 8,* 629–633.

Salazar, L.T.H., Laska, M., and Luna, E.R. (2003). Olfactory sensitivity for aliphatic esters in spider monkeys (*Ateles geoffroyi*). *Behavioral Neuroscience, 117,* 1142–1149.

Salcedo, E., Zhang, C.B., Kronberg, E., and Restrepo, D. (2005). Analysis of training-induced changes in ethyl acetate odor maps using a new computational tool to map the glomerular layer of the olfactory bulb. *Chemical Senses, 30,* 615–626.

Schoon, G.A.A. (1998). A first assessment of the reliability of an improved scent identification line-up. *Journal of Forensic Sciences, 43,* 70–75.

Schoon, G.A.A. and Debruin, J.C. (1994). The ability of dogs to recognize and cross-match human odors. *Forensic Science International, 69,* 111–118.

Sicard, G. and Holley, A. (1984). Receptor cell responses to odorants: similarities and differences among odorants. *Brain Research, 292,* 283–296.

Smith, T.D., Bhatnagar, K.P., Tuladhar, P., and Burrows, A.M. (2004). Distribution of olfactory epithelium in the primate nasal cavity: are microsmia and macrosmia valid morphological concepts? *Anatomical Record Part A, 281A,* 1173–1181.

Sokolic, L., Laing, D.G., and McGregor, I.S. (2007). Asymmetric suppression of components in binary aldehyde mixtures: behavioral studies in the laboratory rat. *Chemical Senses, 32,* 191–199.

Staubli, U., Fraser, D., Faraday, R., and Lynch, G. (1987). Olfaction and the "data" memory system in rats. *Behavioral Neuroscience, 101,* 757–765.

Steen, J.B., Mohus, I., Kvesetberg, T., and Walloe, L. (1996). Olfaction in bird dogs during hunting. *Acta Physiologica Scandinavica, 157,* 115–119.

Steen, J.B. and Wilsson, E. (1990). How do dogs determine the direction of tracks? *Acta Physiologica Scandinavica, 139,* 531–534.

Stevens, D.A. and O'Connell, R.J. (1995). Enhanced sensitivity to androstenone following regular exposure to pemenone. *Chemical Senses, 20,* 413–419.

Stockham, R.A., Slavin, D.L., and Kift, W. (2004). Survivability of human scent. *Forensic Science Communications, 6,* 1–10.

Strausfeld, N.J. and Hildebrand, J.G. (1999). Olfactory systems: common design, uncommon origins? *Current Opinion in Neurobiology 9,* 634–640.

Takahashi, Y.K., Nagayama, S., and Mori, K. (2004). Detection and masking of spoiled food smells by odor maps in the olfactory bulb. *Journal of Neuroscience,* 24, 8690–8694.

Takiguchi, N., Okuhara, K., Kuroda, A., Kato, J., and Ohtake, H. (2008). Performance of mice in discrimination of liquor odors: behavioral evidence for olfactory attention. *Chemical Senses,* 33, 283–290.

Thesen, A., Steen, J.B., and Doving, K.B. (1993). Behaviour of dogs during olfactory tracking. *Journal of Experimental Biology, 180,* 247–251.

Tinbergen, L. (1960). The natural control of insects in pine woods. 1. Factors influencing the intensity of predation by song birds. *Archives Neerlandaises de Zoologie, 13,* 265–343.

Vickers, N.J. (2000). Mechanisms of animal navigation in odor plumes. *Biological Bulletin, 198,* 203–212.

Vickers, N.J., Christensen, T.A., Baker, T.C., and Hildebrand, J.G. (2001). Odour-plume dynamics influence the brain's olfactory code. *Nature, 410,* 466–470.

Waggoner, L.P. et al. (1998). Effects of extraneous odors on canine detection. *SPIE Proceedings.*

Wang, H.W., Wysocki, C.J., and Gold, G.H. (1993). Induction of olfactory receptor sensitivity in mice. *Science, 260,* 998–1000.

Wang, L., Chen, L., and Jacob, T.J. (2003). Evidence for peripheral plasticity in human odour response. *Journal of Physiology, 554,* 236–244.

Wasserzug, L. et al. (2002). Development and canine testing of nonexplosive training aids for tagged C4, NG dynamite, and Semtex-H. OMB 0704-0188, 1-94. CTTSO/TSWG.

Welch, J.B. (1990). A detector dog for screwworms (Diptera: Calliphoridae). *Journal of Economic Entomology, 83,* 1932–1934.

Williams, M. and Johnston, J.M. (2002). Training and maintaining the performance of dogs (*Canis familiaris*) on an increasing number of odor discriminations in a controlled setting. *Applied Animal Behaviour Science, 78,* 55–65.

Williams, M. et al. (1998). SPIE Conference on Enforcement and Security Technologies. Vol. 3575. DePersia, A.T. and Pennella, L.J., Eds. Boston: SPIE, 291–301.

Willis, C.M. et al. (2004). Olfactory detection of human bladder cancer by dogs: proof of principle study. *British Medical Journal* 329, 712A–714A.

Wilson, D.A. (2003). Rapid, experience-induced enhancement in odorant discrimination by anterior piriform cortex neurons. *Journal of Neurophysiology, 90,* 65–72.

Wilson, D.A. and Stevenson, R.J. (2003). Olfactory perceptual learning: the critical role of memory in odor discrimination. *Neuroscience and Biobehavioral Reviews, 27,* 307–328.

Wilson, D.A. and Stevenson, R.J. (2003). The fundamental role of memory in olfactory perception. *Trends in Neuroscience, 26,* 243–247.

Wilson, D.A. and Stevenson, R.J. (2006). *Learning to Smell.* Baltimore: John Hopkins Press.

Wilson, R.I. and Mainen, Z.F. (2006). Early events in olfactory processing. *Annual Review of Neuroscience, 29,* 163–201.

Wiltrout, C., Dogra, S., and Linster, C. (2003). Configurational and nonconfigurational interactions between odorants in binary mixtures. *Behavioral Neuroscience, 117,* 236–245.

Yee, K.K. and Wysocki, C.J. (2001). Odorant exposure increases olfactory sensitivity: olfactory epithelium is implicated. *Physiology and Behavior, 72,* 705–711.

Young, J.M. et al. (2002). Different evolutionary processes shaped the mouse and human olfactory receptor gene families. *Human Molecular Genetics, 11,* 1683.

Zhang, Y., Lu, H., and Bargmann, C.I. (2005). Pathogenic bacteria induce aversive olfactory learning in *Caenorhabditis elegans. Nature, 438,* 179–184.

Zhao, H. and Firestein, S. (1999). Vertebrate odorant receptors. *Cellular and Molecular Life Sciences, 56,* 647–659.

Zou, Z.H. and Buck, L.B. (2006). Combinatorial effects of odorant mixes in olfactory cortex. *Science, 311,* 1477–1481.

Zou, Z.H., Li, F.S., and Buck, L.B. (2005). Odor maps in the olfactory cortex. *Proceedings of the National Academy of Sciences of the United States of America, 102,* 7724–7729.

9 Conservation Dogs

Aimee Hurt and Deborah A. Smith

CONTENTS

INTRODUCTION

Used by researchers, managers, and conservationists, a conservation dog team (specially trained dog and its handler) works to locate a biological target of interest, helping professionals obtain information about target species. Such teams are gaining recognition around the world for their specific efforts. For example, in New Zealand, dogs involved in protected and pest species programs are formally recognized as conservation dogs. In North America, descriptions such as conservation detection dogs, wildlife detection dogs, and scat detection or scat sniffing dogs are used.

Conservation dog can, however, serve as a parent term that encompasses dogs trained to: (1) find feces (scat detection or scat sniffing dogs), (2) match biological-based scents (scent-matching dogs; Kerley and Salkina 2007), (3) find live animals, insects, or plants in natural settings, and (4) sort target from nontarget samples in a laboratory-style line-up (discrimination or comparable species differentiation dogs; Harrison 2006). Dogs trained to find contraband, poached, trafficked, or other illegal plants, animals, or animal components are often recognized as wildlife dogs and work in a law enforcement context (for example, the U.S. Department of Agriculture's beagle brigade and New Zealand's Ministry of Agriculture and Forestry). They are not commonly considered conservation dogs, despite their ability to detect similar biological materials. Regardless of their titles, dogs formally trained in detection methods and used with systematic search tactics offer biologists strong tools for locating wildlife or other biological samples, and collecting or documenting their sign.

This chapter will illustrate numerous uses of conservation dogs and their significant contributions to wildlife research and conservation through descriptions of historical and current uses. It will also provide information on the types of dogs and handlers best suited for this type of work and the careful processes of selection, training, and care required. While the field of conservation detection is not new, many technological fields are newly discovering it as complementary technologies advance and interest increases in noninvasive survey and data gathering techniques that eliminate the need to capture or visually observe targets. By discussing some of the ways that conservation dogs have contributed to science, we aim to provide a greater understanding of the field so scientists can better decide whether conservation dogs may have applications in their work. Most people are aware of narcotic and explosive detection dogs (see Chapter 8 by Goldblatt, Gazit, and Terkel). Conservation dogs, while similar in many ways to the better known detector dogs, have unique characteristics and we hope readers will appreciate their uniqueness.

HISTORY OF CONSERVATION DOGS

The incredible olfactory sensory systems of canines have long amazed people. We have been intrigued by the possibility of training dogs to assist us in a variety of odor detection jobs. The best known examples are dogs that work with police, military, and search-and-rescue forces and successfully use scent to locate objects such as drugs, bombs, weapons, land mines, and people. It may be surprising that for decades dogs have served as aids to biologists in wildlife research, management, and conservation and have proven extremely useful in handling a multitude of tasks in this field.

The first documented use of dogs in a conservation capacity dates back to the 1890s in New Zealand. Dogs were trained to find kiwi (*Apteryx* spp.) and kakapo (*Strigops habroptilus*) birds with the purpose of capturing and relocating them to an island free from introduced mammalian predators (Hill and Hill, 1987). Subsequent conservation work in New Zealand used dogs to assist in finding kiwi, takahe (*Porphyrio mantelli*), waterfowl, and upland game birds (Clegg, 1995). Currently, the New Zealand Department of Conservation has implemented a formal National Conservation Dog Program. Under official guidelines and standards, dogs are used for locating myriad protected native species, searching for introduced pest species, and continuing research on reptile and plant applications (see Browne 2005; Figure 9.1). Conservation dogs have proven invaluable to a number of management programs and have advanced the conservation of kiwi and kakapo 20 years beyond recovery efforts without the help of dogs (Browne and Stafford, 2003).

In North America, studies from the 1930s to 1960s also demonstrated that dogs consistently improved the efforts of biologists to locate wildlife and their sign such as hair, feces (scat), and urine (MacKay et al., 2008b, discusses scat detection dogs and canine detection in conservation). Their powerful odor detection abilities allowed

FIGURE 9.1   In New Zealand, a dog waits patiently while handler and technician attach a transmitter to the leg of a kiwi that the dog just located. Under Department of Conservation guidelines, conservation dogs in New Zealand wear well fitted, comfortable muzzles as an extra safety precaution in the search for delicate live protected species. Photo courtesy of Working Dogs for Conservation Foundation.

dogs to assist wildlife biologists with censuses and brood counts, finding nests, capturing animals for banding or tagging, relocating marked animals, collecting animals as laboratory or museum specimens, and searching for animals dying from natural causes or insecticides (Zwickel, 1969).

Through the 1990s, trained dogs continued to contribute to the collection of information on wildlife species. They were used to successfully locate bird nests, ringed seal (*Phoca hispida*) structures (lairs and breathing holes), and urine marks (Crabtree et al., 1989; Evans and Burn, 1996; Furgal et al., 1996) and improve detection of live cougars (*Puma concolor*), brown tree snakes (*Boiga irregularis*), and box turtles (*Terrapene carolina triunguis*). See Logan et al., 1999; Engeman et al., 1998; Schwartz and Schwartz, 1974; and Chapter 10 of this book. Dogs assisted biologists in locating the scats of specific species of study interest such as black-footed ferrets (*Mustela nigripes*; Dean, 1979 and Winter, 1981), wolves (*Canis lupus*), coyotes (*Canis latrans*), black bears (*Ursus americanus*; Paquet 1982–1989, University of Calgary, unpublished data), and Eurasian lynx (*Lynx lynx*; Breitenmoser and Breitenmoser-Wursten, 1984–1994, IUCN/SSC Cat Specialist Group, unpublished data). The collection and analysis of scat samples over many years has allowed biologists to infer information about habitat use, range size, relative abundance, food habits, and parasitology of wild populations (Putman, 1984; Kohn and Wayne, 1997). The use of dogs to help locate scat samples through scenting held great promise. In

1993, a graduate thesis detailed the potential of dogs specially trained in tracking and scent discrimination to help locate wolves and their sign (scats, tracks, urine) and described the possible functions that dogs could serve in studies of elusive wildlife species (Sturdivan, 1993).

By the late 1990s, the record of dogs used in wildlife and conservation work was extensive, but their importance in contributing to conservation and the frequency of their use in survey work was about to increase dramatically. Developments in molecular genetics made it possible to extract DNA from scats and enhance the abilities of biologists to determine characteristics such as species presence, population size, sex ratio, home range, paternity, and kinship from animal droppings (Kohn and Wayne, 1997; Kohn et al., 1999; Mills et al., 2000). Scats appeared to be bountiful data sources, but finding them continued to be a great challenge. To answer population questions, fecal DNA analysis can require hundreds of scat samples. Finding scats from relatively rare animals and cryptic scats in various habitats is difficult (Smith et al., 2003; Wasser et al., 2004). Using formally trained dogs to provide "a more systematic and efficient approach to scat detection" (MacKay et al., 2008b) provided a solution to the difficult problem of finding valuable wildlife scats.

In the state of Washington, in the 1990s, wildlife researchers and a professional narcotic detection dog trainer formed a team to teach dogs to locate scats of specific species in a more systematic way (MacKay et al., 2008b). Dogs were trained to target only scat, not urine and other signs. Trained in a similar manner to other professional scenting disciplines (narcotic, cadaver, search-and-rescue), dog-and-handler teams searched specific transects for scats of bear and other carnivores in the Okanogan National Forest (Wasser, Davenport, and Parker, unpublished data).

To date, many dogs have been selected, trained, and used in a comparable approach to locate scats of a vast number of species in a great diversity of habitats (MacKay et al., 2008b). Scientific literature discusses studies in which dogs located scats to help answer research and management questions for species such as kit foxes (*Vulpes macrotis mutica*; Smith et al., 2006a and b), gray wolves (Beckmann, 2006), fishers (*Martes pennanti*; Long et al., 2007a), cougars (Beckmann, 2006), grizzly bears (*Ursus arctos*; Wasser et al., 2004; Beckmann, 2006), black bears (Wasser et al., 2004; Beckmann, 2006; Long et al., 2007a), bobcats (*Lynx rufus*; Harrison, 2006; Long et al., 2007a), and right whales (*Eubalaena glacialis*; Rolland et al., 2006). Additional systematic uses of dogs were also tested. Dogs demonstrated equal promise in helping biologists locate bird and bat (Chiroptera) carcasses (Homan et al., 2001; Arnett, 2006), invasive animal species (see Chapter 10), desert tortoises (*Gopherus agassizii*; Cablk and Heaton, 2006), and black-footed ferrets (Reindl-Thompson et al., 2006) and testing scent-matching abilities of dogs with individual Amur tiger (*Panthera tigris altaica*) scats (Kerley and Salkina, 2007).

COMPLEMENTARY TOOLS

Many techniques are available for studying wild species and obtaining critical data for research and conservation. One method is radio tracking—capturing individual animals and placing radio, satellite, or global positioning system transmitters on their bodies (Mech and Barber, 2002). Biologists can then follow these individuals with a receiver and antenna or acquire animal location coordinates via a satellite system. Other methods are noninvasive, in that they "do not require target animals to be directly observed or handled by the surveyor" (MacKay et al., 2008a) and include remote cameras, track stations, track surveys, hair collection, and scat collection (Long et al., 2008).

Cameras offer biologists the ability to photograph animals for positive species identification and even for individual identification. Track plates and scent stations involve setting up specific sites with manmade materials and natural substrates, respectively, and bait scents. Tracks left at sites confirm visits of a particular species and can allow biologists to identify presence and track trends in abundance. Hair collection is often achieved by installing barbed wire in an area treated with a bait scent. When an animal investigates the lure, it rubs against the station and strands of hair or fur are pulled and collected; the strands can then be used in DNA analysis. Finally, scat collections can also be subjected to DNA analysis. These and other methods benefit biologists and are practical and productive tools for obtaining data necessary for developing wildlife conservation and management plans. Each method possesses benefits and shortcomings, and conservation dogs serve as complementary tools that augment our ability to gather vital data.

Conservation dogs trained to detect scat of specific species may allow us to increase the number of scats collected. This is particularly important for scats found in low densities. With the aid of dogs, biologists can locate scats in various terrains and with greater accuracy in identification of species than was possible when locating scats without trained dogs (Smith et al., 2003; Wasser et al., 2004; Rolland et al., 2006; MacKay et al., 2008b; Figure 9.2). Additionally, rather than luring a target animal to a fixed baited area, dogs roam a landscape searching for targets and thus allow collection of fine-scale habitat use data. Furthermore, with no equipment to set up and remove, dogs provide tools for compiling data within a single field session (Long et al., 2007b).

Conservation dogs present disadvantages such as the need for extensive training and cost of their upkeep. Dogs are accompanied by their human handlers who cannot traverse challenging terrain as well as dogs. Variability among dogs is to be expected, although training can mitigate some variability. Overall, dogs are highly mobile sensitive detectors that can maximize some forms of data collection. Their potential is fully utilized when realistic expectations of performance are taken into account (Chapter 8).

UTILITY OF CONSERVATION DOGS

Dogs can and have been used to obtain diverse types of information for research and conservation needs. The first question commonly asked is whether the species of interest is present in the area; the location of scat samples followed by subsequent DNA analysis can confirm that answer. Then, from these scats, a variety of other questions can be answered through further analysis and used to make informed management and conservation decisions for species and populations. For example, in Canada, dogs were used to detect scats of grizzly and black bears in montane forests where numerous scats found in two different years allowed biologists to confirm presence and

FIGURE 9.2 Trained dogs can increase the number of samples located. This dog is surrounded by 47 bear scats located in a single day. Photo courtesy of Working Dogs for Conservation Foundation.

determine which areas were utilized by the two species (Wasser et al. 2004). Analysis of these samples led to identification of land use patterns that may aid in future bear management strategies and reveal other important population information.

In another example, in desert scrub and grasslands in California, dogs were used to find scats of San Joaquin kit foxes on specific lands in the San Joaquin Valley (Smith et al., 2006b). Results of these surveys revealed current information on status and relative abundance of kit foxes within their northern, central, and southern ranges. These data served as the basis of recommendations for future conservation efforts concerning this endangered fox. Additionally, kit fox scats located by dogs in particular population areas were used to reveal information on sex ratio, relatedness, movement patterns, scent marking, and size of home range (Ralls and Smith, 2004; Smith et al., 2006a).

In a temperate forest environment of Vermont, dogs were trained to locate scats from black bears, fishers, and bobcats (Long et al., 2007a). They were very effective at locating scats from these three species, and this survey method proved efficient for collecting detection and nondetection data. Moreover, dogs yielded the highest raw detection rates and probability of detection for each target species along with the greatest number of unique detections than other noninvasive survey methods such as remote cameras and hair snare stations (Long et al., 2007b). Similarly, in New Mexico, dogs used to locate scats allowed bobcat presence to be detected at much higher rates (~10 times) than remote cameras, hair snares, and scent stations combined (Harrison, 2006).

Dogs have also proven their ability to search and locate a great number of scats in a water environment. In a North Atlantic right whale study, scat collection rates using detection dogs on boats were more than four times higher than opportunistic methods (Rolland et al., 2006). The authors reported that for the first time adequate numbers of right whale scats were obtained for statistical analyses and allowed endocrine, genetic, disease, and biotoxin studies to be performed. Currently, dogs are undergoing training to assist biologists in locating scat latrines of North American river otters (*Lontra canadensis*) with the intent of using the dogs to search on land and from boats traveling along river corridors (Alexander, 2005).

The range of species for which dogs can search and questions answered from scat samples found continue to expand at increasingly fast rates. For example, DNA isolated from scats collected from grizzly bear and kit fox populations were identified at individual and sex levels and used to estimate the overall and minimum numbers, respectively of individuals in the study areas (Wasser et al., 2004; Smith et al., 2006a). In the Centennial Mountains along the Idaho and Montana border, dogs were used to locate scats of four species simultaneously (black bears, grizzly bears, wolves, and cougars). The locations of these scat samples will be used to identify specific areas or zones that can support these carnivores at low densities over time (Beckmann, 2006).

Outside North America, work with conservation dogs has also been increasing. In Brazil, dogs are used in and around Emas National Park to study five cryptic, wide-ranging mammal species: maned wolf (*Chrysocyon brachyurus*), jaguar (*Panthera onca*), cougar, giant anteater (*Myrmecophaga tridactyla*), and giant armadillo (*Priodontes maximus*). Information from these scats will be used to evaluate how these "species are distributed throughout a fragmented landscape mosaic" and "to develop models for how landscape change might affect these populations" (Vynne, personal communication).

The vast number of analyses that can be conducted on scat samples to obtain vital information has made the use of dogs to find scats a very important tool in the field of wildlife research. However, dogs have also shown great success in locating wildlife sign other than scats such as live and dead animals. A preliminary evaluation demonstrated that dogs located more bat carcasses than humans during fatality searches conducted at wind energy facilities (Arnett, 2006). Well-trained dogs were also effective in locating live desert tortoises on the surface and in burrows under a range of environmental conditions (Cablk and Heaton, 2006). Trained dogs have also been used to locate three-toed box turtles in Central Missouri for a long-term study on population characteristics, home ranges, and movements (Schwartz and Schwartz, 1974), and more recently to locate the Florida box turtle (*Terrapene carolina bauri*) to determine its role as a seed dispersal agent (Liu et al., 2004).

Additionally, dogs have been evaluated on their abilities to detect black-footed ferret presence in black-tailed prairie dog colonies, and results demonstrated that dogs offered great promise as additional measures for monitoring populations of ferrets and determining presence data (Reindl-Thompson et al., 2006). Furthermore, trained dogs are used to detect snakes in human environments (see Chapter 10). Finally, in New Zealand, evaluation of the ability of dogs to detect reptile scents demonstrated that they were able to detect Cook Strait tuatara (*Sphenodon punctatus*),

Marlborough green gecko (*Naultinus manukanus*), and forest gecko (*Hoplodactylus granulatus*) scents with high success and could potentially be used for further conservation work with these species (Browne, 2005). In general, the use of conservation dogs as data collection tools is proving quite valuable and contributing greatly to the state of knowledge about a variety of animal species worldwide.

Dogs can also be taught to alert to the presence of plant species. In New Zealand, a dog reduced the time to find specimens of the rare wood rose (*Dactylanthus taylorii*) by up to six times (Browne and Stafford, 2003). Another unique study in Montana tested the ability of dog–handler teams versus unaided humans in detecting invasive spotted knapweed plants (*Centaurea biebersteinii* D.C.; Goodwin et al., 2006). The authors reported that dogs matched or surpassed humans in accuracy, detection distance, and search time for knapweed across three seasons of data collection. Furthermore, the dog–handler teams covered more area, possibly resulting in more thorough searches and higher accuracies. The ability of dogs to assist human surveyors in locating rare native plants such as Kincaid's lupine (*Lupinus sulphureus kincaidii*) in Oregon is under evaluation (Vesely, Smith, and Whitelaw, unpublished data; Figure 9.3). Preliminary results revealed that dogs successfully recognize naturally growing native lupine plants and independently and effectively detect the plants, while ignoring all other plant species on the landscape. Hence, the use of trained detection dogs as survey tools may enhance programs to conserve threatened and endangered plants and investigate invasive weeds.

FIGURE 9.3 Detecting rare plants requires a careful search strategy. Here a handler maintains a slow walk along a transect line and the dog works carefully with his nose low to the ground. Other targets may require dogs to work on-leash, others require dogs to range further to cover more area. Photo courtesy of Working Dogs for Conservation Foundation.

CAPACITIES FOR SCENT DETECTION

The canine ability to detect and discriminate odors leads to remarkable versatility for specially trained dogs. While *scent detection* is the perception of odors, *scent discrimination* is the ability to distinguish one odor from another, and dogs excel at both functions (see Chapter 6 by Lit). In general dogs' olfactory abilities are not fully understood, and huge variations exist in answers to questions such as how well do dogs smell? How much better than humans do dogs smell? See Chapters 5 and 8.

A complex interplay surrounds the number of olfactory cells that dogs have and the amount of brain power dedicated to processing information gathered by their scent receptor cells. In addition, the determination of the components a dog perceives in an odor it is trained to detect constitutes a specialized science. Cocaine, a standard scent target for narcotic detection dogs, contains a dominant chemical, methyl benzoate, and therefore we understand that methyl benzoate is the odor that dogs recognize as cocaine (Furton et al., 1997). However, most other detection targets are composed of multiple molecular compounds and a dog uses a subset of these compounds to recognize a target. Without understanding precisely how dogs make decisions about odors, we can nonetheless assess their behavioral responses to odors and measure their detection accuracy. Williams and Johnston (2002) determined that dogs trained to detect ten odors did not show diminished abilities to detect these odors, and, in fact, noted that training time decreased as the dogs were trained to more odors. Also, dogs could detect trained odors 4 months after being trained to the odor with no scent training in the intervening time (Johnston, 1999) without suffering decreases in detection ability.

Similarly, discrimination abilities can be measured. Scent matching is a form of discrimination in which a dog matches one odor to another. Analogous to a crime witness matching his or her mental recollection of a perpetrator to a line-up of suspects, a dog is offered an odor to smell and then selects the matching odor from a line-up of odors (see Chapter 6). Scent matching does not require a dog to be trained to any particular odor, but rather trained to the process and sequence of smelling the target to be matched and then methodically sniffing the line-up to find the match. *Discrimination* also refers to the ability of a dog to recognize the target odor to which it was trained and not confuse it with similar odors.

Conservation dogs have demonstrated detection and discrimination abilities, although few controlled studies have focused on this discipline. One conservation dog mastered 11 trained target odors and recognized a target odor 3½ years after the most recent exposure (Whitelaw, personal communication). Four scat targets have successfully been searched for simultaneously, as indicated by dogs finding samples from all trained species regardless of abundance of each type of scat (Beckmann, 2006). A detection distance of 62.8 m was measured for desert tortoises and probably does not describe the maximum detectable distance of tortoises by dogs (Cablk et al., 2008). Another study estimated a detection distance of 1.93 km for North Atlantic right whale scat in the ocean (Rolland et al., 2006).

Conservation dog discrimination prowess continues to be examined. Smith et al. (2003) describe field accuracy rates of 100% in the presence of nontarget scats of sympatric carnivores such as coyote, striped skunk (*Mephitis mephitis*), and

American badger (*Taxidea taxus*). This is not to say that dogs found 100% of target scats deposited on the landscape. Of the scats selected by the dogs that were subjected to DNA analysis, 100% came from the target species, San Joaquin kit fox. In controlled line-ups, dogs were 100% accurate in selecting San Joaquin kit foxes from related foxes (*Vulpes vulpes*; Smith et al., 2003); 88% accurate in selecting bobcats from sympatric carnivores such as gray foxes (*Urocyon cinereoargenteus*) or kit foxes and coyotes (Harrison, 2006); up to 85% accurate at selecting grizzly scat from black bear scat (Hurt, unpublished data); and 87% accurate at matching individual Amur tiger scats to one another (Kerley and Salkina, 2007). It is also commonly reported that dogs are less accurate at correctly ignoring nontarget scats when no target scats are present in a line-up (Hurt, unpublished data; Smith et al., 2003), and target scats may be missed altogether in a line-up (Harrison, 2006).

TRAINING CONSERVATION DOG TEAMS

With such impressive scenting abilities, dogs clearly exhibit a great deal of natural potential, but careful and consistent training by handlers and trainers is required to achieve this potential and mold dogs into accurate and safe tools that can assist their human partners in conservation. We will discuss the general selection, training, and fielding of one subset—scat detection dogs—as in many respects, this group allows the greatest variety of dogs to be successful.

A field-ready dog–handler team must be able to search a natural area for deposited scats. The handler must recognize the dog's searching and scenting behavior and be able to direct and support the dog while allowing it independence of thought and movement. In turn, the dog must respect the handler's authority and yield to direction while maintaining a general attitude of independence and confidence. The dog must recognize the target scent, locate it, and communicate the find to its handler. It must be motivated to locate the object and willingly search for hours on end with the intent of finding the target while avoiding distractions and enticements such as wildlife, domestic animals, and human activity. Successful teams are created when the proper dogs are selected, partnered with appropriate handlers, given sufficient training, and live structured lives with maintenance training.

DOG SELECTION

People typically wonder whether certain breeds work best as conservation dogs. Generally speaking, all breeds have the inherent olfactory ability for most forms of detection work and no single breed presents a particular advantage. However, a dog must demonstrate qualities that may be more common among the sporting and herding breeds. In early years, most dogs used to assist in research studies were sporting breeds (Zwickel, 1969). Based on particular research and management needs, many of these breeds continue to be utilized by biologists in the field, but more dogs from a variety of pure and mixed breeds that possess certain unique characteristics appropriate for detection are also employed.

Canine drive is generally described as a "motivational characteristic inherent in a dog" (Cablk and Heaton, 2006) or "the propensity of a dog to exhibit a particular pattern of behaviors when faced with particular stimuli" (SWGDOG, 2006). Unlike a behavior that can be created or extinguished, a drive is inherent and can be modestly enhanced or diminished, but not fundamentally altered. Sources differ concerning the numbers and definitions of drives. We use the drive categories described by Brownell and Marsolais (2002). Important characteristics to consider for conservation dogs are social/pack drive, play drive, hunt drive, and prey drive. In addition, other important qualities include work ethic, general morphology, agility and athleticism, general health, and nerve strength/mental fortitude.

Social/pack drive is a "desire to interact with a group" and speaks to a dog's willingness to work with a handler by communicating to the handler and accepting directions and corrections from the handler. As scat detection dogs are fielded without other dogs, they are not required to be exceptionally sociable with other dogs, but the need for sociability may depend on housing arrangements available. A dog who is socially unmotivated may, however, demonstrate too much independence of thought and become hard to manage. A dog with too much social motivation may stay too close to its handler or appear needy or clingy. A moderate level of social/pack drive is required.

Play drive or a "desire to entertain itself or engage in entertaining behavior with others" is required so that a dog will value a toy reward for performing a correct behavior. A dog must endure hundreds of repetitions of an activity to be awarded a short period of play for correctly locating a scent and communicating the find to its handler. Therefore, a dog must be obsessively driven to possess a toy in order to endure the arduous training. An extremely high level of play drive is required (see Figure 9.4). Note that some detection dog trainers prefer to substitute food drive for play drive—the dog works for a food reward instead of a play reward. These divisions in the training community represent fervently held beliefs, but no scientific evidence at present supports play over food rewards or vice versa. Thus the food-versus-play argument remains in the realm of belief and preference. Should food drive be substituted for play drive, an extremely high level of food drive is required.

Hunt drive is the "desire to search" and is required to sustain the dog's motivation to work between opportunities of finding a target and receiving a reward. Generally, most of a scat detection dog's day will be filled with searching as scats within a search area can range from common to infrequent or even nonexistent. An extremely high level of search drive is required.

Prey drive is a "desire to pursue, capture and kill quarry." It is often linked with play and hunt drives. While prey drive is not a requirement for a scat detection dog, it is often present in dogs with the required high play and hunt drives. Low to moderately high levels of prey drive are acceptable. Dogs with low prey drives will be safer in the presence of wildlife, less distracted, and therefore more manageable. Extremely high prey drive is not advised because it presents risks to both dogs and wildlife.

Work ethic or eagerness to work is commonly attributed to people but it is also an important quality for scat detection work. A dog with a high work ethic will begin searching without cues from a handler, continue to work at the cessation of a training exercise, and may appear impatient during periods of rest. Like drives, work ethic may be an innate quality that cannot be taught. However, puppies and adolescent dogs generally do not have the maturity to possess work ethics; maturity is typically

FIGURE 9.4   Dog and handler enjoy playing tug after the dog locates a Kincaid's lupine plant in dense prairie vegetation. The dog's training and work are based on earning the opportunity to enjoy this highly prized reward. Photo courtesy of Working Dogs for Conservation Foundation.

required before a dog can sustain attention and demonstrate its work ethic. Most dogs under 1 year old are too young to focus on training exercises lasting more than a few minutes. Some slow-to-mature breeds may need to be 2 years or older before their attention spans can accommodate the long training required for field preparation and deployment. An extremely strong work ethic is required.

General morphology, agility, physical fitness, and health are other important considerations. For example, brachiocephalic dogs have respiratory challenges that make them unsuitable candidates. Also, scat detection dogs will typically encounter rugged environments. Downfalls, ravines, and other natural obstacles require some jumping and agility skills. Dogs should be of moderate size and have legs long enough to handle various terrains. There is a safety advantage to having a dog light enough to be carried by a handler in the event the animal sustains an injury in the field. Dogs also have varying degrees of temperature tolerance that appear to be individual and unrelated to coat color (skin temperatures are not warmer under a black coat; Chesney, 1997) although related to coat length (see Chapter 13 by Schneider and Slotta-Bachmayr). A heat-tolerant dog will be the most versatile, as overheating reduces search efficiency and can be dangerous for the dog. An extensive veterinary check including blood chemistry analysis and radiographs is necessary to ensure that a dog is in good health and able to enjoy the rigors of training and field work.

Nerve strength or mental fortitude refers to "ability to deal with or adapt to stress producing environmental stimuli." Nerve strength is not a drive, but rather a mix of innate qualities as well as life experience, exposure, and training. Physical challenges such as sharp cacti, rock fields, and inclement weather, and mental challenges such as unfamiliar terrain, traveling, training repetitions, and handler corrections may undermine a dog with poor nerve strength. Dogs that search disaster areas require extremely high nerve strength because they often work in rubble piles and other hazardous environments. They must be able to rest on-site in the midst of bustling activity and generally contend with people, noises, and stressful conditions. Because scat detection takes place in natural environments, dogs do not contend with most man-made stressors and require only moderate nerve strength, although the greater the nerve strength, the more versatile the dog.

While no specific testing protocol has been designed for scat detection dogs, evaluation guidelines for other disciplines may be modified to assess a conservation dog candidate. An assessment such as The Brownell–Marsolais scale offers many exercises that can help select a potential dog. Appropriate candidate dogs are rare. In animal shelters, where the qualities of play and hunt drive can be evaluated, only one of 200 to 300 dogs will exhibit the degree of play and hunt drives necessary to be selected. Of those selected dogs, after beginning training exercises that allow evaluation of the remaining drives and qualities, only 40% complete the training and become field ready (Working Dogs for Conservation Foundation, unpublished data). Typically, this convergence of drives and traits describes an intense, high-energy dog that may be too demanding for living in a home as a pet. For these rare and special animals, a career as a conservation dog provides a rewarding, stimulating, and purposeful life.

SPECIAL CONSIDERATIONS FOR SELECTION

Note that dogs selected for certain types of work within the conservation dog family require certain attributes not fully discussed here. For example, dogs trained to find live protected species require low prey drive so they can remain safe (Cablk and Heaton, 2006). Scent discrimination dogs must stay motivated with simple repetitive tasks to a greater extent than required for a field-going scat dog (Smith et al., 2003).

HANDLER SELECTION

As important as selecting the right dog is selecting the right person as the dog's handler and partner. Approved guidelines by the Scientific Working Group on Dog and Orthogonal Detector Guidelines (SWGDOG) available at www.swgdog.org offer a framework for selecting dogs and handlers along with other evaluation guidelines. Like dogs, some qualities required seem to be innate while others can be learned, but no overall assessment of humans has been developed. SWGDOG guidelines note a combination of training, experience, and personality traits (including work ethic, trainability, teamwork, dedication, judgment) that handlers should possess.

The most explicit description of handler and trainer skills and knowledge probably comes from the Geneva International Centre for Humanitarian Demining (GICHD, 2003). This report identifies 23 tasks and the associated skills and knowledge handlers should have, offers a scoring system, and even groups the tasks into three levels of difficulty. Some qualities they suggest can be learned relatively quickly:

- Ability to observe one's own behavior
- Ability to explain events that happen during training and working situation
- Ability to observe a dog in a training situation

More time and technique are required for candidates to learn other qualities:

- Ability to accept instruction
- Ability to share practical experiences with others
- Interest in learning theory

The hardest set of essential attributes to acquire includes:

- Ability to analyze events as they occur during training
- Ability to set realistic goals for the dog and oneself during a training situation

While the GICHD report offers no specific evaluation exercises, a prospective handler can be evaluated by a professional trainer in the course of training. To some extent, attribute information may be obtained through an interview or by checking professional references.

TRAINING

This training discussion merely constitutes a framework to provide a general understanding of the process. We strongly recommend working with professional trainers who have experience with conservation dogs. The handler skills described above as difficult to attain are typically associated with planning and modifying training and understanding how to evaluate training actions. These skills are attained with experience. While a novice handler/trainer may be able to train a dog to detect an odor, locate its source, and perform an alert, it is highly unlikely that she or he will be equipped to troubleshoot behavioral challenges, customize training plans, or seamlessly transition from training to field exercises. For experienced handler/trainers, we recommend continued education with professional colleagues and evaluation of your team's performance.

Training conservation dogs requires a similar progression to other detection disciplines. A dog must first learn that smelling a particular odor will result in a reward. Then the dog must be taught to perform an alert (e.g. sit) after locating a target. After an unprompted alert is consistently performed, the training focuses on developing search behavior, maintaining fidelity to trained scents, building communication between dog and handler, and maximizing endurance. Throughout the exercises, a dog must be exposed to a diversity of target scents so that it can learn to generalize the odor beyond a few specific scats to all varia that may be encountered in the wild. Training is never finished. Dynamic, thinking animals require regularly scheduled maintenance training and performance changes and glitches are to be expected.

If an exceptional handler and dog are selected and ready to begin training, the team may be generally prepared to begin field work in about 6 weeks. However, several months or more may be required. Professional trainers may elect to train a dog and then pair a handler with the dog for additional weeks of training. No studies currently exist on whether teams trained together perform differently from teams paired after training.

At present, conservation dogs in North America are not subjected to certification requirements or certifying authorities; however, based on the increased use of conservation dogs, this will likely change and would-be conservation dog–handler teams should expect external review. Professional conservation dog trainers utilize internal standards to assess a team's readiness, which at a minimum demonstrate a team's ability to locate a target, have the dog communicate the find to the handler, and perform as a directed, controlled team throughout the search.

CARE OF CONSERVATION DOGS

It is an outdated assumption that working detection dogs must live in kennels, away from family and daily life, although many detection programs have logistical or care and use guidelines best met by quality and careful kenneling. As part of the conservation ethic, we consider it important that conservation dogs live full and active lives in a safe and enriching environment. Both home environments and kennels can offer this experience. Dogs benefit from structure, routine, and boundaries, especially working dogs that obtain their greatest pleasure from training and working and must be in the habit of minding their handlers. Additionally, trained dogs are very valuable, having undergone hundreds of hours of training; they thus become very expensive even if acquisition costs were not high.

Conservation dogs are not pets. Some common activities for pets may pose untenable risks for conservation dogs, such as dog park visits or jarring play (such as leaping for Frisbees). Additionally, rough-housing with owners and unattended free time are examples of social activities typically not appropriate for conservation dogs. These measures are taken to ensure the dog's safety and devotion to its handler and maintain training integrity. As a tradeoff, these dogs enjoy hundreds of hours of attention, training, and togetherness, as well as bonds with their handlers that are unrealized between pet dogs and owners who do not have interdependent relationships.

Due to the high drive, focus, and intensity of these dogs, handlers who maintain their working dogs in their homes describe many life-modifying choices that include crate training, maintaining constant supervision, making special efforts to train dogs for physical condition and scent training, and providing veterinary care, quality food, and supplements that exceed the needs of pet dogs. Would-be conservation dog owners should strongly consider how these care and structure requirements will be accommodated by their lifestyles.

CONSERVATION WORK: A UNIQUE FIELD FOR DETECTION DOGS

Because so many selection criteria and basic training components overlap other detection disciplines, how do conservation dogs vary from other detection dogs? Many conservation dogs are trained to detect noninstinctual targets. Unlike hunting dogs who are bred to seek birds and point and/or retrieve them, scat dogs, plant detection dogs, and scent matching dogs are all trained to detect something of no inherent interest to the dog. Trainers capitalize on the dogs' interest in receiving rewards and couple that with the understanding that detecting an otherwise meaningless target will produce the desired reward.

Conservation dogs will be most versatile if they have low prey drives, yet require relatively minimal nerve strength relative to other detection dog disciplines. In this context *versatile* means the breadth of targets a dog can be trained to detect. A single conservation dog may recognize live animal targets, distinguish scat from many species—carnivores, omnivores, herbivores—as well as plants, insects, parasites, and other animal signs (shed skins, carcasses, eggshells).

Some of these targets can be searched for simultaneously. These wide varieties of targets mean that handlers/trainers must be equipped to customize training and fielding strategies for the target at hand. A single dog may be expected to work on-leash in detailed searches or off-leash and cover many miles a day. Many targets utilize off-leash teams that require additional skills such as alerting when out of sight and at varying distances from the handler, and responding to directional voice commands and corrections—but also require the independence to ignore directional commands issued by handlers when on-scent (see Figure 9.5). Additionally, working a variety of targets means a dog has a variable opportunity for reward. Some targets may not exist in the search area and a dog must be able to work all day without a reward. Other targets are abundant and a dog must obsess over a reward enough to still want it after 50 or more rewarding opportunities during a search.

Typically, conservation dogs maintain longer duty cycles (durations of search time between breaks) than many other detection dogs. Conservation dogs commonly work 2 to 5 hours per day and take handler-directed breaks dependent upon temperature or fatigue. Projects often require dogs to work weeks and sometimes months on end. The Working Dogs for Conservation Foundation typically recommends a baseline of planning for 4 hours of searching per day, 3 days working, 1 day off, 2 days working, 1 day off, and then repeating the cycle. This is based on the subjective assessment that this cycle can be maintained for approximately 1 month while allowing a dog to exhibit similar search vigor at the end of the 4-hour day as he had at the start of the day.

Fatigue, hydration, pad wear, and lameness are taken into account to arrive at this baseline. Although dogs have detected targets in temperatures up to 29.9°C (Nussear et al., 2008), heat-related detection limitations have been suspected at 23°C (Smith et al., 2003). Generally, dogs must take more breaks as temperature warms and thus a search loses some efficiency. Reported linear distances covered in 1 day range from 2 km (Long et al., 2007b) to 10 to 14 km for handlers (Nussear et al. 2008) while dogs cover greater distances than handlers.

FIGURE 9.5 Conservation dogs may work on- or off-leash, but must remain under the direction of handlers at all times. Training, type of target, personality, and terrain determine the distance a dog will range from its handler. Photo courtesy of Working Dogs for Conservation Foundation.

FUTURE OF CONSERVATION DOGS

A survey of peer-reviewed articles clearly shows an increase in the reported use of conservation dogs in scientific research and conservation efforts. We anticipate these contributions to grow as dogs continue to search for new types of targets and assist scientists in answering study questions not yet explored. Additionally, the science of conservation dogs is poised for more exploration. Examinations using physiological parameters such as heart rate and circulating stress hormones to measure the strains placed on dogs (see the application to avalanche search dogs in Chapter 13) may better determine work-and-rest schedules.

Other training and fielding techniques could be maximized by assessing performance limitations on abilities to search for scents simultaneously, determining how handlers influence performance, and whether a food or toy reward offers a performance advantage. The study of conservation dogs offers conservationists, trainers, behaviorists, ecologists, and other scientists a unique portal into understanding wildlife, wild lands, and dogs.

REFERENCES

Alexander, B. (2005). Volunteer wildlife detector-dog training for assisted detection of river otter latrines. Waco: Texas A&M University, ethology internship evaluation.

Arnett, E.B. (2006). A preliminary evaluation on the use of dogs to recover bat fatalities at wind energy facilities. *Wildlife Society Bulletin, 34,* 1440–1445.

Beckmann, J.P. (2006). Carnivore conservation and search dogs: the value of a novel, non-invasive technique in the Greater Yellowstone ecosystem. In Wondrak-Biel, A., Ed. *Greater Yellowstone Public Lands: A Century of Discovery: Hard Lessons and Bright Prospects.* Proceedings of 8th Biennial Scientific Conference on Greater Yellowstone Ecosystem, Yellowstone National Park, Wyoming, 28–34.

Browne, C.M. (2005). The use of dogs to detect New Zealand reptile scents (M.S. thesis). New Zealand: Massey University.

Browne, C. and Stafford, K. (2003). The use of dogs in conservation work in New Zealand. *New Zealand Veterinary Association Companion Animal Society Newsletter, 14,* 58–59.

Brownell, D.A. and Marsolais, M. (2002). The Brownell–Marsolais scale: a proposal for the qualitative evaluation of SAR/disaster K9 candidates. *Advanced Rescue Technology,* 57–67.

Cablk, M.E. and Heaton, J.S. (2006). Accuracy and reliability of dogs in surveying for desert tortoise (*Gopherus agassizii*). *Ecological Applications, 16,* 1926–1935.

Cablk, M., Sagebiel, J., Heaton, J., and Valentin, C. (2008). Olfaction-based detection distance: a quantitative analysis of how far away dogs recognize tortoise odor and follow it to source. *Sensors, 8,* 2208–2222.

Chesney, C. (1997). The microclimate of the canine coat: the effects of heating on coat and skin temperature and relative humidity. *Veterinary Dermatology, 8,* 183–190.

Clegg, S. (1995). *Certification of Dogs Used in Department of Conservation Protected Species Management Programmes.* Wellington: Threatened Species Unit, New Zealand Department of Conservation.

Crabtree, R.L., Burton, F.G., Garland, T.R., Cataldo, D.A., and Rickard, W.H. (1989). Slow release radioisotope implants as individual markers for carnivores. *Journal of Wildlife Management, 53,* 949–954.

Dean, E.E. (1979). *Training of Dogs to Detect Black-Footed Ferrets.* Albuquerque, NM: Southwestern Research Institute.

Engeman, R.M., Rodriquez, D.V., Linnell, M.A., and Pitzler, M.E. (1998). A review of the case histories of the brown tree snakes (*Boiga irregularis*) located by detector dogs on Guam. *International Biodeterioration and Biodegradation, 42,* 161–165.

Evans, M.R. and Burn, J.L. (1996). An experimental analysis of mate choice in the wren: a monomorphic, polygynous passerine. *Behavioral Ecology, 7,* 101–108.

Furgal, C.M., Innes, S., and Kovacs, K.M. (1996). Characteristics of ringed seal, *Phoca hispida,* subnivean structures and breeding habitat and their effects on predation. *Canadian Journal of Zoology, 74,* 858–874.

Furton, K., Hsu, Y., Luo, T., Alvarez, N., and Lagos, P. (1997). Novel sample preparation methods and field testing procedures used to determine the chemical basis of cocaine detection by canines. *Proceedings SPIE* 2941, 56, DOI:10.1117/12.266313.

GICHD. (2003). *Mine Detection Dogs: Training, Operations, and Odour Detection.* Geneva: International Centre for Humanitarian Demining.

Goodwin, K., Jacobs, J., Weaver, D., and Engel, R. (2006). Detecting rare spotted knapweed (*Centaurea biebersteinii* DC.) plants using trained canines. Bozeman: Montana State University.

Harrison, R.L. (2006). A comparison of survey methods for detecting bobcats. *Wildlife Society Bulletin, 34,* 548–552.

Hill, S. and Hill, J. (1987). *Richard Henry of Resolution Island.* Dunedin, NZ: John McIndoe.

Homan, H.J., Linz, G., and Peer, B.D. (2001). Dogs increase recovery of passerine carcasses in dense vegetation. *Wildlife Society Bulletin, 29,* 292–296.

Johnston, J.M. (1999). *Canine Detection Capabilities: Operational Implications of Recent R&D Findings.* Auburn, AL: Auburn University.

Kerley, L.L. and Salkina, G.P. (2007). Using scent-matching dogs to identify individual Amur tigers from scat. *Wildlife Society Bulletin*, *71*, 1341–1356.

Kohn, M.H. and Wayne, R.K. (1997). Facts from feces revisited. *Trends in Ecology and Evolution*, *12*, 223–227.

Kohn, M.H., York, E.C., Kamradt, D.A., Haught, G., Sauvajot, R.M., and Wayne, R.K. (1999). Estimating population size by genotyping faeces. *Proceecings of the Zoological Society of London*, 266, 657–663.

Liu, H., Platt, S.G., and Borg, C.K. (2004). Seed dispersal by the Florida box turtle (*Terrapene carolina bauri*) in pine rockland forests of the lower Florida Keys. *Oecologia*, 138, 539–546.

Logan, K.A., Sweanor, L.L., Smith, J.F., and Hornocker, M.G. (1999). Capturing pumas with foothold snares. *Wildlife Society Bulletin*, *27*, 201–208.

Long, R.A., Donovan, T.M., MacKay, P., Zielinski, W.J., and Buzas, J.S. (2007a). Effectiveness of scat detection dogs for detecting forest carnivores. *Journal of Wildlife Management*, *71*, 2007–2017.

Long, R.A., Donovan, T.M., MacKay, P., Zielinski, W.J., and Buzas, J.S. (2007b). Comparing scat detection dogs, cameras, and hair snares for surveying carnivores. *Journal of Wildlife Management*, *71*, 2018–2025.

Long, R.A., MacKay, P., Zielinski, W.J., and Ray, J.C., Eds. (2008). *Noninvasive Survey Methods for Carnivores*. Washington, D.C.: Island Press.

MacKay, P., Smith, D.A., Long, R.A., and Parker, M. (2008*b*). Scat detection dogs. In Long, R.A. et al., Eds. *Noninvasive Survey Methods for Carnivores*. Washington, D.C.: Island Press, 183–222.

MacKay, P., Zielinski, W.J., Long, R.A., and Ray, J.C. (2008*a*). Noninvasive research and carnivore conservation. In Long, R.A. et al., Eds. *Noninvasive Survey Methods for Carnivores*. Washington, D.C.: Island Press, 1–7.

Mech, L.D. and Barber, S.M. (2002). A critique of wildlife radio-tracking and its use in national parks. Minnesota: Biological Resources Division, U.S. Geological Survey and University of Minnesota.

Mills, L.S., Citta, J.J., Lair, K.P., Schwartz, M.K., and Tallmon, D.A. (2000). Estimating animal abundance using noninvasive DNA sampling: promise and pitfalls. *Ecological Applications*, *10*, 283–294.

Nussear, K., Esque, T., Heaton, J., Cablk, M., Drake, K., Valentin, C., Yee, J., and Medica, P. (2008). Are wildlife detector dogs or people better at finding tortoises? *Journal of Herpetological Conservation and Biology*, *3*, 103–115.

Putnam, R.J. (1984). Facts from faeces. *Mammal Review*, *14*, 79–97.

Ralls, K. and Smith, D.A. (2004). Latrine use by San Joaquin kit foxes (*Vulpes macrotis mutica*) and coyotes (*Canis latrans*). *Western North American Naturalist*, *64*, 544–547.

Reindl-Thompson, S.A., Shivik, J.A., Whitelaw, A., Hurt, A., and Higgins, K.F. (2006). Efficacy of scent dogs in detecting black-footed ferrets at a reintroduction site in South Dakota. *Wildlife Society Bulletin*, 34, 1435–1439.

Rolland, R.M., Hamilton, P.K., Kraus, S.D., Davenport, B., Gillett, R.M., and Wasser, S.K. (2006). Faecal sampling using detection dogs to study reproduction and health in North Atlantic right whales (*Eubalaena glacialis*). *Journal of Cetacean Research and Management*, *8*, 121–125.

Schwartz, C.W. and Schwartz, E.R. (1974). The three-toed box turtle in central Missouri: its population, home range and movements. Jefferson City: Missouri Department of Conservation.

Scientific Working Group on Dog and Orthogonal detector Guidelines. (2006) *SWGDOG SC1abcd-Terminology*. Retrieved July 18, 2008 from: http://www.swgdog.org/

Smith, D.A., Ralls, K., Hurt, A., Adams, B., Parker, M., Davenport, B., Smith M.C., and Maldonado, J.E. (2003). Detection and accuracy rates of dogs trained to find scats of San Joaquin kit foxes (*Vulpes macrotis mutica*). *Animal Conservation*, *6*, 339–346.

Smith, D.A., Ralls, K., Hurt, A., Adams, B., Parker, M., and Maldonado, J.E. (2006a). Assessing reliability of microsatellite genotypes form kit fox faecal samples using genetic and GIS analyses. *Molecular Ecology*, *15*, 387–406.

Smith, D.A., Ralls, K., Cypher, B.L., Clark Jr., H.O., Kelly, P.A., Williams, D.F., and Maldonado, J.E. (2006b). Relative abundance of endangered San Joaquin kit foxes (*Vulpes macrotis mutica*) based on scat-detection dog surveys. *Southwestern Naturalist*, *51*, 210–219.

Sturdivan, C. (1993). Locating wolves with a dog trained in tracking and scent discrimination (M.S. thesis). Washington: Evergreen State College.

Wasser, S.K., Davenport, B., Ramage, E.R., Hunt K.E., Parker, M., Clarke, C., and Stenhouse, G. (2004). Scat detection dogs in wildlife research and management: application to grizzly and black bears in the Yellowhead ecosystem, Alberta, Canada. *Canadian Journal of Zoology*, *82*, 475–492.

Williams, M. and Johnston, J. (2002). Training and maintaining the performance of dogs (*Canis familiaris*) on an increasing number of odor discriminations in a controlled setting. *Applied Animal Behaviour Science*, *78*, 55–65.

Winter, W. (1981). *Black-Footed Ferret Search Dogs*. Albuquerque, NM: Southwestern Research Institute.

Zwickel, F.C. (1969). Use of dogs in wildlife management. In Giles, R.H., Ed. *Wildlife Management Techniques*, 3rd ed. Washington, D.C.: Wildlife Society, 319–324.

10 Working Dogs
The Last Line of Defense for Preventing Dispersal of Brown Treesnakes from Guam

Daniel S. Vice, Richard M. Engeman,
Marc A. Hall, and Craig S. Clark

CONTENTS

INTRODUCTION

The inadvertent introduction of the brown treesnake (*Boiga irregularis*) to Guam resulted in unprecedented losses to the island's fragile ecology and economy. A primary management objective is preventing the spread of brown treesnakes to other locations via Guam's transportation network. To achieve this, snake populations are suppressed in and around port and cargo staging areas through an integrated wildlife damage management approach, with the last line of defense for preventing their entrance into the outbound cargo flow inspection by trained snake detector dogs. The efficacy of working dogs on Guam has been evaluated from a variety of aspects.

Most snakes found in the course of dog inspections have been in immediate positions for transport to locations potentially vulnerable to their invasion; snakes typically removed from the cargo flow are generally smaller than snakes removed using other population control measures. The efficacy of the dog teams at finding randomly hidden snakes (not known to handlers) has remained at 62 to 70% as long as such unannounced plants are parts of regular training. Detector dog teams are limited resources relative to the volume of cargo and transportation flow from Guam. Thus, a thorough understanding of the transportation network is required to prioritize and apply detector dogs for maximal impact in reducing brown treesnake exportation risk. The dog handling teams used in conjunction with other control methods as part of a comprehensive containment program have been highly effective at preventing the transport of brown treesnakes from Guam to other vulnerable locations.

BROWN TREESNAKE IMPACTS ON GUAM

The brown treesnake (*Boiga irregularis*) on Guam is an extreme example of the effects an introduced predator can exert on native insular faunal populations. This snake, native to the northern and eastern coasts of Australia, eastern Indonesia, New Guinea and the Solomon Islands, likely was brought to Guam accidentally through post-World War II shipments of war materiels from New Guinea (Rodda et al., 1992). By the 1970s, native bird populations were absent from all but the northern third of Guam.

Disease and pesticides were first speculated to be responsible for the losses of avifauna (Grue, 1985; Savidge, 1987; Savidge et al. 1992), but predation by the arboreal and nocturnal brown treesnake ultimately was identified as the cause of the birds' disappearances (Savidge, 1987). Guam's wildlife evolved a resilience to the often dramatic habitat changes regularly inflicted by typhoons (and also by World War II; Engbring and Pratt, 1985), but native birds and other potential prey species on Guam had not evolved in the presence of predators such as the brown treesnake.

Of the 12 native species of forest birds on Guam, only the Mariana crow (*Corvus kubaryi*), the Mariana grey swiftlet (*Aerodramu vanikorensis bartschi*), and the Micronesian starling (*Aplonis opaca*) survive in the wild (Wiles et al., 2003; Clark and Vice, 2001). Guam's crow population is extinct. The birds remaining in the wild are offspring of individuals translocated from Rota (National Research Council, 1997; Wiles et al., 2003). Two species, the Guam rail (*Gallirallus owstoni*) and the Micronesian kingfisher (*Halcyon cinnamomina*), were taken into captive breeding programs, with reintroductions of Guam rails initiated in the late 1990s (Anderson et al., 1998; Vice et al., 2001).

Recruitment in Mariana fruit bat (*Pteropus mariannus*) populations, already limited by human harvest, has essentially ceased as a result of predation by the brown treesnake (Wiles, 1987a and b; Wiles et al., 1995). Similarly, several indigenous or endemic species of lizards became extinct or endangered primarily due to brown treesnake predation (Rodda and Fritts, 1992), with only one of the 12 native lizard species appearing in similar density on Guam as on nearby snake-free islands (Rodda and Fritts, 1992).

Brown treesnakes are agricultural pests through depredations on chickens, pigeons, caged songbirds, newborn pigs, kittens, and puppies (Fritts and McCoid, 1991). The arboreal snake also climbs utility poles and wires, causing frequent electrical power failures when their bodies connect live and grounded wires. The outages result in millions of dollars of losses from damaged power equipment, electrical appliances and machines, as well as repair costs and losses of productivity (Fritts et al., 1987). Moreover, the brown treesnake is mildly venomous and readily enters dwellings at night. Many victims have been bitten during sleep. The brown treesnake is rear-fanged and must chew to envenomate its victims, making it primarily a health hazard to infants and small children who are less able to defend themselves from its bite and its constriction. A number of life-threatening snake bite incidents involving children have occurred on Guam (Fritts et al., 1990; Fritts et al., 1994), although no human fatalities are known from brown treesnake bites.

THREAT OF DISPERSAL FROM GUAM

As global commerce increases, a concurrent, exponential increase in accidental transportations of species outside their native ranges has been widely documented (Mack et al., 2000). The brown treesnake presents an acute and chronic threat to areas beyond Guam because it is well suited for transport to and establishment at other locations (Fritts et al., 1999). The range of the brown treesnake on Guam encompasses the entire island, occupying virtually every habitat across urban and rural areas. High densities of snakes can be found in small forested patches in developed areas, landscaped areas adjacent to habitations and other buildings, and military and commercial port areas. Brown treesnakes are highly mobile, agile climbers that seek refuge from heat and light during the daylight.

Many types of cargo, shipping containers, and air and sea transport vessels offer daytime refuge. These elements, coupled with Guam's position as a focal point for commercial and military shipments of cargo and passengers throughout the Pacific, present a significant threat for further dispersal of brown treesnakes (Vice et al., 2003; Vice and Vice, 2004). Brown treesnakes associated with military or civilian transportation from Guam have been sighted on virtually every major island in the tropical western Pacific, including Oahu in Hawaii, Saipan, Tinian and Rota of the Commonwealth of the Northern Mariana Islands (CNMI), Kwajalein, Pohnpei, Chuuk, and Diego Garcia in the Indian Ocean, and Okinawa in the Ryukyu Islands of Japan. Brown treesnakes have also been found on the North American mainland and as far away as Rota, Spain (McCoid et al., 1994; Fritts et al., 1999). An incipient population likely now exists on Saipan, as credible sightings and captures now total over 75 in the past 10 years (McCoid et al., 1994). Even a single dispersal event may yield ecological disaster for a recipient location, considering that the Guam brown treesnake population may have originated from a single female snake (Rawlings et al., 1998).

Not surprisingly, the brown treesnake management objective that has received the most effort and attention to date is to deter their further dispersal beyond Guam (Engeman and Vice, 2001a). Federal control efforts were implemented in 1993 to address this objective (Vice and Pitzler, 2002). The primary areas on Guam targeted for snake control include the commercial and naval wharves, associated warehouses and outdoor cargo staging sites around Apra Harbor, the Won Pat International Airport and its cargo staging facilities, the flight line, warehouses and outdoor cargo staging facilities at Andersen Air Force Base (AAFB), commercial packers and shippers (distributed around central Guam), and military housing areas (high turnover of personnel at military bases daily presents a large amount of cargo associated with household moves). The areas subjected to control have continued to evolve along with greater definition of the cargo traffic flows within and from Guam (Vice et al., 2003).

DETECTOR DOG INSPECTIONS

Snake population suppression techniques such as trapping (Engeman and Linnell, 1998; Engeman and Vice 2001a and b; Vice et al., 2005), oral toxicants (Savarie et al., 2001), and spotlight searches of fences (Engeman et al., 1999; Engeman and Vice 2001a and b) effectively reduce snake populations locally, but snakes occasionally circumvent primary control measures and stow in outbound cargo. To minimize this risk, trained detector dogs are used to search, locate, and remove brown treesnakes prior to the departures of outbound military and commercial cargo and transportation vessels from the island. Each team is comprised of a handler and a unique detector dog assigned to that handler. Jack Russell terriers serve as the breed of choice due to their energetic and aggressive nature and ease of handling in cargo and confined spaces (see Figure 10.1). A variety of commercial and military locations are inspected, with 24-hour availability of handlers and their dogs for conducting inspections.

EVALUATION OF DETECTOR DOG EFFICACY

Efficacy based on risks posed by discovered snakes — Examination of the records of brown treesnakes detected during dog inspections revealed that 80% of the snakes found by the dogs were at high risk for export, with Hawaii, followed by the Micronesian islands, the most frequently identified potential destinations (Engeman et al., 1998a). A subsequent study corroborated these results, also showing snakes located by detector dogs were in positions for immediate export from Guam to vulnerable locations (Vice and Vice, 2004). Typically, snakes found during dog inspections averaged smaller in snout–vent length than snakes captured in traps or from spotlight searches of fences (Vice and Vice, 2004). These smaller snakes are more likely to evade other control measures and also less likely to be spotted by cargo or transportation workers.

Natural disasters such as the typhoons that frequently strike Guam exert substantial impacts on snake habitat and movements, result in increased cargo flow for the recovery process, and damage the traps and fences used in control efforts or force their removal from the environment. This combination of impacts increases the likelihood of the snakes' entry into the cargo flow and elevates the importance of detector dog inspections following such events (Vice and Engeman, 2000). Additionally, stochastic events such as typhoons impact surrounding islands and may dramatically affect transportation volume, as recovery support that originates on Guam or transits the island further increases the risk of brown treesnake dispersal. Vice and Engeman (2000) observed an increase in the rate of brown treesnake discoveries by detector dogs in the 10 weeks following the highly destructive supertyphoon Paka. While that storm, one of the most powerful ever recorded in the world, rendered many brown

FIGURE 10.1   Snake-detecting Jack Russell terrier.

treesnake control technologies inoperable, the detectors dogs were able to resume duties the day after the storm (Vice and Engeman, 2000).

Efficacy based on locating planted snakes — The efficacy of the teams of handlers and their dogs for locating stowed brown treesnakes was investigated by planting live brown treesnakes (in escape-proof containers) in cargo without the knowledge of the handlers responsible for inspecting the cargo (Engeman et al., 1998b; Engeman et al., 2002). See Figure 10.2. When an observer attended the inspection to watch procedures, 80% of the planted snakes were located. Otherwise, 70% of the planted snakes were discovered, but only after such plantings had become routine. Prior to that, efficacy was nearly 50% less. The reasons dog teams missed some planted snakes were attributed to insufficient search patterns by the handlers or the failures of handlers to detect indications from the dogs that snakes were present. The interaction between dog and handler is complex and it is impossible to precisely determine in the latter situation whether (1) the dog did not detect the snake, (2) the dog detected the snake but did not respond, or (3) the handler did not recognize a response by the dog. Continued testing has found efficacy to remain around two-thirds for finding brown treesnakes planted in cargo and fewer missed snakes caused by insufficient search patterns (Engeman et al., 2002). The same study also indicated efficacy was higher for daytime inspections indoors than outdoors.

These studies indicated that discontinuation of the random trials of the dog teams with planted snakes likely would lead to decreased attentiveness to inspection

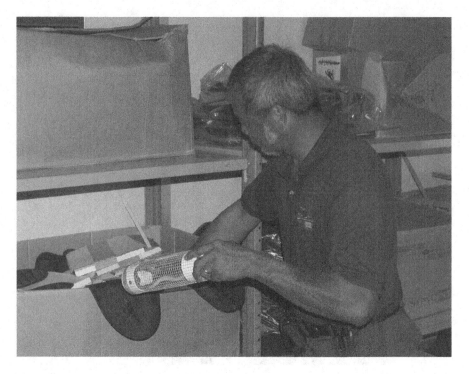

FIGURE 10.2   Planted snake.

procedures and a subsequent decrease in efficacy. Beyond that, finding planted snakes instills confidence in the dogs from their handlers. Similarly, facility workers and managers at sites where inspections occurred have expressed greater confidence and interest in the abilities of the handlers and dogs, leading to more proactive snake control efforts by employees at regularly inspected facilities (Engeman et al., 1998b).

In recent years, the detector dog program has transitioned through aggressive (biting) to passive (sitting) response training protocol to an active (scratching) response. Since the transition to the active response protocol, the Guam program has also instituted a permanent quality assurance program using snakes planted in cargo in a manner consistent with earlier efficacy evaluations. Although not applied in a scientifically rigorous manner, this project has produced detection rates consistently over 75%. This effort is used to enhance and maintain dog team efficacy and build employee morale via a reward system for high achieving canine teams.

DEFINING AND PRIORITIZING EFFORTS

The use of the dog teams to investigate cargo is the result of cooperative arrangements and coordination with agencies, organizations, and companies transporting cargo from Guam. A thorough understanding of cargo transport from Guam is necessary to effectively apply the dogs (and other control methods) as part of a comprehensive containment program. Cargo inspections on Guam are prioritized according to several risk factors, such as probability of establishment in recipient location, type of movement, and size of movement.

Although desirable, it is not logistically feasible under current program structure to search all outbound cargo. To augment existing containment activities on Guam, some recipient locations have established their own detector dog programs. Hawaii conducts detector dog inspections of inbound cargo from Guam using trained beagles. The dogs are available for commercial flights from Guam and they are cross-trained to also detect agricultural products (Kaichi, 1998). Searches of inbound cargo from Guam with trained detector dogs have been conducted for several years on Saipan, Tinian, and Rota in the CNMI. No live brown treesnakes have been located by detector dogs on Hawaii or the CNMI.

The use of dogs to inspect cargo also points to some policy issues (Immamura, 1999) related to training issues such as standards for methods and efficacy across programs and inspection quality control in the face of task monotony. Economic issues relating to vessel delays due to inspection times, refusal of entry for unsearched cargo, costs of certification requirements, expanded search times following a positive dog response, and protocols for handling cargo after a positive response was exhibited but no snake found must be resolved in an acceptable manner. Resolution of such policy issues will insure the efficacy and harmonious coordination of detector dog programs with cargo facilities and the cargo handling process. These points become more acute at recipient locations where the probability of locating a snake appears extremely low.

CARGO CERTIFICATION PROCESSES

In most quarantine programs, recipient locations ultimately shoulder the responsibility for ensuring that goods and vessels arriving at their borders are pest-free. Containment of the brown treesnake on Guam presents a unique quarantine situation in which recipient locations depend upon the prevention of snake incursions into cargo prior to departure from Guam. Quarantine efforts at the origination site create challenges in documenting (certifying) the completion of adequate canine inspections, particularly for cargo.

Commercial surface cargo shipments are consolidated at more than 30 different warehouses around Guam, with contents typically placed inside containers over the course of one or more business days. Upon completion of consolidation, containers are customs-sealed and transported to their respective civilian or military shipping ports. To validate the inspection of all contents on a single manifest, canine teams must be on-site for the entire loading process that may last 4 or more hours. The length of this process renders efficient certification of surface commodities virtually impossible based on the current manner in which cargo is loaded. Managers on Guam are working with local businesses, port operators, and other pertinent organizations to develop procedures for certification that are more conducive to canine inspection requirements.

Commercial air freight presents even more substantial challenges in certification of inspection. Typically, air freight is consolidated into unsecured containers that are closed via Velcro or snapped curtains. No customs seals are applied to the containers (airway bills typically have customs information attached). Shippers and airlines frequently add commodities to a container after initial loading on a space-available basis. Again, program managers on Guam work closely with commercial aviation companies to minimize opportunities for missed inspections via this pathway.

TRAINING AND EVALUATION STANDARDS

A critical component of effective canine detection is consistent, relevant training. All canine handling teams on Guam undergo periodic proficiency training and are tested annually; formal testing consists of written and a canine handling components. The annual testing requirement coupled with the transition to active response protocols has enhanced the efficacy of handling teams, as reflected by increased snake detection rates in the overall quality assurance program. Further improvements in inspection procedures and detection rates can be attributed to the current training regime. Dog handlers must achieve better understanding of canine responses to effectively detect target odors.

DISCUSSION

The integrated wildlife damage management methods for snake removal (Engeman and Vice, 2001), of which detector dog inspection is a vital component, have been highly effective on Guam. As effective as the methods may be individually, they must be carefully applied using the available information on their application or their efficacy will suffer. Also, their use must be integrated to maximize efficiency of the methods and ensure that the scenarios by which a brown treesnake could evade the controls and depart the island are minimized. Continued improvements in understanding snake survivability in transit from Guam, a thorough understanding of cargo flows from Guam (Vice et al. 2003, Perry and Vice, 1998), and public awareness of the need to cooperate with snake control efforts allow more precision and efficiency in the application of detector dogs, a relatively high investment technology in limited supply.

REFERENCES

Anderson, R.D., Beauprez, G.M., and Searle, A.D. (1998). Creation of a snake-free area on Guam using trapping and barrier technology. *Brown Treesnake Research Symposium.* Honolulu, HI. 10.

Clark, C.S. and Vice, D.S. (2001). Protecting the endangered Vanikoro swiftlet from brown treesnakes. *Brown Treesnake 2001 Research and Management.* Andersen Air Force Base, Guam. 9–10.

Engbring, J. and Pratt, H.D. (1985). Endangered birds in Micronesia: their history, status, and future prospects. In Temple, S.A., Ed. *Bird Conservation,* Vol. 2. Madison: University of Wisconsin Press, 71–105.

Engeman, R.M. and Linnell, M.A. (1998). Trapping strategies for deterring the spread of brown treesnakes (*Boiga irregularis*) from Guam. *Pacific Conservation Biology.* 4, 348–353.

Engeman, R.M., Linnell, M.A., Aguon, P., Manibusan, A., Sayama, S., and Techaira, A. (1999). Implications of brown treesnake captures from fences. *Wildlife Research, 26,* 111–116.

Engeman, R.M., Rodriguez, D.V., Linnell, M.A., and Pitzler, M.E. (1998a). A review of the case histories of brown treesnakes (*Boiga irregularis*) located by detector dogs on Guam. *International Biodegradation and Biodeterioration, 42,* 161–165.

Engeman, R.M. and Vice, D.S. (2001a). Objectives and integrated approaches for the control of brown treesnakes. *Integrated Pest Management Reviews, 6,* 59–76.

Engeman, R.M. and Vice, D.S. (2001b). A direct comparison of trapping and spotlight searches for capturing brown treesnakes on Guam. *Pacific Conservation Biology. 7,* 4–8.

Engeman, R.M., Vice, D.S., Rodriguez, D.V., Gruver, K.S. Santos, W.S., and Pitzler, M.E. (1998b). Effectiveness of detector dogs for locating brown treesnakes in cargo. *Pacific Conservation Biology, 4,* 348–353.

Engeman, R.M., Vice, D.S., York, D., and Gruver, K.S. (2002). Sustained evaluation of the effectiveness of detector dogs for locating brown treesnakes in cargo outbound from Guam. *International Biodegradation and Biodeterioration, 49,* 101–106.

Fritts, T. H. and McCoid, M.J. (1991). Predation by the brown treesnake (*Boiga irregularis*) on poultry and other domesticated animals on Guam. *Snake, 23,* 75–80.

Fritts, T.H., McCoid, M.J., and Gomez, D.M. (1999). Dispersal of snakes to extralimital islands: incidents of the brown treesnake (*Boiga irregularis*) dispersing to islands in ships and aircraft. In Rodda, G. et al., Eds. *Problem Snake Management: Habu and Brown Treesnake*. Ithaca: Cornell University Press, 209–223.

Fritts, T. H., McCoid, M.J., and Haddock, R.L. (1990). Risks to infants on Guam from bites of the brown treesnake (*Boiga irregularis*). *American Journal of Tropical Medicine and Hygiene, 42,* 607–611.

Fritts, T.H., McCoid, M.J., and Haddock, R.L. (1994). Symptoms and circumstances associated with bites by the brown treesnake (Colubridae: *Boiga irregularis*) on Guam. *Journal of Herpetology, 28,* 27–33.

Fritts, T.H., Scott, N.J., and Savidge, J.A. (1987). Activity of the arboreal brown treesnake (*Boiga irregularis*) on Guam as determined by electrical outages. *Snake, 19,* 51–58.

Grue, C.E. (1985). Pesticides and the decline of Guam's native birds. *Nature, 316,* 301.

Immamura, C.K. (1999). A preliminary examination of public policy issues in the use of canine detection of brown treesnakes. In Rodda, G. et al., Eds. *Problem Snake Management: Habu and Brown Treesnake*. Ithaca: Cornell University Press, 353–362.

Kaichi, L. (1998). An overview of the state of Hawaii's detector dog program. *Brown Treesnake Research Symposium*. Honolulu, 23.

Mack, R.N., Simberloff, D., Lonsdale, W.M., Evans, H., Clout, M., and Bazzaz, F. (2000). Biotic invasions: causes, epidemiology, global consequences and control. *Issues in Ecology, 5,* 1–20.

McCoid, M.J., Fritts, T.H., and Campbell, E.W., III. (1994). A brown treesnake (Colubridae: *Boiga irregularis*) sighting in Texas. *Texas Journal of Science, 46,* 365–368.

National Research Council. (1997). *The Scientific Bases for the Preservation of the Mariana Crow*. Washington: National Academy Press.

Perry, G. and Vice, D.S. (1998). Evaluating the risk of brown treesnake dispersal in surface and air shipping: lessons from thermal research. *Brown Treesnake Research Symposium*. Honolulu, 30.

Rawlings, L.R., Whittier, J., Mason, R.T., and Donnellan, S.C. (1998). Phylogenetic analysis of the brown treesnake, *Boiga irregularis*, particularly relating to a *population on Guam*. Brown Treesnake Research Symposium. Honolulu, 31.

Rodda, G.H. and Fritts, T.H. (1992). The impact of the introduction of the colubrid snake *Boiga irregularis* on Guam's lizards. *Journal of Herpetology,* 26, 166–174.

Rodda, G.H., Fritts, T.H., and Conry, P.J. (1992). Origin and population growth of the brown treesnake, *Boiga irregularis*, on Guam. *Pacific Science, 46,* 46–57.

Savarie, P.J., Shivik, J.A., White, G.C., Hurley, J.C., and Clark, L. (2001). Use of acetamino-
phen for large-scale control of brown treesnakes. *Journal of Wildlife Management, 65,*
356–365.

Savidge, J.A. (1987). Extinction of an island forest avifauna by an introduced snake. *Ecology,*
68, 660–668.

Savidge. J.A., Sileo, L., and Siegfried, L.M. (1992). Was disease involved in the decimation of
Guam's avifauna? *Journal of Wildlife Diseases, 28,* 206–214.

Vice, D.L., Beck, R., Aguon, C.F., and Medina, S. (2001). Recovery of native bird species on
Guam. *Brown Treesnake 2001 Research and Management.* Andersen Air Force Base,
Guam, 24–25.

Vice, D.S. and Engeman, R.M. (2000). Brown treesnake discoveries during detector dog
inspections following Supertyphoon Paka. *Micronesica, 33,* 105–110.

Vice, D.S., Engeman, R.M., and Vice, D.L. (2005). A comparison of three trap designs for
capturing brown treesnakes on Guam. *Wildlife Research, 32,* 355–359.

Vice, D.S., Linnell, M.A., and Pitzler, M.E. (2003). *Summary of Guam's Outbound Cargo
Handling Process: Preventing the Spread of the Brown Treesnake.* Working Draft
Report, USDA/APHIS/Wildlife Services, Barragada, Guam.

Vice, D.S. and Pitzler, M.E. (2002). Brown treesnake control: economy of scales. In Clark,
L., Ed. *Human Conflicts with Wildlife: Economic Considerations.* Fort Collins, CO,
127–131.

Vice, D.S. and Vice, D.L. (2004). Characteristics of brown treesnakes (*Boiga irregularis*)
removed from Guam's transportation network. *Pacific Conservation Biology, 10,*
216–220.

Wiles, G.J. (1987a). The status of fruit bats on Guam. *Pacific Science. 41,* 148–157.

Wiles, G.J. (1987b). Current research and future management of Marianas fruit bats (Chiroptera:
Pteropodidae) on Guam. *Australian Mammalogy, 10,* 93–95.

Wiles, G.J., Aguon, C.F., Davis, G.W., and Grout, D.J. (1995). The status and distribution of
endangered animals and plants in northern Guam. *Micronesica, 28,* 31–49.

Wiles, G.J, Bart, J., Beck, R.E., and Aguon, C.F. (2003). Impacts of the brown treesnake: patterns of decline and species persistence in Guam's avifauna. *Conservation Biology, 17,*
1350–1360.

11 Canine Augmentation Technology for Urban Search and Rescue

Alexander Ferworn

CONTENTS

INTRODUCTION

This chapter describes the role dogs play in responses to urban disasters. Many people are aware of fire dogs and rescue dogs but few understand the profound difference that a disaster dog makes in the course of searching for people trapped in the rubble left in the wake of a building collapse. It is common practice for emergency first response organizations (EFROs) around the world to use dogs for search and rescue (SAR). Their keen senses of smell, focus, and tenacity allow dogs to detect and locate people who are lost or trapped. Working over vast distances and in large areas, these dogs can often find missing people or rule out an area from further search.

It can be argued that the tragic events of, and subsequent responses to, the Oklahoma City (1995; Figure 11.1) and World Trade Center (2001) terrorist attacks caused many EFROs to become keenly aware that SAR in an urban area after a natural (earthquake, hurricane, tornado, etc.) or man-made (terrorist or military attack) disaster is different from other forms of SAR and require new skills, techniques, and equipment. The need to standardize, certify, and deploy more canine urban search and rescue (USAR or US&R) teams was one of the inevitable results of this growing awareness. This chapter seeks to illuminate the roles of USAR dogs as members of teams that search rubble to save lives.

URBAN DISASTERS

USAR involves the mobilization of resources required to safely and quickly locate, extricate, and bring to safety trapped victims from partially or totally collapsed

FIGURE 11.1 VA-TF2 in front of The Alfred P. Murrah Federal Building in downtown Oklahoma City in 1995; 168 people died and more than 800 were injured. The search involved 11 FEMA Task Forces and dozens of canine teams. (Photo courtesy of Teresa MacPherson.)

structures. It usually involves multiple organizations that provide first responses. USAR is a multidisciplined endeavor involving fire fighters, police, managers, medical personnel, heavy equipment operators, and other related services (Perry, 2003). Urban disasters share common characteristics regardless of how they occur (Solway, 1999):

- They often result from collapses of one or more human-occupied buildings and are caused by single events.
- A common origin causes one or more disasters.
- Little warning of the event is usually available.
- Most direct losses of life and property occur during or shortly after the event.
- Exposure to the event is involuntary.
- The results are so intense that some form of specialized emergency response is necessary as local resources may be overwhelmed in the aftermath.

While SAR operations generally take place in large areas like forests, bodies of water, and mountain ranges, USARs often take place in much more intimate settings (Figure 11.2). While an urban disaster within a sprawling urban area like Mexico City in 1985 (10,000 estimated dead, thousands of collapsed structures), Kobe in 1995 (6,500 estimated dead, 103,521 buildings destroyed), or Sichuan Province in China in 2008 (over 80,000 estimated dead, thousands of collapsed buildings) can be widespread, individual USAR efforts tend to focus on relatively small areas in

FIGURE 11.2   Scene at simulated structural collapse 36 hours after shoring operations commenced to stabilize the structure and gain access to victims trapped inside.

which closely located buildings suffer damage resulting in partial or complete structural collapse.

The structural collapses of multistory buildings often lead to massed casualties from the direct effects of the collapse and possible entombments of many more individuals within new structures or voids formed by collapsing and settling structural elements (McGuigan and Friedman, 2006). Often these voids, while providing temporary safe havens for those trapped within them, have no safe exits. Voids are unmapped and potentially unstable as building materials continue to settle or are further disturbed by new events such as aftershocks from earthquakes, secondary explosions, and movements of equipment on the rubble above them.

Those who live through a structural collapse can expect their everyday existence to be turned upside down. The experience will be accompanied by terrible rending noises, material upheavals and, finally, darkness. Confusion, panic, and uncertainty are common feelings (Bal and Jensen, 2007). Those lucky enough to survive a collapse may experience long hours of anticipation followed by disappointment and despair as they begin to believe no one will find them (Roy et al., 2002).

UTILIZING CANINE SENSE OF SMELL

The ability of dogs to detect a wide range of scents in very small concentrations, including explosives (Ford and Reeve, 2007; Chapter 8 by Goldblatt, Gazit, and Terkel), drugs (Acree, 1976), and even cancer (Pickel et al., 2004) is well known. It is not surprising that this ability can be used to find humans. Human scent is a combination of skin rafts shed in the tens of thousands every minute and volatile chemicals released by our bodies and our apparel (Cablk et al., 2008).

Probably the best-known use of dogs to find people by their scent is tracking. A tracking dog is presented with an article of clothing carrying a missing person's scent and brought to a start point where the person was last seen. This is assumed to be the area of the person's freshest known scent. The dog then follows the scent trail. A handler follows and controls the dog via a tracking harness and lead worn by the dog.

While dogs used for USAR also follow scents, some important differences surround their use. USAR dogs are not presented with scented articles because such articles are probably not available. In a process known as Air Scenting, dogs use the air in an area of interest to find scents of humans who cannot be directly seen. A USAR dog works off-lead, and usually without a collar or harness because it does not follow a track; it searches for scent wherever it can find it. A dog must be free to move around in order to find a human scent. If possible, a handler will attempt to place a dog downwind of the search area to make the task easier. The dog will sample many areas in an attempt to find the largest scent pool and follow it to its largest concentration of scent. Another important difference from tracking is that a USAR dog will not proceed to search from a starting location. Essentially, the dog finds the most likely location of a trapped human through movement and olfactory ability.

SAR dogs employed to search in the aftermath of an urban disaster are called urban SAR (USAR) dogs, US&R dogs, or disaster dogs, canines, or K-9s. This category is relatively new and specialized training is required of both dogs and handlers.

A disaster dog is any canine that has received training in USAR and works urban disasters to find people trapped in the debris of collapsed structures such as office towers, homes, and other occupied structures.

Regional and national organizations have developed certification processes to allow canine teams to prove that they have acquired certain skill sets and can act as USAR canine teams within the context of an emergency task force.* In the United States, the certification process is administered by the Department of Homeland Security (DHS) Federal Emergency Management Agency (FEMA) Canine Subgroup (FEMA, 2008). Similar standards also exist in Canada and other countries. The processes, skills, and procedures for certification constantly undergo revision to improve the "bottom line"—more people found faster.

CANINE USAR PROCESS

The nose of a dog may be the first and best hope of entombed individuals caught in urban disasters. Canine USAR teams consist of a human handler who has been trained as an emergency first responder and canine handler and one or more dogs. The dogs have undergone specialized agility training related to moving around, within, and over unfamiliar and often dangerous environments. They are also conditioned to associate the scent of a hidden live human with a reward. They learn that they must find people they cannot see to earn their rewards.

Often, the canine team is the first element of the EFRO to enter a restricted disaster site (Figure 11.3). The handler and dog will be transported to the search site by whatever means are available. Many canine teams employ specially designed mobile kennels or modified vehicles to carry all the equipment and supplies required to sustain their activities at the scene of a disaster. Table 11.1 lists items carried by emergency task forces in the United States. However, many canine handlers carry extra equipment and supplies related to the safety and comfort of their dogs, including additional first aid supplies, specialized canine apparel, food, and toys. It is not uncommon for volunteer canine handlers to invest thousands of dollars and many hours of work in the wellbeing, performance, and comfort of their dogs.

Teams may need to be moved to a disaster scene via safety harnesses and rope systems handled by other first responders (Figure 11.4). In the first hours of a search, movement of the canine teams into the disaster area is of prime importance. As soon as the search site is reached, handlers will prepare the dogs for the search by giving them the cues they learned in training. Each handler will remove the dog's collar and lead and point the dog in approximately the right direction and verbally encourage it to start searching. Theses dogs are trained to search off-lead and away from their handlers who may not be able to follow them into the search area. It is important to remember that handler and dog work as a team. While a dog searches for scent, the handler will direct the dog by voice commands and hand signals to move to areas that may have been identified by other first responders who may be more familiar with local conditions and have clues as to where people may have been trapped.

---

* In the U.S., FEMA task forces consist of approximately responders along with their personal and special equipment. Each roster includes four canine team positions.

FIGURE 11.3 Mobile kennel truck.

TABLE 11.1
Minimum Equipment for Canine Team

Canine Search Specialist Equipment

Container, food
Harness, tie-down (for air transport)
Kennel, portable, collapsible
Lead, tie-out
Lifting harness
Pad, resting
Stake, tie-out
Shower, solar
Decontamination kit

Source: 2003–2004 Task Force Equipment Cache List, National
Urban Search and Rescue Response System, Federal
Emergency Management Agency. http://www.fema.
gov/emergency/usr/equipment.shtm.

FIGURE 11.4   Dog and handler are moved in a harness. (Photo courtesy of The SOOTODAY. com.)

Because canine teams are normally some of the first rescue groups to reach a disaster, the structures to be searched may not have been assessed as safe for humans to enter. However, a canine search may be allowed to proceed at the discretion of a handler even if difficulties may prevent the handler from following the dog. A structure may be too difficult for a human to enter but a dog may be small and agile enough to gain entry and continue searching.

Much has been written about the amazing canine sense of smell (Brisbin and Austad, 1991; Schoon and De Bruin, 1994; Schoon, 1996; Mesloh et al., 2002; Pickel et al., 2004; Ford and Reeve, 2007; Cablk et al., 2008). Essentially, canine smell is the "gold standard" for olfaction-based detection systems. A dog will use its sense of smell to find scent plume from live humans trapped in rubble (Figure 11.5). It will move to the area with the strongest scent and attempt to get as close to the source of the scent as possible. The scent may be carried through the rubble and disperse in many ways; determining where a scent originates is often quite challenging.

When a dog decides it can get no closer, it will begin a sustained and loud series of barks to indicate the presence of a live human. The location indicated by the dog will be marked, and the dog will be rewarded with a play period. If more casualties are suspected in the area, the dog may be redeployed multiple times or several canine teams may work in rotation.

Barking is considered a sign of life and a good indication of the location of a trapped person. It often determines a point from which a rescue may be attempted because the same plume of scent detected may reveal a way to reach the people

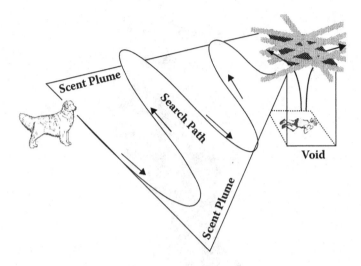

FIGURE 11.5 Following a scent plume to a live human. The diagram has been significantly simplified because a human scent will normally move in various ways through rubble where it will be found by a dog.

trapped in the rubble. After a live human is found and the location marked, the job of the canine team is normally complete.

The length of time it takes to find a person in rubble depends on many factors including the amount and direction of airflow through the rubble, the type of debris searched, the type of collapse, air leakage from the structure, and the experience of the dog. No one may have survived entombment in the rubble. In this case, handlers will often engage volunteers to hide in the rubble so that the dog can find them. Finding people represents a game and the dog's interest in the game must be encouraged, maintained, and reinforced. The paradox for a handler is to remain positive and enthusiastic for the dog even when the result of the search is inconclusive or worse.

USAR CANINE SELECTION AND SCREENING

Successful USAR dogs tend to come from certain breeds. In the case of FEMA-certified teams, most USAR dogs are members of five basic breeds (Table 11.2 and Figures 11.6 through 11.10). Not all dogs become good disaster dogs. The USAR selection process is intended to identify appropriate dogs that will accept training in the least amount of time and present the highest probability of achieving certification within the jurisdiction in which the team will work.

SELECTING PUPPIES

The selection of a particular dog may be based on factors not directly related to the task of finding people and may include the preferences of the handler. While final screening of a candidate usually occurs when the dog is at least a year old, candidate selection can take place in the puppy stage. Since puppies are

TABLE 11.2
FEMA-Qualified USAR Canines by Breed

Breed	Percentage
Labrador Retriever	54.5%
Border Collie	11.0%
German Shepherd	10.5%
Golden Retriever	9.9%
Belgian Malinois	7.3%

Source: DHS/FEMA National Response System Canine
Search Specialist Roster (May 2008).

FIGURE 11.6   Belgian Malinois.

FIGURE 11.7   Border Collie.

relatively easy to find and litters can be large, greater choices of breeds and sexes may be possible. Since puppies are usually not exposed to training, they present few issues related to poor training. Because they are very young, it is feasible to introduce them quickly to a controlled environment that closely resembles that of an urban disaster, using rubble piles and wrecked buildings as play and training areas.

USAR dogs should not be aggressive. Aggressive tendencies can be circumvented and positive behaviors can be established in a young dog. Furthermore, a trainer has more time to condition the dog to become familiar with and work within the USAR dog reward system.

However, puppies may not be the ideal candidates. For example, it is impossible to test whether puppies possess the characteristics of a good USAR dog because so many of their early characteristics change as they mature. Their total training will

FIGURE 11.8   German Shepherd.

also take much longer because intensive training may have to wait a year or more until they mature. Another factor is a longer period during which physical problems may develop and disqualify a dog after considerable time and effort have already been invested.

SELECTING ADULTS

Selecting young adult dogs mitigates some of the problems of puppy selection as the dog has grown into its current characteristics and the behaviors it exhibits are likely to be the ones that it will continue to exhibit—making a prediction of successful training much easier to make.

When the dog is older it has more experience and is likely to take less time to train. In addition, any physical incapacity or problems will likely be exhibited already. However, older dogs are not without their problems and risks. The number

FIGURE 11.9   Golden Retriever.

of available dogs will be limited and their early history may not be known and could be problematic. They may have been socialized in ways that are undesirable including the possibility of hidden aggressive tendencies that may be difficult to curb.

The sources for adult dogs vary from direct purchase from breeders, rescued dogs from pounds, or service and charitable organizations that may have access to dogs.* Some organizations specialize in certain dog breeds such as the various Labrador Retriever rescue organizations.**

SCREENING CRITERIA

Final screening normally occurs when a dog is at least a year old. The screening process is normally undertaken in an area that is unfamiliar to the dog to ensure that the

---

* Humane societies often have suitable dogs as do organizations such as 1-800-save-a-pet.com.
** http://www.lrr.org.

FIGURE 11.10   Labrador Retriever.

behaviors it exhibits are indicative of how it will normally behave in an unfamiliar environment. USAR canines are generally screened for four qualities:

1. Drive—the innate impulse spurring a dog to action
2. Demeanor or behavior in unfamiliar conditions and situations—often called nerve strength
3. Sociability around other dogs and people
4. Physical abnormalities

Drive is simply the amount of emphasis a dog places on performing certain behaviors (van der Waaij et al., 2008). Drive can be used by a handler to improve physical fitness and serve as an indicator of the ability of a dog to focus on a task. The drive of a dog to perform certain behaviors can be used to train it to perform useful functions for people. The drive to follow and retrieve thrown objects is a common example. A number of innate drives are considered highly desirable in USAR dogs. The drive to play with and possess toys indicates that a dog will stay focused and adapt well to a play-for-work reward system. During testing, dogs are observed as they play with toys and their interest is assessed. Dogs that have high play drives will play vigorously with toys and may present the toys to whoever plays with them in attempts to solicit even more play. Figure 11.11 depicts common dog toys.

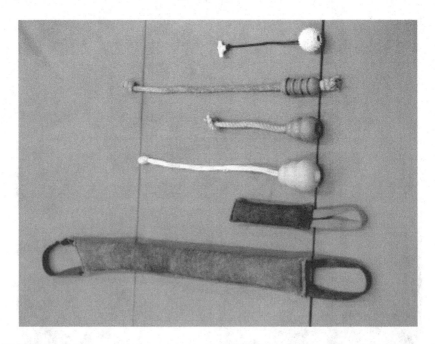

FIGURE 11.11 Common dog toys used as rewards allow a dog to grab with its teeth and pull. Toys vary and include Kongs, ropes, balls, and variants.

A dog's prey drive can be seen when it chases and retrieves thrown toys (Figure 11.12). High prey drive dogs will stay focused on toys and become increasingly enthusiastic as the play continues. Hunting drive is assessed by hiding or throwing a toy into rubble, pallets, or other difficult terrain and observing the response. An assessor can determine how well a dog hunts for a toy and can see whether the dog will hunt by using its nose—clearly a desirable quality in a dog meant to use scent to find people.

An important selection criterion is nerve strength, also referred to as calm demeanor. This is loosely defined as the ability of a dog to tolerate a great number of distractions of varying intensities from external stimuli without responding defensively. This ability is very important because a dog must stay focused on the task of finding people even when surrounded by the many distractions present at a disaster scene. Nerve strength is critical because a dog will work in an inherently unfamiliar, noisy, and dangerous environment—full of hazards, people, and other dogs. A dog must be tolerant of unfamiliar and possible unstable or slippery surfaces. It must be insensitive to height because many surfaces on a rubble pile may be reachable only across beams or may be elevated above other visible surfaces.

Beerda and colleagues (1999 and 2000) reported the stress that some dogs experience because they are sensitive to certain physical and social phenomena around them. Clearly, a dog must be tolerant of close or confined spaces and loud noises that often accompany work at a disaster site. It will encounter situations that expose it to

FIGURE 11.12   Testing for good play drive.

the elements it must be able to tolerate. It must recover quickly if startled and become accustomed to repeated exposure without reacting.

In addition to challenges of physical environments, a dog will work in unfamiliar social situations while deployed. It must ignore or make friendly overtures to the people and canines encountered. Situations will inevitably include handling by unfamiliar people and the nearby presence of unfamiliar dogs for long periods. Nerve strength is especially important for the trapped people who may be found. The dog must be perceived as a hope for rescue, not a threat. The dog must be physically able to perform searches in demanding environments; it must be physically sound. Several breeds of dogs develop hip and elbow problems. Many breeds have tendencies to develop specific ailments. A veterinary examination, tests, and x-rays are required before a dog can be considered for further training.

Many other factors must be considered when selecting a suitable candidate dog. At least one screening tool was proposed by Brownell and Marsolais (2000) to provide objective performance measures in the screening process. Screening does not guarantee success but it selects high probability candidate dogs that exhibit the same characteristics of dogs that have served as good disaster dogs.

USAR CANINE HANDLERS

Generally speaking, USAR canine handlers share a number of characteristics including familiarity with, tolerance of, and fondness for dogs. Handlers must be physically fit because their dogs can move very quickly over broken ground and they must keep up with the animals. A handler must be able to transport his or her dog quickly and efficiently to a work site. In an emergency, a handler may have to carry a dog to procure medical assistance.

The USAR process involves many people in different roles. Handlers are motivated to work in team environments, but they should be able to work independently with their dogs in a search environment that may be extremely dangerous—the best method for searching may be ambiguous. Handlers must have strong commitment to their task since many hours of training and practice are required. They must be reliable because human lives are at risk when their services are requested. Handlers must be trustworthy because they may be required to search crime scenes.

Handlers have varied perspectives on their relationships with their dogs and the jobs they do together. While it is difficult to categorize all types of canine handlers they generally fall into two groups, depending on who owns the dog and whether they serve as volunteers or paid employees. In jurisdictions where canine teams work within a police or fire establishment, the handler cares for the dog but the establishment owns it. The relationship between dog and handler tends to be a partnership. The handler likes the dog but sees it as a tool for finding people. Typically, the state pays all or most of the not inconsiderable expenses of maintaining the dog. They may include housing in the form of a kennel, food, and regular veterinary care. Better-equipped teams may be issued with specially modified vehicles that act as mobile kennels.

USAR canines may also be cross-trained to acquire other skills such as explosive detection or cadaver search to make the investment in the team more cost effective and to keep the dog and handler engaged in related activities.

Many canine teams consist of one or more dogs and a volunteer who owns the dogs and undergoes various levels of training and certification with them. In this situation, canine handlers and/or owners must pay all or most of the costs associated with their animals. These handlers often feel close bonds with the dogs. Their dogs may often fill secondary roles as pets, companions, and even as therapy dogs. In all cases, the dogs probably view their handlers as dominant members of their packs.

CANINE COMMUNICATION

Dogs communicate through innate responses and learned signals. Much can be learned about a dog's state simply by observing its body position and activity. Dogs follow a complex set of behaviors related to their social positions relative to other dogs and their physical and mental states. However, dogs can also learn to communicate through barking and pose. This makes them ideal for USAR work as they can "tell" a handler what is going on when they find something.

TRAINING USAR DOGS

It is always dangerous to anthropomorphize an animal's motivation but a useful heuristic for understanding the innate behaviors of a dog and how they can be modified through conditioning to assist humans is related to self-interest (see discussion of anthropomorphism in Chapter 1). Essentially, dogs will attempt do what they consider best for themselves. While it is difficult to know what a dog thinks, its job is not so much about work as it is about play—elaborate and controlled play. Since the dog appears to want more play, we assume it thinks play is fun. Finding people is a game a dog can learn to play. Through a series of enjoyable activities performed in conjunction with a handler, perspective disaster dogs are introduced to unfamiliar environments and situations that they learn to associate with fun.

Games and orchestrated play build on a foundation of innate canine agility, and natural sensing—dominated by smell—that makes a disaster dog an effective tool for finding live people. Prospective disaster dogs are gradually introduced to increasingly complex tasks starting with movement over and through unstable, multilevel environments like rubble piles (Figure 11.13). This is done through special agility training employing various obstacles on courses designed to familiarize a dog with movements and techniques it may never have used. For example, many obstacles are intended to acquaint a dog with controlling and using its hind legs to climb onto and walk over narrow, high, or unstable surfaces (Figures 11.14 and 11.15). The obstacles also introduce the dog to the confined spaces of tunnels and the need to jump between surfaces (Figures 11.16 through 11.18).

Later the dog will be taught to bark for a reward. This may start with a bark barrel. The dog is rewarded with play for finding and barking at a person—its quarry (Figure 11.19). Initially the quarry will be visible to the dog and later will hide in a barrel, hidden from the dog's view. The goal is to encourage the dog to find hidden people through the use of scent. The bark barrel is a means of accomplishing this goal. The bark barrel will be moved to a rubble environment as the dog becomes more familiar with the game and eventually the barrel is removed and quarry will be hidden in the rubble.

Teaching the right lesson is always a concern. Prospective USAR dogs tend to be intelligent. Care must be taken to thwart the dog's natural self-interest. For example,

FIGURE 11.13   Negotiating elevated plank on agility course.

FIGURE 11.14 Spaced footing obstacle.

if a quarry is always placed in the same hole in a training rubble pile, the dog will soon learn that all it must do to gain a reward is walk to each hole and give a quick sniff rather than expending effort on a proper search of the area. This problem can be challenging for handlers as the number of good training sites is limited and, if they are not careful, their dogs may learn where all the hiding holes are.

FIGURE 11.15 Teeter totter.

FIGURE 11.16   Tunnels.

The dog is also trained to take direction from the handler who will help guide its search. This is done via a series of hand signals, body movements, and voice commands. This is a very important skill as it allows the handler to modify the search pattern while the dog is searching. This may be necessary if a handler receives more information from emergency responders or sees that the path the dog is taking will be fruitless or too dangerous for the dog to traverse.

During a 15- to 20-week training period, a dog bonds with its handler (normally in a few days) and is conditioned to search for hidden live humans using their scents and bark continuously when they are found. The bonding period is important because

FIGURE 11.17   Wooden pallet piles.

FIGURE 11.18   Surface of fencing.

FIGURE 11.19   Bark barrel training.

the dog and handler form a team and each must learn the personality and characteristics of the other so they can work together well.

CANINE SEARCH

To an observer, a searching dog may look like it is solving an extremely complex optimization problem using its nose. As scent pools form around a hidden human, air movement picks up the scent and shifts it around. Some is blown downwind and some forms other scent pools in sheltered areas of a pile of rubble (Figure 11.20). Running a criss-cross or box pattern across a scent plume, a dog samples the available scent and makes decisions about the most fruitful path to follow. However, the dog is not merely following scent. Experienced dogs will pick their paths to facilitate their movements. In other words, they do not follow a path that provides the shortest distance to the source of the scent. They seem to pick paths that allow them to detect the human scent and facilitate their movements. In fact, a dog usually follows the same task sequence a searching robot might follow if such a robot were available. Of course, the dog performs these tasks at high speed with little human intervention.

A dog can often pick up the distinct scent of a human buried many meters inside rubble. The scent moves and can be misleading as it filters through the material in the rubble, is caught by the wind, or artificially pools in still areas. The dog takes all these factors into account and, more often than not, heads for the right area. This performance is unmatched by any other device and is the source of envy for any would-be inventor seeking to replace the disaster dog.

FIGURE 11.20   Canine teams from OPP and Toronto Police Services conduct training on rubble pile.

CONTINUOUS TRAINING

The basic skills required of a USAR team can be learned in several months, but must be continually practiced on a weekly basis. Refresher training is mandatory for canine teams attached to USAR task forces. Many handlers will train at every opportunity. Training is also a mechanism for introducing new skills and new scents. Many canine handlers attempt to vary individuals who act as quarries (Figure 11.21). Their reasoning is that people smell different and their dogs should become familiar with various human scents. Handlers will work with their dogs to inure them against distractions such as dead animals, food, or clothing.

IMPROVING CANINE SEARCH: SITUATIONAL AWARENESS

One major problem in disaster response is determining what is happening at key locations and times. This is the concept of situational awareness. To a large extent,

FIGURE 11.21   Female quarry found buried in drywall board by Dare, a disaster dog.

this concept has been borrowed from more theoretically mature fields such as aircraft control and human–robot interaction (Demchak et al., 2006; O'Brien and O'Hare, 2007; Roman et al., 2008). Situational awareness has been defined as the perception of elements in the environment within a volume of time and space, the comprehension of their meaning, and the projection of their status in the near future (Endsley, 1988). Briefly, this means developing an accurate understanding of what is going on and deciding what to do about it.

Clearly, one area of critical importance in developing a sense of awareness is building an understanding of the situation of a trapped person. This may be more difficult than one might think even when a barking disaster dog indicates a location. When a canine handler and other human responders are precluded from reaching trapped people, they cannot gather useful information about the situations of the people found other than their approximate locations relative to the dog. Even this information may be impossible to ascertain in cases where the dog has moved through or past structural elements that block the view of the rescue team. Even when the dog does not bark, useful information may be missed because the dog cannot communicate it. Situations like this were reported at the site of The World Trade Center disaster in 2001.

In one situation, a handler sent her dog into a void, as requested. She watched her dog make his way through the debris and then disappear from sight. As time passed, the handler's concern grew. Even when she called her dog, she got no response. Finally, a firefighter crouching near the entrance to the void reported that he could see the dog making his way back over the debris. Somehow, the dog found its way out of the wreckage through a distant opening and returned unhurt (ARDA, 2002).

Situational awareness can be examined from many perspectives including that of a trapped individual found by a dog. While the barking dog brings hope that help is near and may soon arrive, the dog cannot convey information. The handler will eventually recall the dog and the person will remain trapped with no indication of when rescue might arrive—hours or days later. This situation can produce great stress and anxiety and possible psychiatric disorders (North, 1999; Hagh-Shenas et al., 2006) along with the physical trauma caused by the collapse.

THE DOG'S VIEW

Barking does not supply enough information to make decisions about how a rescue might be attempted. One obvious solution is to provide a dog a way of telling its handler more about what it experiences while searching. A potential solution is to attach some kind of camera to record the events of the dog's journey. Attempts to attach video recording cameras to disaster dogs working the World Trade Center rubble were reported by Heckedorn (2006). Cameras were attached in several ways. Wireless cameras were attached to dog collars at the same disaster, as reported by Sandia National Laboratories (German, 2002; Report, 2003). Various other camera arrangements have been suggested for related applications (Thomson, 2005).

EXPERIMENTAL USAR CANINE PROGRAM

The Canine Augmentation Technology (CAT) project is a multidisciplinary research effort centered at the Network-Centric Applied Research Team (N-CART)* laboratory in the Department of Computer Science at Ryerson University. The results of this work have been reported by Ferworn et al. (2006 and 2007). The primary goal of the project is to improve the performance of trained USAR canine teams. Note that the goal is to improve team performance as a whole—the performances of the humans must be improved; the dogs actually do very well on their own. They move through rubble, detect people, and indicate where they are. Problems occur when the handler cannot go where the dog goes and therefore does not share the dog's experience. The project was initiated in 2005 and has continued to evolve and expand. In its present form, it consists of five separate subcomponents based around the central theme of improving the situational awareness of the first responders who follow the USAR dogs:

- Canine augmentation technology (CAT)
- Canine remote deployment system (CRDS)
- Canine pose estimation (CPE)
- Canine work apparel (CWA)
- Canine brain function (CBF)

The CAT project attracted the active participation of the Ontario Provincial Police (OPP) almost from inception and the Toronto Heavy Urban Search and Rescue (HUSAR) organization within the City of Toronto Fire Services since 2006. These organizations form the bulk of Canada Task Force 3 (CAN TF-3)—one of five national USAR task forces.

CANINE AUGMENTATION TECHNOLOGY

CAT is a wireless video, audio, telemetry, and sensing system worn by USAR canines that actively search for survivors in areas where their handlers cannot follow. The system has been refined and radically altered as more is discovered about what can actually be sensed, how dogs react to the device, and how dog and system can better work together. Initially, it was believed that the best place to put a camera on a disaster dog was on its head. The reasoning was that a dog generally points its head in the direction of travel and travels in the direction that will lead it to trapped people. The assumption was that the dog's head would point to the trapped people when it arrived at the site where they were buried.

The prototype (CAT I) employed analog video and audio transmission with an effective range of several hundred feet and a harness system that strapped the camera, microphone, and transmitter to the dogs (Figure 11.22). CAT I was tested in various configurations as suggested by existing canine applications and described by Haug et al. (2002). Initial testing proved encouraging as the dogs head is in a

---

* http://ncart.scs.ryerson.ca.

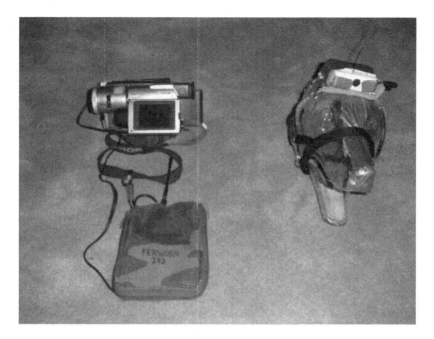

FIGURE 11.22   Early prototype CAT with modified Halti collar and protective housing made from cough drop box.

forward position when it performs certain tasks. For example, a dog tracking human scent will face the human when it finds one. We confirmed this and recorded various images similar to the one in Figure 11.23 of the face of a quarry engaged in rewarding the dog after she was found in a copse of trees.

Our harness designs concentrated on keeping the camera stable on the dog's head and ensuring that the dog could not remove the harness by scratching or rubbing. However, as the harnesses became more secure, they also disturbed the dogs and distracted them from searches. Clearly, performance suffered as the dogs concentrated on removing the harnesses rather than searching. Our testing shifted from the tracking task to searching rubble and the results worsened. Very few camera images related to quarry. We assumed that the problem was with the harness.

We continued testing with a variant of the head harness based on commercially available canine goggles* (Figure 11.24). We believed the dogs would accept the goggles because they were far less restrictive than the harnesses. Initially, this proved true. However, after many trials, it became apparent that the dogs—while accepting the goggles when their handlers were near—universally rejected the goggles after only a few sessions with them. The handlers began to suspect that the goggles interfered with their dogs' peripheral vision. However, what caused our rejection of canine goggles was the failure of the video they recorded (with some exceptions) to reveal much of interest to rescuers.

---

* http://www.dog-goes.com/.

FIGURE 11.23   View of quarry from head-mounted camera system attached to tracking dog.

We concluded that while a dog's head often points in the direction of travel when it uses its eyes, the carriage of its head was unpredictable when the animal used its nose in USAR work. We found that the dogs spent a lot of time looking down or around but rarely looked straight ahead and almost never at the quarry except when receiving a toy as a reward. This is interesting in that it is apparent from some parts of the video that dogs have good abilities to coordinate their movements to take advantage of information they receive only visually. This is strikingly evident in Figure 11.25. The image was taken with a CAT camera (canine goggle variety) on a USAR dog pursuing a Kong toy at approximately 20 kph. However, when dogs employed their noses, visual information appeared to be far less important to the decision making process.

Based on our experience with canine goggles, we devised a body harness that would accommodate a twin camera system (one on each side of the dog), a light system providing both visible and infrared light, and computing and network components. The decision to use twin cameras was not made lightly. While twin cameras provide opportunities to record more information, they also occupy more space on the dog and impose increased power and other requirements. Active electronics generate heat and heat burden must be directed away from the dog because a dog has very few ways to cool itself.

Radio communication at most urban disaster sites is problematic. Reinforcing steel rods, concrete, and other materials inside collapsed structures inhibit reliable radio communications. To mitigate this problem, we designed CAT II to act as a WiFi node in a redundant mesh network to allow multiple paths for data transmission. This type of

FIGURE 11.24   Dare wearing CAT I dog goggles in snow.

network is becoming increasingly available and access nodes can be quickly deployed around a search area. As a dog moves through a search area, the CAT system will connect, disconnect, and reconnect to different peripheral access nodes as the dog travels into and out of transmission range. As the access nodes also connect with each other, they can relay CAT data to a receiving node, usually where the handler was when the dog was released to search. We worked closely with Anvil Technologies* whose network access nodes provided considerable flexibility in our testing.

Efficacy of CAT II

The CAT system has been deployed in two large-scale, realistic, structural collapse exercises (Figure 11.26). The system worked in both cases and quarries simulating trapped people were found and documented. However, the CAT system is challenged

* http://www.anviltech.biz/index.shtml.

FIGURE 11.25 Canine visual tracking of a Kong (lower right) at 20 kph.

FIGURE 11.26 Moose, a USAR canine, searching with CAT II harness.

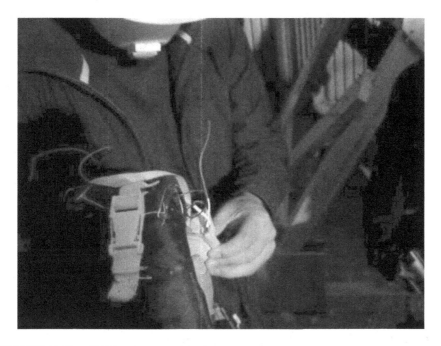

FIGURE 11.27   CAT II eviscerated by rebar.

by agility in that a dog often scrapes against or rubs on structural elements that caused considerable damage to our prototypes (Figure 11.27). In its first deployment in June 2006 in the wreckage of a simulated collapsed parking structure and hotel in Toronto, the system transmitted images of structural elements of the parking garage that the dog passed during its search. Figure 11.28 shows damage at the top of the pillar. Coincidentally, structural engineers independently investigated this damage and selected this pillar for reinforcement with additional shoring that can be seen in the figure.

At its second deployment, the CAT system captured a single image of a crouched individual (Figure 11.29) It was later determined that this crouched figure was, in fact, a quarry placed in the rubble who stood up and was bent over when the dog walked by. The dog was accustomed to not receiving rewards if he barked at people standing in rubble and thus learned not to bark when it saw standing humans.

CANINE REMOTE DEPLOYMENT SYSTEM (CRDS)

The CRDS is a remotely operated release mechanism worn by a disaster dog and triggered by a handler or assistant using a wireless handset (Figure 11.30). When the handler hears a bark indication from the dog, she can release a bag (underdog) by pressing a button on the transmitter. The underdog may contain medical supplies, a radio, food, water, or other sensors (Figure 11.31). Various tests have indicated that the underdog can usually be placed within a foot of where the dog indicates a human scent. The CRDS works effectively as a tool to begin the rescue process even before

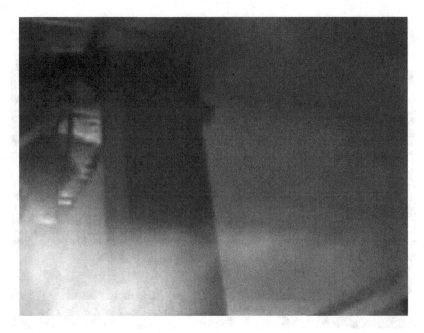

FIGURE 11.28   Damaged pillar in disaster captured by CAT.

FIGURE 11.29   Crouched quarry found by CAT and ignored by dog (outline added).

FIGURE 11.30   Dog wearing CRDS and underdog.

FIGURE 11.31   Underdog left behind.

rescuers arrive. The first prototypes were available for a 2006 structural collapse exercise in which the system was demonstrated. A year later, the improved system was used at a similar exercise and achieved over 20 successful operational drops. The system is currently undergoing an extended trial by the USAR canine teams of PERT.

THREE-DOG PROTOCOL

On occasion it is useful to use the CRDS in combination with the CAT system in what has come to be known as the three-dog protocol (3DP). When someone is suspected of being trapped within a collapsed structure that cannot be entered by humans, a disaster dog is sent to conduct a search (first dog). If the dog gives a bark indication and cannot be seen by its handler, it is recalled. Because the dog cannot be seen, trapped people cannot be seen and their locations cannot be identified. A second dog wearing a CRDS with a bright orange underdog is then sent to the site. The underdog usually contains a radio for communication with the trapped person but may also be empty. When the second dog gives a bark indication, the CRDS is activated and the underdog is released, leaving a bright orange indication in the vicinity of the person's scent. A third dog wearing a CAT system is then deployed to again find the trapped person. Someone viewing the CAT video feed can use the bright orange bag as a reference point to spot the likely location of the person. This protocol was used in a November 2007 exercise in which the Figure 11.32 image was captured.

In the upper right corner of the image (circle), the underdog can be seen. Slightly to the left and down a reflective strip of a rescuer acting as a quarry can be seen hidden under a layer of drywall. The video feed was recorded from the dog and processed offline after the search with a laptop computer. Including this post-processing step, the entire search with video "scrubbing" produced a usable image in less than 25 minutes.

CANINE POSE ESTIMATION (CPE)

The goal of CPE is to allow a handler to know what a dog is doing even when he or she cannot see the dog. Through a set of accelerometers, a wireless signaling system, signal analysis, and various algorithms, the position of the dog will be indicated to the handler or an assistant on a display (Figure 11.33). This is important because a dog's body position often indicates many parameters including camera angles for images received from CAT. CPE also increases the potential communication vocabulary between dog and handler. For example, cadaver dogs are often trained to sit when they discover a dead human. CPE makes it possible to see the dog sit and thus alert the handler.

CANINE WORK APPAREL (CWA)

The pet and service animal apparel trade is a multimillion dollar business in the United States, but no available apparel article provides a base for mounting delicate hardware components on a dog. CWA is an effort to design and test prototype canine harnesses that fulfill the requirements to provide freedom of movement, comfort,

FIGURE 11.32 Simulated patient can be seen (box) and to left of orange underdog (circle) at top right.

and safety for the dog (Figure 11.34). In addition, the harness must provide a stable platform for the sensing components, the release mechanism of a CRDS, and other components that constitute parts of the CAT ensemble.

Other reasons dictate designing of CWA. In the United States, the Canine Search Specialist Certification Process (FEMA, 2008) directs handlers to remove collars and vests from dogs during testing. This has created the impression that a dog

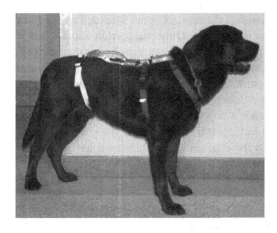

FIGURE 11.33 Canine with CPE apparatus.

FIGURE 11.34 CAT attached to CWA.

should never wear apparel while searching. Other jurisdictions such as New Zealand (NZUS&R, 2004) imposed similar requirements. The reason apparel is removed before a dog traverses rubble is the potential danger to the dog. The apparel may become snagged on a piece of exposed rebar or other protrusion that would hold the dog beyond the handler's reach. To avoid this problem, all dogs search "nude."

CWA for USAR is designed to eliminate the possibility of pinning a dog to a metal bar or other object by designing it to tear away when snagged. However, unlike other tear-away systems, the apparel is not intended to shed completely away, but will remain on the dog's body. Only the portion caught will tear away and thus release the dog. This strategy increases the chance that the equipment will return with the dog without endangering the dog. This can be important for systems like CAT. If wireless communication proves impossible, an on-dog recording may prove very useful for review after a dog returns from a search.

Figure 11.35 shows a dog wearing a CAT CWA. A detached metallic clasp can be seen on the left foreleg near the bottom of the image. The clasps are designed to keep the apparel on the dog, detach when force is applied, and reattach easily with the help of a handler.

CANINE BRAIN FUNCTION (CBF)

We have found that it is possible to sense what a USAR dog experiences by measuring its physiology during an activity. We use multichannel near-infrared brain

FIGURE 11.35 CWA with CAT and detached clasp.

spectroscopy (Boecker, 2007; Helton et al., 2007; Herrmann et al., 2008). This type of sensor is capable of measuring brain oxygenation by illuminating brain tissue within the skull and detecting its reflection.

We believe this is a fruitful area for further investigation. Many canine handlers believe they will be able to determine aspects of the mental state of a USAR canine. This may allow them to determine whether the animal is actually working on finding people (see Chapter 7 by Helton). Clearly, other potential advantages include the ability to differentiate various canine mental states and apply them to the sensing task. For example, explosive detection dogs will indicate the presence of an explosive but perhaps measurement of physiological response will enable handlers to determine what explosive is actually present.

The ability to assess mental state has additional implications for canine training. If it becomes possible to determine CBF changes during training sessions, it may become possible to determine exactly when a dog requires a rest break, additional motivation through the introduction of a play period, or termination of a training period after optimum benefit has been obtained.

We conducted several proof-of-concept experiments replicating part of Pavlov's famous work (1927) on conditioned response. A canine handler was asked to hold a sensor on a dog's head for several minutes until the dog became familiar with the device and became calm. After the calming period, the sensor output was recorded. Another participant would then show the dog its favorite toy. It was possible to detect a change in the blood oxygenation shortly after the toy was presented and before the dog exhibited any physical reaction (Figure 11.36).

FUTURE OF DISASTER DOGS

The use of disaster dogs to find people trapped in rubble is perhaps one of the most poignant examples of the deep and beneficial relationship between humans and dogs that contributes to the safety of our society. Canine teams undergo rigorous training, testing, retraining, and retesting as a type of insurance policy against disasters that we hope we will never need.

Undoubtedly, canine handlers have several motivations for their work but the role of the dog should never be underestimated. Working for nothing more than a chew toy, a dog performs critical life-saving tasks by employing its natural sense of smell. It is not surprising that the relationships between dogs and handlers in canine teams

FIGURE 11.36 Kong is presented to USAR canine while its brain function is measured.

have worked so well that they continue to defy all attempts to replace them. Disaster dogs constitute a relatively new class of working canines. With the impetus of recent history, new techniques and training methods continue to be developed and modified as more experience is gained with urban disasters. Standardization, practice, and refinement have made the canine team into a formidable search tool that can be deployed quickly with a high probability of success.

Perhaps it is fitting that—instead of replacing disaster dogs with technology— progress focuses on adding technology to the dogs. By equipping dogs with non-intrusive and inherently safe sensing, communication, and computing systems we hope to improve their performance even more. Strides have been made to assist teams in working remotely to find people faster and immediately help those who are found. Technology like CAT has shown the potential to provide incident commanders with timely and precise information about activities at the forward edge of a USAR effort—improving situational awareness about key locations where only a dog can go so that people may live.

ACKNOWLEDGMENTS

I would like to acknowledge the valuable contributions of several people to this work: canine handlers Constable Kevin Barnum (OPP/CAN TF-3), Constable Mike Dallaire (OPP/CAN TF-3), Constable Denis Harkness (OPP/CAN TF-3), Staff Sergeant Wayde Jacklin (OPP Canine Coordinator), Professor Lucia Dell'Agnese (Ryerson University, School of Fashion), Devin Ostrom (Ryerson University, Department of Mechanical Engineering), Jimmy Tran (Ryerson University, Department of Computer Science), Cristina Ribeiro (The University of Guelph, Department of Computing and Information Science), Salah Sharieh (Ryerson University, Department of Computer Science), Teresa MacPherson (FEMA Canine Sub-group and VA-TF1), and Sonja Heritage (FEMA VA-TF1).

REFERENCES

Acree, V.D. (1976). Customs tailored enforcement techniques: trouble for federal lawbreakers. *FBI Law Enforcement Bulletin, 45*(1): 16–20.

ARDA (2002). *Search and Rescue Dogs: Training the K-9 Hero.* New York: John Wiley & Sons.

Bal, A. and Jensen, B. (2007). Post-traumatic stress disorder symptom clusters in Turkish child and adolescent trauma survivors. *European Child and Adolescent Psychiatry 16*(7), 449–457.

Beerda, B. et al. (1999). Chronic stress in dogs subjected to social and spatial restriction. I. Behavioral responses. *Physiology and Behavior, 66*(2), 233–242.

Beerda, B. et al. (2000). Behavioural and hormonal indicators of enduring environmental stress in dogs. *Animal Welfare, 9*(1), 49–62.

Boecker, M. et al. (2007). Prefrontal brain activation during stop-signal response inhibition: an event-related functional near-infrared spectroscopy study. *Behavioural Brain Research, 176*(2), 259–266.

Brisbin, I.L., Jr. and Austad, S.N. (1991). Testing the individual odour theory of canine olfaction. *Animal Behaviour, 42*(1), 63–69.

Brownell, D. and Marsolais, M. (2000). The Brownell–Marsolais Scale: A Proposal for the Quantitative Evaluation of SAR/Disaster K9 Candidates.

Cablk, M.E. et al. (2008). Olfaction-based detection distance: a quantitative analysis of how far away dogs recognize tortoise odor and follow it to source. *Sensors, 8*(4), 2208–2222.

Demchak, B. et al. (2006). Situational awareness during mass-casualty events: command and control. *AMIA Annual Symposium Proceedings*, 905.

Endsley, M.R. (1988). Situation awareness global assessment technique (SAGAT). *Proceedings of National Aerospace and Electronics Conference*. New York: IEEE.

FEMA. (2008). Canine Search Specialist Certification Process. National Urban Search and Rescue Response System, Department of Homeland Security, Federal Emergency Management Agency, 38.

Ferworn, A. et al. (2006). Urban search and rescue with canine augmentation technology. *IEEE International Conference on Systems Engineering*, Los Angeles, CA.

Ferworn, A. et al. (2007). Canines as rescue robot surrogates. *Second ACM/IEEE International Conference on Human Robot Interaction*, Washington, D.C.

Ferworn, A. et al. (2007). Canine as robot in directed search. *IEEE International Conference on Systems Engineering*, San Antonio, TX.

Ferworn, A. et al. (2007). Rubble search with canine augmentation technology. *IEEE International Conference on Systems Engineering*, San Antonio, TX.

Ford, A.R. and Reeve, S.W. (2007). Sensing and characterization of explosive vapors near 700 cm^{-1}. *Proceedings of SPIE (International Society for Optical Engineering)*.

German, J. (2002). Sandia explores K-9 collar camera kits for hostage rescue, emergency response. Albuquerque, NM: Sandia National Laboratories.

Hagh-Shenas, H. et al. (2006). Post-traumatic stress disorder among survivors of Bam earthquake 40 days after the event. *Eastern Mediterranean Health Journal 12* (Suppl. 2).

Haug, L.I. et al. (2002). Comparison of dogs' reactions to four different head collars. *Applied Animal Behaviour Science, 79*(1), 53–61.

Heckendorn, F. (2006). A rescuer's perspective on the World Trade Center disaster recovery. *Journal of Failure Analysis and Prevention, 6*(5), 9–16.

Helton, W.S., Hollander, T.D., Tripp, L.D., Parsons, K., Warm, J.S., Matthews, G., and Dember, W.N. (2007). The abbreviated vigilance task and cerebral hemodynamics. *Journal of Clinical and Experimental Neuropsychology, 29*, 545–552.

Herrmann, M.J. et al. (2008). Brain activation for alertness measured with functional near infrared spectroscopy (fNIRS). *Psychophysiology, 45*(3), 480–486.

McGuigan, K. and Friedman, J. (2006). Canine-firefighter teams search for survivors. *Fire Engineering, 159*(5), 197–198.

Mesloh, C. et al. (2002). Scent as forensic evidence and its relationship to the law enforcement canine. *Journal of Forensic Identification, 52*(2), 169–182.

North, C.S. et al. (1999). Psychiatric disorders among survivors of the Oklahoma City bombing. *Journal of the American Medical Association, 282*(8), 755–762.

NZUS&R. (2004). Best Practice Guideline, Canine Readiness Evaluation Process: Advanced Level.

O'Brien, K.S. and O'Hare, D. (2007). Situational awareness ability and cognitive skills training in a complex real-world task. *Ergonomics 50*(7), 1064–1091.

Pavlov, I.P. (1927). *Conditioned Reflexes: An Investigation of the Physiological Activity of the Cerebral Cortex*. Oxford: Oxford University Press.

Perry, R. (2003). Incident management systems in disaster management. *Disaster Prevention and Management, 12*(5), 405–412.

Pickel, D. et al. (2004). Evidence for canine olfactory detection of melanoma. *Applied Animal Behaviour Science, 89*(1–2), 107–116.

Report, Annual. (2002–2003). Albuquerque, NM: Sandia National Laboratories, 27.

Roman, R. et al. (2008). Situation awareness mechanisms for wireless sensor networks. *IEEE Communications Magazine, 46*(4), 102–107.

Roy, N. et al. (2002). The Gujarat earthquake (2001) experience in a seismically unprepared area: community hospital medical response. *Prehospital and DisasterMedicine, 17*(4), 186–195.

Schoon, G.A.A. (1996). Scent identification lineups by dogs (*Canis familiaris*): experimental design and forensic application. *Applied Animal Behaviour Science, 49*(3), 257–267.

Schoon, G.A.A. and De Bruin, J.C. (1994). The ability of dogs to recognize and cross-match human odours. *Forensic Science International, 69*(2), 111–118.

Solway, L. (1999). Socioeconomic perspectives of developing country megacities vulnerable to flood and landslide hazards. In Margottini, C., Ed. *Floods and Landslides: Integrated Risk Assessment.* Heidelberg: Springer, 245–277.

Thomson, I. (2005). Police unleash dogcam crime busters, vnunetwork.

van der Waaij, E.H. et al. (2008). Genetic analysis of results of a Swedish behavior test on German Shepherd dogs and Labrador Retrievers. *Journal of Animal Science*, in press.

12 Physiological Demands and Adaptations of Working Dogs

Michael S. Davis

CONTENTS

OVERVIEW

Physical activity places metabolic demands on working dogs above their basal metabolic rates, and successful adaptation to routine physical activity is the result of both long-term and short-term changes in physiology that permit augmentation and reallocation of finite physiological resources. Conversely, failure to adapt to the physiological demands of exercise is associated not only with fatigue and failure, but also with specific exercise-induced injuries and disease that may compromise physical performance and, in severe instances, the health and wellbeing of a working dog. This chapter will highlight the desirable adaptations that facilitate physical activity, the consequences of these adaptations on other physiological functions, and the common conditions related to failure to adapt to exercise.

SHORT-TERM AND LONG-TERM ADAPTATIONS THAT FACILITATE EXERCISE PERFORMANCE

Physical exercise is, at its simplest, the conversion of chemical energy to mechanical energy by muscle. This simple task is facilitated by a variety of organs, tissues, cells, and proteins in a coordinated process that dynamically allocates finite physiological resources to achieve muscular work, but without excessively impairing the short-term or long-term viability of the exercising subject. This chapter will outline the basic process by which physical exercise is achieved, highlight the factors that often limit the capacity for physical exercise in dogs, identify the mechanisms by which the capacity for physical exercise is increased, and describe some of the unique concerns and adverse consequences specific to working dogs.

DEMANDS OF MUSCULAR WORK

The process of extracting energy from complex macromolecules by biological tissues can be viewed as a whole as tissue metabolism. Skeletal muscle is perhaps the tissue with the greatest range of metabolism. Resting muscle consumes relatively little energy, whereas maximally exercising muscle can increase its metabolism 20-fold or more. The processes collectively known as metabolism include substrate delivery, conversion of chemical energy to mechanical energy, and removal of waste products. Although these processes are highly coordinated and interrelated, it is useful to consider them individually before integrating them into the whole process of muscle metabolism.

SUBSTRATES

The basic substrates for the conversion of chemical energy to mechanical energy are oxygen and a carbon–carbon bond. The latter requirement can be fulfilled by a variety of macromolecules including carbohydrates, fats, and proteins. Each macromolecule has its own specific chemistry, dictating different enzymatic processes that provide access to the energy in the carbon–carbon bonds, but ultimately most of the energy is extracted by converting the macromolecules into multiple 2-carbon acetyl groups that are then processed through a series of enzymes (the tricarboxylic acid or Krebs cycle and the electron transport chain) and combined with oxygen to produce the key energy "currency" within cells: adenosine triphosphate (ATP). These terminal processes occur within the mitochondria of cells.

Depending on the compound and enzymatic processes, additional energy can be captured in the process of rendering the macromolecules down to acetyl groups, directly or through the creation of reducing equivalents (intermediate molecules that provide short-term storage of energy captured while breaking down macromolecules) that are also eventually combined with oxygen to produce ATP. ATP is the molecule that is used for virtually all energy-consuming processes in cell including muscle contraction.

DELIVERY

The desired capacity for metabolic activity in muscle far exceeds the capacity to store metabolic substrates within muscle. Therefore, the storage and delivery of these substrates is an important feature in establishing and maintaining the capacity for muscular work. The storage of oxygen is deceptively simple, in that it is stored in the atmosphere that the dog breathes. Although obviously abundant in a physical sense, oxygen gas cannot be moved by any body system against its chemical gradient (established as the pressure of the gas), and thus an ample gradient must be present in the atmosphere to permit the flow of oxygen into muscle.

At high altitudes or, more rarely, in confined spaces in which a finite amount of oxygen can be consumed, the drop in the pressure of oxygen has a cascading downstream effect of impairing oxygen delivery to muscle cells and thus impairing the conversion of chemical energy to mechanical energy. Delivery of oxygen from the atmosphere to the muscle is facilitated by (1) the lungs that permit oxygen in the gas phase to dissolve into the red blood cells that chemically store oxygen using hemoglobin to increase the amount of oxygen in the blood, and (2) the cardiovascular system that moves the oxygenated blood from the lungs to the muscle.

To a large extent, the greater the capacity of these systems, the greater the capacity for delivery of oxygen to muscle. However, it is important to note that because these systems act in a serial fashion, increasing the capacity of one without increasing the capacity of the others will have minimal effect on overall oxygen delivery. In other words, increases in the number of red blood cells will not produce a profound increase in oxygen delivery without corresponding increases in absorption of oxygen or cardiovascular transport. Conversely, an impairment in any one of these features can be expected to exert an adverse effect on oxygen delivery, i.e., a loss of red blood cells is likely to decrease overall capacity for oxygen delivery even if the lungs and cardiovascular systems are completely normal.

Storage and delivery of macromolecules suitable for providing chemical energy to muscles is more complex than the storage and delivery of oxygen, and vary for different macromolecule classes. The storage of carbohydrates is achieved using glycogen (a complex of simple sugar molecules called glucose) stores in the muscle and the liver. Delivery of glycogen stored in the muscle is simple. Since it is already present in muscle, the only step required is to break it down. However, in general the body treats this store as an energy source of last resort, to be used only when all other processes fail to meet the demands of the muscle. This regulatory strategy is discussed further below.

Glycogen stored in the liver must be broken down in the liver into individual glucose units before it can leave the liver. It is transported to muscles using the same transport system as oxygen (the blood and cardiovascular system). Fats are stored in muscle tissues as well, although typically outside cells in the form of fat droplets rather than within cells. Fats are also stored in the liver and adipose (fat) tissue. Release of these stores is regulated but somewhat less regimented compared to glycogen, in that stored fat can be released as a complex of multiple fatty acids or as individual fatty acids. Like glucose, the fats and fatty acids are transported to the muscle via blood. However, in order to be absorbed by muscle cells, they must be in the form of fatty acids; if they were not released as fatty acids, they must be broken down to fatty acids outside the muscle cells.

Proteins are not stored as energy sources per se, for the simple reason that they are poor and inefficient energy sources, and are generally more valuable to the body as functional elements of metabolism. However, during periods of high energy demand relative to energy supply, the body will break down functional proteins into component amino acids. Some, but not all, amino acids can be readily metabolized into 2-carbon acetyl groups and used as energy sources. However, since quality protein intake is frequently a limiting factor in the working dog diet and in certain cases using protein as energy actually consumes energy, the use of protein, particularly proteins serving important functions in the body, is an inefficient means of meeting energy demands.

Another, and often unappreciated, site of energy storage is the gastrointestinal tract in the form of a meal consumed prior to exercise. With the proper composition and timing of feeding and exercise, a meal inside the gastrointestinal tract can become the primary source of fuel for muscular work. This aspect of energy storage is best exemplified by endurance athletes such as racing sled dogs. Built like human endurance athletes, racing sled dogs have relatively little body fat. Using a highly digestible, high fat diet, the dogs that race in multiday events burn energy as they consume it, making them highly efficient working animals (see Figure 12.1). This

FIGURE 12.1   Fat snacks used to fuel exercise in racing sled dogs. Slightly larger than golf balls, the fat snacks consist of pure rendered animal fat. They are provided during brief rests to maintain the energy supplies required for the dogs to continue running.

strategy requires a well functioning gastrointestinal tract and a healthy appetite. A meal missed for any reason has an immediate and negative effect on an animal's capacity for work.

MYOLOGY

Muscle work is the direct result of energy consumption by the contractile protein filaments of muscles. Actin and myosin are long, complex proteins that span the length of a muscle fiber. Actin contains multiple repeating sequences of amino acids that present binding sites to the head of the myosin filament, but are normally covered by small troponin molecules. Contraction begins with the release of calcium ions into the muscle cytoplasm. The calcium ions cause the troponin to shift, exposing the actin binding sites.

The myosin head has two primary conformations: extended and contracted. The contracted conformation is the native conformation (that exists outside a living system). The myosin head is converted to the extended conformation when the energy from ATP is transferred to the myosin head. Once in the extended position, conversion back to the contracted position occurs when the binding sites on actin are exposed, allowing the energy to be released and the myosin head to contract. On a macroscopic scale, shortening of the muscle occurs due to the rapid and repeated ratcheting of the myosin head along the actin filament.

A single muscle contains numerous individual cells, with potentially different metabolic capabilities. A basic criterion for classification is the type of myosin within a cell (Type I or Type II). Type II myosin is further subdivided into Types IIA, IIB, and IIX. Type I myosin has a very strong tendency toward the contracted configuration and therefore provides greater contractile power per filament. However, it is slow to convert to the extended configuration with the binding of ATP, and thus cannot cycle through a repeated ratcheting process quickly. Because of this, Type I fibers are frequently designated "slow-twitch" fibers. In most species, Type II fibers are the opposites of Type I: not as much power generated by contraction of the myosin head, but capable of rapidly ratcheting through a series of conformational changes. Type II fibers are thus referred to as "fast-twitch" fibers. However, in dogs, Type IIA and Type IIB fibers are capable of greater force than Type I fibers, while retaining their capacity for rapid cycling.

As a general pattern, Type I fibers also contain large numbers of mitochondria and thus relatively greater capacity for completely metabolizing macromolecular substrates into energy and terminal wastes. For this reason, Type I fibers are frequently termed oxidative (due to the large capacity to consume oxygen in the mitochondria) and fatigue-resistant (due to the ability to produce large quantities of ATP). In contrast, Type IIX fibers contain relatively fewer mitochondria but large amounts of glycogen, permitting very rapid, but very short-lived production of energy even in the absence of oxygen. For this reason, Type II fibers are frequently referred to as glycolytic (due to the tendency to fuel contraction by the incomplete metabolism of glycogen) and fatigable (due to the limited amount of contraction performed before fatigue). The ideal fiber (possessing the most robust capacities without shortcomings) is Type IIA. This type possesses the capacity for rapid contraction along with high oxidative capacity.

Different muscles require different combinations of fibers, depending on their predominant activities. For example, the retractor bulbi muscle (that pulls the eyeball back into the orbit as a protective measure) requires extremely fast responses to protect the eye, but little capacity for endurance since it rarely is required to fire repeatedly within a short period. Thus, the retractor bulbi is composed almost entirely of Type IIB and Type IIX fibers (Toniolo et al., 2007). Conversely, fatigability would be a serious detriment to a muscle that fills a basic life support function such as the diaphragm, which explains why this muscle is comprised entirely of fatigue-resistant Type I and Type IIA fibers. Muscles of locomotion fall somewhere between these two extremes, with postural muscles (that are under constant tone to maintain posture) tending toward a predominance of Type IIA fibers. Muscles used for quick, short-term responses that may require little power (such as muscles that extend the digits) tend toward Type IIB fibers. Most muscles of propulsion such as the caudal thigh and upper arm fall into both categories, and may contain 15 to 20% Type IIB fibers, with the balance Type I or Type IIA.

WASTES

The process of converting chemical energy to mechanical energy yields waste products that must be removed from muscles and ultimately from the body for exercise to continue. The principal chemical wastes of complete substrate oxidation are carbon dioxide and water. Carbon dioxide is a gas that is removed essentially using the reverse process of oxygen delivery; it is dissolved in blood, transported by the cardiovascular system to the lungs, and exhaled. The differences in physical chemical properties of oxygen and carbon dioxide are very relevant. As a molecule that can not only dissolve in water but also react with water, carbon dioxide is easily eliminated from muscle. Excess carbon dioxide is also very easily cleared from the body through the lungs. Thus, carbon dioxide buildup is rarely a cause of impaired exercise, as long as the subject continues to breathe. Because carbon dioxide is a critical component of the body's systems for buffering pH, excessive elimination of carbon dioxide is undesirable and will cause dangerously high alkalinity in body fluids. Elimination of water as a metabolic waste product is rarely a concern, since the body has an absolute requirement for water at rest, and this requirement increases during exercise due to the demands of heat dissipation.

The list of chemical waste products expands if exercise is supported energetically by anaerobic processes or by significant amounts of protein. Anaerobic metabolism to support exercise occurs when rate of initial breakdown of glucose exceeds the capacity of the mitochondria to fully oxidize glucose to carbon dioxide. Although normal extracellular fluid contains approximately 85 mg of glucose per milliliter of fluid, uptake of this glucose into muscle is rarely rapid enough to exceed oxygen delivery and utilization by mitochondria. Thus when extramuscular glucose is the sole source of glucose for energy metabolism, anaerobic metabolism rarely occurs. However, during periods of extreme demand, intramuscular glycogen can be broken down and produce glucose faster than can be fully utilized by the muscle.

The initial breakdown of glucose can produce up to 6 moles of ATP per mole of glucose without the need for oxygen. The carbon-based end product of this reaction is pyruvate instead of carbon dioxide. If pyruvate is not further metabolized to carbon dioxide, as is the case with aerobic metabolism, it is converted to lactic acid—a potent organic acid that can lower pH and impair cellular function if allowed to accumulate. The body has numerous mechanisms by which lactic acid can be removed or converted back to pyruvate, but these processes have low capacity and thus accumulation of lactic acid is a distinct possibility. If protein is used as an energy source, the increased rate of ammonia production (due to deamination of amino acids in the liver) results in increased amounts of urea that must be eliminated in the urine (McKenzie et al., 2007). The concentration of urea in the blood is frequently used as a screening test for kidney disease in dogs and other mammals, but exercise-induced increases in blood urea should not be confused with kidney disease.

A final end product of exercise metabolism, and one that is often overlooked, is heat. All reactions involved in the conversion of chemical energy to mechanical energy proceed down a thermodynamic gradient, and thus release energy in the form of heat. In fact, virtually all energy is eventually released as heat; only in rare circumstances is energy stored as kinetic energy (climbing and remaining elevated). In most mammals, a large amount of heat dissipation is accomplished through sweating. However, dogs lack sweat glands except for their footpads. Evaporative cooling is restricted to a dog's respiratory tract where thermoregulatory requirements compete with ventilatory requirements. A compromise is reached through panting—a relatively inefficient means of dissipating the heat of muscular work because it requires muscular work and therefore increases the amount of heat to be dissipated. Furthermore, certain working dog activities, particularly those involving olfactory function, may be compromised by the need for panting. Overheating is a serious and common consequence of muscular work in dogs, and will be discussed later in this chapter.

REGULATION AND RAPID ADAPTATION

In order to facilitate a 20-fold increase in metabolism, it is necessary to have the capacity to increase each individual process 20-fold. On the other hand, maintaining that magnitude of delivery, conversion, and removal when it is not necessary (i.e., during periods of rest) would be extremely wasteful of body resources. The regulatory processes can be broadly divided into processes that regulate the movements of oxygen and carbon dioxide, processes that regulate the movement and storage of macromolecules used for energy, and thermoregulation.

Efficient movement of oxygen and carbon dioxide is achieved through control of the respiratory and cardiovascular systems. The amounts of oxygen and carbon dioxide in the blood stream are monitored individually by the brain, which increases the rate and depth of breathing based on decreasing amounts of oxygen or increasing amounts of carbon dioxide. Typically, a slight lag in timing occurs between the onset of exercise and changes in the blood concentrations of these gases due to exercise, so breathing is not immediately stimulated. However, the combination of high sensor sensitivity and a robust capacity for ventilation makes it possible for dogs to closely control the blood concentrations of both oxygen and carbon dioxide, so that only in extreme circumstances do the concentrations of these compounds deviate from normal resting values.

Rapid adaptation of the cardiovascular system to changing exercise intensities (and by extension, energy demands of muscles) is achieved on multiple levels. At the level of working muscles, blood flow through the muscles is regulated by many individual valves within the capillaries. These valves open and close regularly, but the durations of their open conformations can be affected by metabolic signals from muscles. A decrease in the local oxygen concentration (pH) or an increase in local carbon dioxide or potassium (all consistent with contraction or increased metabolism) will result in greater percentages of the valves in the open position at any given time, thus increasing blood flow to muscle cells. The increase in blood flow will increase delivery of oxygen, and increase removal of metabolic wastes (like carbon dioxide and organic acids).

The regulation of delivery of macromolecular fuel for muscle contraction is a much more complex process than the regulation of oxygen and carbon dioxide. The increased level of complexity arises from two properties of the macromolecules: (1) they are not freely available, and (2) they do not readily diffuse across cellular membranes. The former is addressed through storage and release of the molecules within muscles and other body tissues. The latter is addressed through specific transporters that facilitate the movement of the molecules across muscle cell membranes.

A limited amount of energy is stored in muscle, and thus very little exercise can be performed without movement of energy sources into the muscle. Stored ATP can fuel only a few muscle contractions. Intermediate energy storage molecules including intracellular glucose, fatty acids, and their intermediate breakdown products can provide 15 to 30 seconds of maximal contraction. Any further work must be supported by breakdown of intramuscular macromolecular stores or transport of glucose and/or fatty acids from the bloodstream to the muscles.

The processes that permit increased availability of macromolecular energy to muscle cells are critical adaptations that permit exercise. When a muscle cell is stimulated to contract, small amounts of calcium are released into the cell. The primary role of the calcium ions is to change the conformation of secondary proteins associated with the actin and myosin, thereby exposing the reactive sites of these molecules and allowing contraction to occur. A second role for the calcium ions is to stimulate the movement of transporters from the interior of a cell onto its surface. These transporters, specific for glucose and fatty acid, increase the movement of these energy sources into cells, thus providing the energy needed for subsequent contractions. In the absence of these transporters on cell surfaces, very little glucose and fatty acid can be transported into cells, making movement of these transporters a critical feature of sustained exercise.

Breakdown of intramuscular stored macromolecules (fat and glycogen) occurs when the energy state of a muscle cell begins to drop, despite the recruitment of glucose and fatty acid transporters. When the amount of de-energized ATP increases within a cell, selected enzymes are activated that break down intramuscular glycogen, releasing glucose within the cell for use as energy.

A similar process occurs with respect to intramuscular fat, producing fatty acids. Greater decreases in the energy state of a muscle activate still more enzymes that recruit additional biological machinery for absorbing and breaking down macromolecules for energy, as well as activating genes to stimulate the production of more energy-harvesting enzymes and transporters. This sequence of priorities is important in two ways: (1) by placing a priority on use of whole body stores (as opposed to internal stores), a muscle maximizes its overall exercise capacity, and (2) to increase the exercise capacity of a muscle, it is necessary to work it hard enough to create an energy deficit, thus stimulating cells to produce greater capacity. This latter aspect is critical to the long-term adaptation of a dog to exercise, and will be discussed later in this chapter.

The concentrations of glucose and fatty acids circulating in the bloodstream must be maintained as exercise-stimulated transport into the muscle increases. Maintenance of blood glucose and fatty acid concentrations is achieved by hormonally mediated release of these compounds from storage sites within the body. Glycogen and fat are the primary storage forms of glucose and fatty acids, respectively. When these nutrients become available in excess of demand, such as immediately following ingestion of a meal, the body releases insulin to facilitate storage of the excess energy nutrients.

During exercise, most dogs will experience decreases in circulating insulin concentration and increases in circulating glucagon concentration. The combination of these hormonal changes will permit the release of fatty acids from fat stores (insulin drop) and stimulate the release of glucose from the liver—due to breakdown of hepatic glycogen or conversion of certain amino acids to glucose (glucagon rise). Under resting conditions, insulin is needed to activate macromolecular substrate transport into muscle cells. However, during exercise the lack of insulin is compensated by the previously described effect of contraction on substrate transport.

Glucocorticoids may also play a role in maintenance of energy supply to muscles by stimulating protein breakdown and liberation of amino acids for gluconeogenesis. In other species, catecholamines such as epinephrine play dominant roles in the release of glucose from the liver. However, this pathway appears to be less active in dogs. Studies of dogs performing long-duration moderate exercise did not reveal evidence of systemic release of catecholamines (Durocher et al., 2007).

CONDITIONING ADAPTATIONS

SPECIFICITY OF CONDITIONING

A critical principle of athletic conditioning, regardless of species or discipline, is that conditioning is activity-specific. Throughout this chapter and most texts on exercise physiology, conditioning is defined as the physiological adaptation to increase exercise capacity or the activity that produces such physiological adaptations. Training is the mental process of learning a particular activity, behavior, or response.

The body seeks efficiency in the use of energy and nutrients. Because the physiological changes that comprise effective conditioning require both energy and nutrients, the body will not increase its capacity for exercise unless such increase is clearly needed. In other words, some degree of physical or metabolic stress is required to stimulate conditioning. The stress need not be enormous, and in fact too much stress can result in damage to stressed tissues, resulting in injury and impeding conditioning. It is also important to recognize that a tissue that is not stressed will not adapt. Thus, if conditioning does not closely mimic the desired activity, some or all of the tissues needed to perform the desired activity will not be conditioned to a higher level of performance and will thus act as limiting factors in exercise performance.

CARDIOVASCULAR ADAPTATIONS

Cardiovascular adaptations are the most prominent adaptations resulting from effective conditioning and generally contribute the most to increased exercise capacity. Cardiovascular adaptations fall into three general categories: heart, capillaries, and blood volume. The primary direct effect of conditioning on the heart is cardiac hypertrophy or enlargement of the heart. The heart has more muscle and its chambers are also larger. The combination of these two changes results in much more effective pumping of blood, a greater capacity for cardiac output during exercise, and a much slower heart rate at rest.

The normal resting heart rate of a dog is 80 to 100 beats per minute, but it is not unusual to find conditioned working dogs with heart rates below 60 beats per minute. Very well conditioned (and very relaxed) working dogs may even present with intermittent dropped heartbeats. Their heart rates slow so much that the ventricles that provide the force to propel the blood into the arteries do not contract with every electrical impulse. In any other type of dog, this would appear as a serious cardiac abnormality that warranted further examination. In well conditioned working dogs, the slowed heart rates may be normal. In fact, decreasing heart rate in response to conditioning is a predictable enough response that many trainers use the resting heart rates of their dogs to gauge the effectiveness of their conditioning regimens.

Increased blood volume is a second, very common change associated with exercise conditioning. The mechanisms of blood volume expansion are not entirely known, but overall blood volume expansion is the result of fluid retention by the kidneys through conservation of sodium. To the extent this action is driven by release of aldosterone, sodium retention may come at the expense of potassium conservation, potentially leading to a whole-body deficiency of potassium. Potassium deficiency is a potential contributing cause to some exercise-related disease syndromes, to be discussed later in this chapter. Despite greater blood volume, blood pressure does not necessarily increase because veins expand to accommodate the additional fluid. When needed, the veins can constrict, increasing the pressure in that segment of the circulation and facilitating faster filling of the heart. It is important to note that although the amount of water in the circulation increases, the concentrations of various components of blood may drop due to dilution. This is most obvious in the conditioning-associated decrease in the concentration of red blood cells. Different reference ranges for working dogs are necessary when traditional diagnostic tools are used to examine these dogs (Davis et al., 2008).

The collective effect of a larger, more powerful heart and a larger blood volume is the capacity for greater cardiac output, and by extension, greater delivery of substrates used to power muscles and greater capacity to remove wastes from muscles. In fact, modification of the cardiovascular system to exercise can be the sole adaptation responsible for increased aerobic capacity due to conditioning. However, other systems may also be stressed by conditioning and respond with increased functional capacity. Adaptation of these systems may be just as critical to enhanced exercise performance, particularly if the capacity for a given type of exercise is not limited by aerobic capacity.

Increased density of capillaries in muscles is a third area of cardiovascular adaptation to exercise and one that is the least studied. Most scientists assume, based on mathematical models, that muscle develops more capillaries to more effectively transport gases, substrates, and heat during exercise, and some studies support these beliefs. However, the technical difficulties inherent in these studies limits their numbers. From a practical view, the importance of increased capillary density in the muscles of exercising subjects is best illustrated by the artificial effects of some drugs on muscles. If muscle mass is increased through drug use but is not accompanied by increased capillary density, the muscle effectively becomes excess weight and overall athletic performance decreases even if the subject appears "muscled up."

MUSCLE ADAPTATIONS

The primary muscle adaptation resulting from physical conditioning and permitting greater exercise capacity than in unconditioned subjects is synthesis of additional metabolic machinery that converts macromolecular energy sources to ATP. The specific signals and sequences that convert muscle stress to improved muscle capacity are not completely known, but most evidence points to activation of a central energy sensing molecule in muscles called AMP kinase (AMPK). This enzyme is activated by increased concentrations of de-energized ATP, a clear signal that a muscle is not producing energy sufficiently.

In addition to short-term adaptations to improve substrate availability, AMPK appears (indirectly) to activate genes that code for the various enzymes, transporters, and support proteins for converting glucose and fatty acids to energy. Collectively, this results (at least in other species) in a greater percentage of muscle tissue with high oxidative capacity. Based on the inherently high metabolic capacities of dogs in general, it is not a safe assumption that the normal amount of muscle metabolic capacity would be insufficient and need to be augmented. However, recent pilot data from our laboratory documented training-induced increases in the capacity for macromolecular substrate uptake in dogs. Whether this adaptation is due to increased abundance of transporters or altered function of existing transporters is unknown.

It is generally accepted that a well conditioned dog has greater muscle mass than an unconditioned dog, but the characteristics of that muscle mass make as much difference to exercise performance as the overall amount of muscle. Aerobic conditioning may cause changes in overall fiber types, with a general shift from fast twitch, low oxidative fibers (Types IIX and IIB) to fast twitch, high oxidative fibers (Type IIA). In other species, prolonged aerobic training can also cause an increase in Type I fibers, but this has not been conclusively demonstrated in dogs.

It is important to remember that muscle fiber type is most clearly identified by antigenically distinct isotypes of the myosin head proteins of the contractile apparatus.

Whether a collective increase in a particular fiber type is due to hypertrophy of pre-existing fibers, greater numbers of those fibers, or internal switching of the myosin produced by existing fibers, is unclear. Biologically, the simplest and most likely explanation is the selective hypertrophy of pre-existing fibers. This type of modification of overall muscle fiber content is probably the most rapid and sufficient to explain the experimental observations. It is important to note that alteration of fiber types can be independent of, and is probably less important than, changes in available substrate transport and oxidation machinery when overall exercise performance is considered.

Skeletal Adaptations

Adaptation of the skeleton and connective tissue is a logical outcome of exercise conditioning. In other species, stress on the bones, tendons, and ligaments in the legs results in remodeling of these tissues to produce stronger structures capable of undergoing greater stress without injury. Direct evidence to support this change in working dogs is lacking, but the assumption is reasonable. The most important consideration is that, compared to adaptation of the cardiovascular systems and muscles, adaptation of the skeletal system requires less duration of stress (shorter workouts), but more elapsed time for the metabolically slower skeletal tissues to remodel (2 to 4 months versus 4 to 6 weeks). Thus, during initial conditioning, it is common for metabolic adaptations to precede skeletal adaptations, leading to an athlete who is sufficiently fit to overstress and injure his or her skeletal structures.

DISEASES OF WORKING DOGS

Heat Exhaustion and Heat Stroke

The comparatively limited ability of a dog to dissipate excess body heat combined with the prodigious metabolic capacity of a canine athlete makes heat-related disease a primary concern for working dogs. Evaporative cooling (release of body heat by the evaporation of water from body surfaces) is the most potent means of shedding excess heat. As previously described, the limited evaporative cooling of dogs arises from their respiratory tracts. The skins of most other mammals provide evaporative cooling via sweat glands in addition to respiratory evaporation. Thus, dogs are prone to generating metabolic heat faster than they can eliminate it, resulting in increased body temperature.

High body temperature is abnormal, but whether it is pathological (indicates disease) depends on the situation. An important consideration is the differences among hyperthermia due to exercise, hyperthermia due to ambient conditions, and fever (increased body temperature due to resetting of its thermostat by inflammatory mediators). Hyperthermia is typical in strenuously exercising dogs, and a healthy dog can tolerate substantial exercise-induced increases in body temperature for short periods. Fever, on the other hand, is by definition a sign of disease. The normal body temperature for a dog is 100.5 to 101.5°F (38 to 38.6°C). In the author's experience, it is not unusual for healthy, well conditioned racing sled dogs to enter checkpoints with temperatures of 105°F. Rectal temperatures of up to 107°F have been reported in normal retrieving dogs. Conditioning appears to affect a dog's ability to tolerate high body temperatures. The same degree of hyperthermia in an unconditioned dog arising from exercise or extreme ambient conditions (being locked in an enclosed car) invariably leads to serious systemic disease. The mechanism for this apparent protective effect of conditioning remains unclear.

Differentiating a dog that is hot from one suffering from a heat-related disease can be an exercise in subtlety. In both cases, the dogs will demonstrate short-term adaptations that maximize their ability to shed excess heat: rapid panting, reddened oral mucous membranes, swollen tongues. Probably the most important subtle difference is that dogs approaching critical levels of heat build-up will often appear weak or unsteady; they will seem more clumsy than usual. This can appear as a weaving or swaying action as a dog runs. In dogs performing precise movements, the unsteadiness can appear as tripping or misplaced steps. More advanced cases of heat-related disease are easier to recognize. Depression, seizures, and shock are common consequences of severe overheating.

Treatment of overheating depends upon the cause and severity, but in all cases requires a reversal of the imbalance between heat generation and dissipation. The obvious method for reducing heat generation is to reduce the intensity of exercise. This may be the only measure required if the situation has not progressed to systemic disease and the capacity for eliminating excess heat is substantial (e.g., very cold ambient conditions). The options for increasing heat dissipation are limited, but nevertheless should be employed if available.

Because dogs do not rely on sweating for dissipation of heat from the skin, their skins are relatively poorly supplied by blood vessels (blood transports heat from the muscles to wherever it will be excreted). Thus, although spraying a dog with water will increase the amount of heat lost through the skin, the effectiveness of this method is far less than in animals with sweat glands and a well developed ability to move heat to the skin surface. If large amounts of water are not available, the strategic use of cold water packs on areas with minimal insulation (e.g., the groin) or areas carrying blood to critical areas of the body (e.g., the underside of the neck where the vessels supplying the brain are located) can accelerate heat loss and protect the most vital tissues. Any dog that shows signs of systemic compromise should be cooled immediately and examined or monitored by a veterinarian, even if the animal appears to regain normality.

Life-threatening complications such as disseminated intravascular coagulation and multiple organ failure can occur hours after apparent recovery from severe heat-related disease. Although many dog handlers believe that a dog becomes predisposed to subsequent heat-related problems after having suffered from an initial episode, no scientific evidence supports this belief. Rather, it is likely that the predisposition that appears after an initial episode was present earlier and possibly contributed to the initial episode. Since proper conditioning appears to be protective from the deleterious effects of overheating (but not the overheating itself), it is reasonable to employ additional conditioning of working dogs that previously suffered from heat-related disease in an effort to raise their tolerance to this stressor.

EXERCISE-INDUCED COLLAPSE IN LABRADOR RETRIEVERS

A syndrome of exercise-induced collapse (EIC) has been described in field trial-bred Labrador Retrievers. Dogs suffering from this syndrome typically are extremely fit and muscular, with pronounced retrieving drive. At rest, dogs with EIC are completely normal, and can tolerate mild to moderate exercise. After 5 to 15 minutes of strenuous exercise, affected dogs develop clinical signs of weakness, lack of coordination, and collapse. During episodes, the dog's muscles are flaccid, but usually the animal remains mentally alert (although many appear confused, perhaps due to the unexplained inability to continue retrieving).

No characteristic laboratory abnormalities indicate EIC. Most dogs display the typical alterations described for any dog performing strenuous exercise. Episodes can last from 10 to 30 minutes, after which the dogs appear completely normal; fatal episodes of EIC have been noted in isolated reports. It is important to distinguish EIC from heat stroke since they present similarly: progressive weakness and collapse and body temperatures up to 107°F. Although dogs with heat stroke will eventually have much higher concentrations of serum CPK than dogs with EIC, this change occurs hours after the initial onset and is therefore not of value in determining initial treatment. The most reliable means of distinguishing EIC and heat stroke in Labrador Retrievers (EIC has not been described in other breeds) is to assess the speed and completeness of recovery. Dogs with heat stroke remain depressed, weak, and recumbent for hours, even after treatment.

The etiology of EIC is a genetic defect in a protein associated with nerve impulse transmission. The mutation is passed on as a recessive trait, but with very high penetrance, meaning that although only dogs with both genes mutated can be affected, virtually all these dogs will show clinical signs. Dogs with only one defective gene will not be affected, but can pass that gene on to half of their offspring. Interestingly, similar mutations have been described in other organisms and the resulting defective protein is temperature-sensitive. At normal temperatures, the protein functions normally, but becomes dysfunctional at higher temperatures. Whether this is the case in dogs with EIC is unknown, but is supported by the available data.

Affected dogs do not have progressive disease. Treatment is generally unrewarding. In some cases, owners have reported some success of treatment using central nervous system depressants such as phenobarbital, but the effect of such drugs on retrieving ability is likely substantial. In other cases, dogs have been reported to "grow out" of EIC. Whether this is due to better conditioning, changes in management, or less excitability (or a combination thereof) is unknown. A genetic test for the EIC mutation is available from the University of Minnesota Veterinary Diagnostic Laboratory. Based on the strong inheritance of the defect, affected dogs should not be bred. It is well within the ability of the Labrador Retriever community to completely eliminate this genetic defect, and therefore the condition, within a few generations of dogs.

GASTRIC ULCERS

Gastritis and gastric ulceration are primarily described in sled dogs participating in endurance activities, but recent studies of this group of dogs suggest that most working dogs may have at least some risk of developing exercise-associated gastro-intestinal disease. Clinically significant gastritis or gastric ulcers can be found in 50 to 60% of endurance sled dogs despite the complete absence of clinical signs of such lesions (Davis et al., 2003; see Figure 12.2). Although the initial studies were conducted after completion of races lasting as long as 10 days, more controlled studies have demonstrated that gastric lesions can be induced in as little as a single day of moderate intensity exercise (Davis et al., 2006). Thus, a variety of working dogs may be at risk for this condition. Because these lesions are not clinically apparent in most dogs (the exceptions are the few severely affected dogs that vomit blood), gastroscopy is the only means by which the true prevalence of this condition can be determined in working dogs.

The clinical signs of exercise-induced gastrointestinal disease may appear only during or immediately following acute bouts of work, but the underlying pathophysiology appears to begin during physical conditioning. Trained rested sled dogs typically have no visible stomach lesions, but functionally and microscopically their gastrointestinal tracts are not normal. Trained dogs reveal moderate microscopic evidence of inflammation in the wall of the gastrointestinal tract. Although the intensity of this abnormality tends to decrease with seasonal rest, it does not completely resolve.

Tests of gastrointestinal permeability have shown that athletic training tends to increase protein leakage into the lower gastrointestinal tract, and increased permeability develops in more proximal segments during moderate exercise (Davis et al., 2005; Davis et al., 2006). These observations have led to the prevailing hypothesis that exercise-induced gastrointestinal disease is the result of chronic exercise-induced hyperpermeability of the gastrointestinal mucosa, with resulting leakage of gastrointestinal lumen contents into the walls of the affected segments. This initiates an inflammatory response that becomes chronic with repeated episodes of hyperpermeability and leakage, and the chronic inflammatory condition lowers the threshold for abrupt failure of the mucosal barrier during acute exercise.

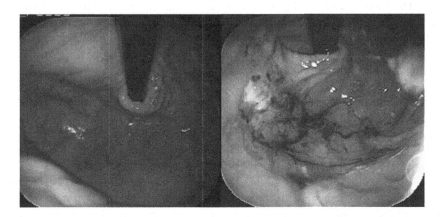

FIGURE 12.2   Endoscopic images of the stomachs of two dogs. Left: normal stomach. Right: severe gastric ulcer. Neither dog showed clinical signs of illness at the time of examination.

Studies of the treatment and prevention of exercise-induced gastric disease in dogs have focused on the acute effects of strenuous exercise, but have not addressed the underlying pathophysiology. Treatment is straightforward, as it is for the majority of exercise-related diseases: cessation of exercise. In the case of occult gastritis and gastric ulceration, cessation of exercise with no additional treatment results in resolution of the endoscopic evidence of disease within 4 days. As a practical matter, the author recommends specific treatment including gastric acid suppression and gastroprotectant medications for any dog that demonstrates outward signs of disease. Gastric acid suppression is the principal approach used for prevention of exercise-induced gastric disease. Prophylactic dosing of proton pump inhibitors and histamine receptor antagonists both demonstrated efficacy in blinded, placebo-controlled studies. The safety of long-term use of these drugs including their possible use during conditioning to prevent subclinical gastric inflammation has not been documented.

RHABDOMYOLYSIS

Exertional rhabdomyolysis is a syndrome of muscle necrosis associated with strenuous exercise. This syndrome is relatively common in most mammalian athletes including humans, dogs, and horses. However, the cause is unknown, and it is likely that multiple factors, including the intensity of exercise relative to the level of conditioning, genetic predispositions, and metabolic factors contribute to variable degrees in most cases of exertional rhabdomyolysis. Dogs with exertional rhabdomyolysis exhibit stiff, painful muscles and an abnormal gait. Any muscle group can be affected. In severe cases, the proteins released from the damaged muscle cells into the blood can appear in the urine and the urine will be dark reddish-brown. These proteins can be toxic to the kidneys, and secondary renal disease is common in cases of rhabdomyolysis that result in discolored urine.

The most common laboratory finding used for the diagnosis of rhabdomyolysis is increased serum creatine phosphokinase (CPK) activity. This enzyme is highly associated with skeletal muscle but is also present in heart and nervous tissue. Increased serum potassium can also occur in severe cases as intracellular potassium is released into the blood stream. An increase in CPK is a very sensitive indicator of muscle damage, and thus values should be interpreted with caution to avoid falsely diagnosing exertional rhabdomyolysis. Modest increases (two to four times normal) are common after strenuous exercise in clinically normal dogs (McKenzie et al., 2007). Serum CPK can also increase following intramuscular injections, muscle bruising, and even prolonged recumbency (in horses, but reasonably possible in dogs as well). Most clinicians feel that a 10-fold increase over normal is necessary to diagnose diffuse muscle damage characteristic of exertional rhabdomyolysis. In severe cases, serum CPK values can reach as high as 1000 times normal.

The mechanism by which the level of conditioning may affect the occurrence of exertional rhabdomyolysis is through the selection of energy substrates. When a dog exercises at a higher intensity than that to which it is accustomed, muscle perfusion and extramuscular substrate transport may be insufficient to meet the energy demands of the muscles. Rapid glycogen breakdown and anaerobic metabolism of large amounts of glucose results in the production of lactic acid and if blood flow is insufficient to remove this metabolite, the muscle can become dangerously acidic. It is believed that this can result in muscle cell damage and death. However, this mechanism has not been clearly proven, and certainly it would be erroneous to presume that any level of glycogen depletion and anaerobic metabolism will lead to exertional rhabdomyolysis.

The primary metabolic factors implicated in the development of exertional rhabdomyolysis in dogs are oxidative stress and potassium deficiency. Oxidative stress occurs as a routine side effect of aerobic metabolism, and typically the rate of oxidant production increases as metabolic rate increases. Oxidative stress is caused by the production of various reactive oxygen molecules, and the deleterious effects result when these reactive oxygen molecules damage proteins and cell membranes. It stands to reason that the structures most likely to be damaged are those in closest proximity to the production of the radicals (i.e., muscles). Although muscles contain extensive protective antioxidant systems in anticipation of this stress, spillover of oxidative stress into the systemic circulation has been demonstrated in working dogs. Using the most extensively studied examples (racing sled dogs), Piercy et al. (2000 and 2001) clearly demonstrated oxidative stress in the form of depletion of circulating α-tocopherol (vitamin E, a critical antioxidant). However, supplementation of dietary α-tocopherol failed to attenuate the exercise-induced oxidative stress or exercise-induced increases in CPK. In fact, dogs withdrawn from competition due to signs of exertional rhabdomyolysis had higher serum concentrations of α-tocopherol than dogs withdrawn for other reasons. Although it is generally acknowledged that the incidence of exertional rhabdomyolysis in this population has decreased coincident with the advent of extremely high levels of dietary α-tocopherol supplementation, there is no clear mechanistic association between these observations.

Potassium deficiency has been linked to exertional rhabdomyolysis experimentally in dogs, but the verification of this mechanism in clinical situations is difficult. The proposed mechanism is inadequate exercise-induced hyperperfusion of muscle since the release of potassium from contracting muscle is one of the metabolic signals resulting in increased perfusion to working muscle. Exercise can lead to potassium depletion through the efforts of the body to increase blood volume by renal exchange of sodium for potassium (with the subsequent loss of potassium in urine).

Clinically, potassium deficiency is difficult to detect because most potassium in the body is within cells and not the free fluid that is normally sampled. Nevertheless, progressive loss of potassium has been documented in working dogs, lending credibility to this mechanism of exertional rhabdomyolysis. Replenishment of body potassium in the face of a whole-body deficiency is difficult because intravenous supplementation is poorly tolerated and oral supplementation is difficult due to the bitter taste. Effective prevention is based on maintaining appropriate dietary intake to prevent deficiency. Genetic causes of exertional rhabdomyolysis have been identified in horses, and although none have been identified in dogs, the possibility remains if familial tendencies are recognized.

REFERENCES

Davis, M.S., Davis, W.C., Ensign, W.Y., Hinchcliff, K.W., Holbrook, T.C., and Williamson, K.K. (2008). Effects of training and strenuous exercise on hematologic values and peripheral blood leukocyte subsets in racing sled dogs. *Journal of the American Veterinary Medical Association, 232,* 873–878.

Davis, M.S., Willard, M., Nelson, S., Mandsager, R.E., McKiernan, B.C., Mansell, J.K., and Lehenbauer, T.W. (2003). Prevalence of gastric lesions in racing Alaskan sled dogs. *Journal of Veterinary Internal Medicine 17,* 311–314.

Davis, M.S., Willard, M., Williamson, K.K., Royer, C.M., Payton, M.E., Steiner, J.M., Hinchcliff, K.W., McKenzie, E.C., and Nelson, S.L. (2006). The temporal relationship between gastrointestinal protein loss, gastric ulceration or erosion, and strenuous exercise in racing Alaskan sled dogs. *Journal of Veterinary Internal Medicine, 20,* 835–839.

Davis, M.S., Willard, M.D., Williamson, K.K., Steiner, J.M., and Williams D.A. (2005). Sustained strenuous exercise increases intestinal permeability in racing Alaskan sled dogs. *Journal of Veterinary Internal Medicine, 19,* 34–39.

Durocher, L.L., Hinchcliff, K.W., Williamson, K.K., McKenzie, E.C., Holbrook, T.C., Willard, M., Royer, C.M., and Davis, M.S. (2007). Effect of strenuous exercise on urine concentrations of homovanillic acid, cortisol, and vanillylmandelic acid in sled dogs. *American Journal of Veterinary Research, 68,* 107–111.

McKenzie, E.C., Jose-Cunilleras, E., Hinchcliff, K.W., Holbrook, T.C., Royer, C., Payton, M.E., Williamson, K., Nelson, S., Willard, M.D., and Davis, M.S. (2007). Serum chemistry alterations in Alaskan sled dogs during five successive days of prolonged endurance exercise. *Journal of the American Veterinary Medical Association, 230,* 1486–1492.

Piercy, R.J., Hinchcliff, K.W., Disilvestro, R.A., Reinhart, G.A., Baskin, C.R., Hayek, M.G., Burr, J.R., and Swenson, R.A. (2000). Effect of dietary supplements containing antioxidants on attenuation of muscle damage in exercising sled dogs. *American Journal of Veterinary Research, 61,* 1438–1445.

Piercy, R.J., Hinchcliff, K.W., Morley, P.S., Disilvestro, R.A., Reinhart, G.A., Nelson, S.L., Schmidt, K.E., and Craig, A.M. (2001). Vitamin E and exertional rhabdomyolysis during endurance sled dog racing. *Neuromuscular Disorders, 11,* 278–286.

Toniolo, L., Maccatrozzo, L., Patruno, M., Pavan, E., Caliaro, F., Rossi, R., Rinaldi, C., Canepari, M., Reggiani, C., and Mascarello, F. (2007). Fiber types in canine muscles: myosin isoform expression and functional characterization. *American Journal of Physiology Cellular Physiology, 292,* C1915–C1926.

13 Physical and Mental Stress of SAR Dogs during Search Work

Michaela Schneider and Leopold Slotta-Bachmayr

CONTENTS

INTRODUCTION

Search and rescue (SAR) dogs have been used successfully for several decades to search for persons gone missing, buried by avalanches, or trapped under rubble. Despite the development of state-of-the-art technical search instruments, the dog continues to be the fastest tool particularly in open or inaccessible terrain. The most important characteristic of a search dog is its sense of smell, which is employed not only during searches for missing persons but also for locating explosives, illegal drugs, and weapons (Almey and Nicklin, 1996).

SAR dogs are often involved in extended search activities when on duty. Search activities in alpine terrain in addition involve particular requirements of mobility and stamina. Additionally, they must undergo transport by helicopter to a site of action, stay at high elevations, and/or share confined spaces with strangers or other dogs. These mental and physical stressors can disturb the physical balance and impair the dogs' well-being (Immelmann, 1982; Schilder, 1992; Kolb, 1993; Ladewig, 1994; Beerda et al., 1996; Murphy et al., 1997; Beerda et al., 1997; Ewert, 1998; Grandjean et al., 1998).

The ideal search dog has a strong sense of smell, enjoys working, is healthy and agile, exhibits stable behavior day after day, and finds affirmation through praise. In reality, however, a gap exists between those theoretical requirements and the performance rendered by search dogs. In particular, health and levels of agitation are often impaired by the underlying conditions described above and fall short of the theoretical requirements (Rooney et al., 2004). The state of health of a dog, its ability to cope with stress, and the bond between handler and dog are major parameters that determine the potential of a dog and thus its success at searching (Murphy et al., 1997).

STRAIN ON SAR DOGS

Only a small number of scientific studies focus on the strain to which SAR dogs are exposed during search work (Grandjean et al., 1998; Shivik, 2002; Slotta-Bachmayr, 2003). These studies deal mainly with the effects of environment, altitude, or nutrition on performance. Some studies analyzed the physical and mental strain drug-sniffing and explosive-sniffing dogs undergo during work (Strasser et al., 1993; Gazit and Terkel, 2003). The results of these studies, however, cannot be directly applied to SAR dogs because SAR dogs search larger territories and are thus subject to greater physical strain due to their running activity. For that reason, various tests were conducted on trained and fit SAR dogs within the scope of a long-term study. The goal of these tests was to obtain an objective measurement of the strain on SAR dogs during search work by means of physiological measures, detect strain factors, and identify possibilities for decreasing the strain imposed on them during search missions and training.

The first two studies conducted by Köhler (2004) and Wust (2006) investigated the physical and mental strain on alpine SAR dogs during wide area searches in the mountains and during avalanche searches at high altitudes. To obtain standardized data, the dogs performed two search operations during all tests, each lasting about 20 minutes and conducted on the same terrain. Between the two search runs, the dogs were given a 20-minute break. The second search run was followed by a 60-minute recovery phase. In summer, the dogs searched for a person hidden in alpine territory during each search run, indicating the find by barking. In winter the dogs had to locate a person buried about 1 m deep in the snow and indicate success by pawing the ground.

In a third study (Wilhelm, 2007), the strain on SAR dogs was investigated during a rubble search spanning several days. The dogs had to perform a total of four rubble searches daily, each lasting 20 minutes, over 3 days, so that the net time of searching totaled 240 minutes over the 3 days. To simulate a strain situation as close to reality as possible (imitating a rubble search after a disaster) the search exercise was carried out on Tritolwerk in Austria, a large exercise ground for disaster training. In addition, the dogs spent the nights with their handlers in tents. The following factors proved important parameters influencing the strain on the dogs:

- Season, particularly outside temperature
- Type of terrain
- Altitude of search area
- Duration of search operation
- Individual characteristics (age, weight, fur structure)

STRAIN PARAMETERS

THERMOREGULATION

Canine rectal body temperature is generally between 37.5 and 39.0°C, slightly lower in medium-sized or large dogs. During physical activity, body temperature increases since muscle activity generates heat. However, mental stress may cause the body temperature to rise as well. Body temperature must be regulated within a narrow range even under physical or mental stress as a temperature exceeding 40°C results in damage to tissue cell, along with a danger of circulatory failure at a body temperature above 41°C.

The ability to regulate body temperature depends mainly on the ambient temperature, in particular, during physical strain. With rising ambient temperature, mammals dissipate more heat by increased sweating and enhanced perfusion of the blood vessels in the skin. Dogs have sweat glands only on the soles of their paws. Therefore, heat dissipation by means of panting and increased perfusion of skin vessels are the most important cooling mechanisms. Producing about 60% of heat dissipation, panting handles a larger share than conduction via the body surface (Young et al., 1959; Phillips et al., 1981; Bjotvedt et al., 1984; Hornicke, 1987; Ilkiw et al., 1989; Hinchcliff et al., 1993; Haupt, 1997; Matwichuk et al., 1999; Jessen, 2000; Kolb and Seehawer, 2002).

The SAR dog studies demonstrated that the body temperatures of the dogs increased to 39.2°C and up to 39.8°C following search work in alpine terrain in summer. During a 3-day rubble search their temperatures reached an average of 39.3°C. In the winter, their temperatures increased to an average of 39.0°C after avalanche search. Dog body temperature thus rose as high as 39.0°C and above even after a mere 20 minutes of search activity. In medium-sized to large dogs, this temperature is considered a fever (see Figure 13.1).

These studies also showed that additional search runs of the day did not lead to further increases in body temperature. The 20-minute break between the two search runs proved sufficient for temperatures to return to baseline only in the avalanche dogs. The body temperatures of the SAR dogs were already elevated when they entered into the second search run during the operations in summer (see Figure 13.1). Despite that fact, the renewed running and searching effort in the second run did not produce stronger increases in body temperature. This effect was also observed during rubble searches over 3 days. Even though work loads were doubled in that study, body temperatures remained within an even range after four daily search runs lasting 20 minutes each (see Figure 13.2).

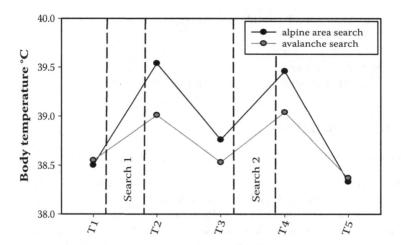

FIGURE 13.1 Body temperatures of search and rescue dogs during search work in alpine terrain in summer and avalanche work in winter. T1 = measurement before first search run; T2 = immediately after first 20-minute search run; T3 = after 20-minute break; T4 = immediately after second 20-minute search run; T5 = after 60-minute recovery.

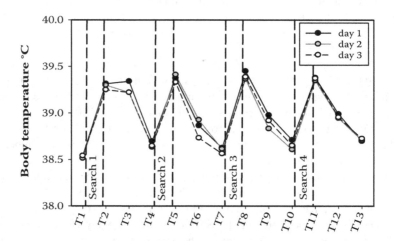

FIGURE 13.2 Body temperatures of search and rescue dogs during four rubble search operations conducted over three consecutive days. T1 = before first search run; T2 = immediately after first 20-minute search run; T3 = after 20-minute break; T4 = after 40-minute break; T5 = immediately after second 20-minute search run; T6 = after 20-minute break; T7 = after 40-minute break; T8 = immediately after third 20-minute search run; T9 = after 20-minute break. T10 = after 40-minute break; T11 = immediately after fourth 20-minute search run; T12 = after 20-minute break; T13 = after 40-minute break.

The increases in body temperature through the search work were found to depend on the activity levels of the dogs, the outside temperature, and the dogs' fur structures. The higher the dogs' level of activity during the search operations, the higher was the temperature increase after search. This can be explained by the increased muscle activity of the more active dogs and the related increased generation of body heat. Outside temperature influences body temperature as the heat generated during physical work must be dissipated to the environment. In the presence of high ambient temperatures and under direct sunlight, the level of heat dissipation via the body surface is lower as heat dissipation is less effective with decreasing temperature difference between body surface and environment. If outside temperatures are high, dogs at the same activity level face increased risk of getting too hot compared to dogs working outside in low temperatures. Low ambient temperatures in connection with a strong wind or precipitation (rain or snow) can significantly increase the percentage of heat dissipation via the body surface. Increased heat loss via the body surface in colder temperatures initially results in enhanced working efficiency because body temperatures rise less and result in extended stamina.

This effect was also observed during a study on the influence of altitude on alpine SAR dogs. It was only during an exercise between 2500 and 2700 m altitude that the body temperature was lower after both search runs (see Figure 13.3). In all five search exercises, the outside temperature was similar at the various altitudes, ranging between 12 and 15°C. This explains why the outside temperature cannot have been the decisive factor responsible for the significantly lower body temperature increase at high altitudes.

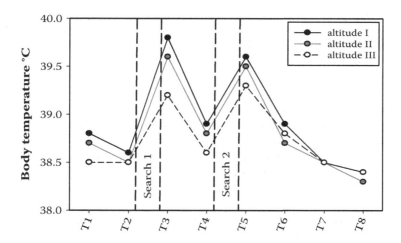

FIGURE 13.3 Body temperatures of search and rescue dogs during alpine rubble searches at three different altitudes (I: 500 to 700 m; II: 1500 to 1700 m; III: 2500 to 2700 m). T1 = measurement 15 minutes before first search run; T2 = immediately before first search run; T3 = immediately after first 20-minute search run; T4 = after 20-minute break; T5 = immediately after second 20-minute search run; T6 = after 20-minute recovery; T7 = after 40-minute recovery; T8 = after 60-minute recovery.

Humidity also did not differ significantly among the various altitudes. Between 2500 and 2700 m, however, a strong wind was blowing. It exerted a cooling effect on the dogs by rumpling their insulating fur, enabling more heat to be dissipated into the environment. Hörnicken (1987) and Jessen (2005) already demonstrated that wind velocities of 27 km/h may double the level of heat dissipation into the environment.

HEART RATE AND ACTIVITY

Measuring the dogs' heart rates enabled the determination of both mental and physical strain levels. The resting heart rate of a dog is 70 to 160 beats per minute (bpm), with small dogs having slightly higher rates than large dogs (Sporri, 1987; Engeland et al., 1990; Beerda et al., 1997; 1998; Harmeyer, 2000). Under physical strain, a dog's heart rate increases during the initial 20 to 60 seconds and may double or even triple under maximum strain. The heart rates of exercised dogs are generally lower than those of unexercised dogs. In addition, exercised dogs recover from strain more quickly. The time span until the return of the heart rate to baseline can be 8 to 60 minutes (Franklin et al., 1959; Smulyan et al., 1965; Mackintosh et al., 1983; Ilkiw et al., 1989; Sneddon et al., 1989; Vincent and Leahy, 1997; Beerda et al., 1998; Matwichuck et al., 1999; Engelhardt, 2000; Kolb et al., 2002).

However, the heart rate also changes under mental strain. Dogs react with increased heart rates to events associated with positive (joyfully anticipated work) and negative expectations (punishment). The heart rate in that context is also dependent on temperament. In dogs characterized as sensitive to stress, heart rates measured were invariably higher than in dogs classified as even-tempered (Beerda et al., 1998). Because working heart rate is substantially influenced by both mental and physical strains, both parameters will be discussed.

The resting heart rates of the SAR dogs participating in the three studies ranged from 88 to 108 bpm and thus fell within the resting range cited in the literature. Observation of working heart rates demonstrated that the dogs exhibited average rates of 156 to 167 bpm during alpine area and rubble search, and 144 to 151 bpm during the 3-day rubble search. The heart rates during avalanche searches were slightly higher, reaching an average of 174 bpm. SAR dogs therefore operate at a heart rate optimal for activities requiring stamina, which enables them to work untiringly even over extended periods.

The study of alpine area and rubble search as well as avalanche search yielded contrasting changes in heart rates and activity levels during two 20-minute search runs performed on one day. While the heart rates were no different in the two search runs, the activity levels of the SAR dogs in the second search run dropped significantly (see Figure 13.4). This decreased activity in the second search run may have been indicative of a loss in motivation or a sign of fatigue. Because the physical strain on the dogs in the second search run (measured by heart rate and body temperature) was on the same level as in the first run, the lower activity suggests fatigue. The dogs ran at their optimum heart rates in the second search run as well, but they already reached that frequency at a lower level of motion activity, which is indicative of higher physical strain in the second run. This increased strain in the second search run was more marked in the older (more than 7 years) and heavier or larger

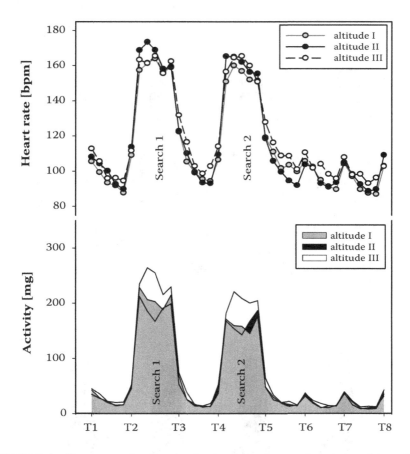

FIGURE 13.4 Heart rate and activity levels of search and rescue dogs during alpine rubble searches at three altitudes (I: 500 to 700 m; II: 1500 to 1700 m; III: 2500 to 2700 m). T1 = measurement 15 minutes before first search run; T2 = immediately before first search run; T3 = immediately after first 20-minute search run; T4 = after 20-minute break. T5 = immediately after second 20-minute search run; T6 = after 20-minute recovery; T7 = after 40-minute recovery; T8 = after 60-minute recovery.

dogs as the drops in their activity levels were significantly greater than in younger and lighter animals.

During the 3-day rubble search, however, other results were noted. Both heart rate and activity remained on much the same level during the four daily search exercises (see Figure 13.5). Only during the fourth search operation was a slight drop in the heart rate registered, with activity remaining on the same level. However, significant differences were observed between the study days. The heart rates reached the highest values on day 1, dropped significantly on day 2, and slightly increased again on day 3. On the other hand, activity on days 1 and 3 was on a similarly low level, reaching the highest values on day 2 by a significant margin. These contradictory changes in heart rate and activity can be explained. Many dogs had problems with the very difficult terrain on day 1 and moved cautiously and uncertainly on the

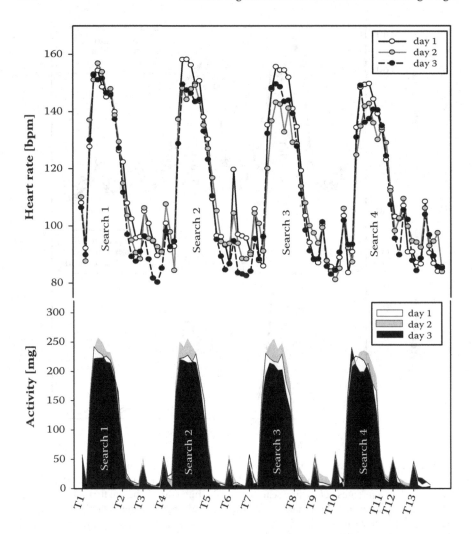

FIGURE 13.5 Heart rate and activity levels of search and rescue dogs during four rubble search operations conducted over three consecutive days. T1 = before first search run; T2 = immediately after first 20-minute search run; T3 = after 20-minute break; T4 = after 40-minute break; T5 = immediately after second 20-minute search run; T6 = after 20-minute break; T7 = after 40-minute break; T8 = immediately after third 20-minute search run; T9 = after 20-minute break; T10 = after 40-minute break; T11 = immediately after fourth 20-minute search run; T 12 = after 20-minute break; T13 = after 40-minute break.

rubble. The high heart rates on day 1 were therefore caused by the mental strain to which the dogs were exposed during the search. By day 2, the dogs were accustomed to the terrain and moved with considerably more security and speed. Based on lower stress levels of the dogs during search activity, their hear rates dropped despite significantly higher levels of activity. On day 3, 63% of the dogs exhibited marked signs of fatigue, and activity during searching decreased again considerably. Heart rates, on the other hand, remained on much the same level on day 3 as the strain intensified significantly due to physical and mental fatigue. The dogs' fatigue on day 3 was also reflected in the time they took to locate victims. On average, victim locations were indicated 4.3 minutes after initiation of the search. It took the most time to locate victims on day 3 even though the dogs by then were very familiar with the terrain.

ALTITUDE

The influence of the altitude of the search area was manifest most by the strain on the dogs' cardiovascular systems. It is well known that the physical performances of humans and animals are influenced by altitude because oxygen supplies deteriorate at high altitudes due to lower partial oxygen pressure. To ensure adequate tissue oxygenation, the heart rate increases. Grandjean et al. (1998) had already conducted studies on SAR dogs in altitudes of 4800 m and 5980 m, and detected increased heart rates during rest and under strain. No studies, however, indicate whether the lower altitudes experienced during alpine SAR dog operations in Europe influence the strains exerted on the dogs.

During the first study on SAR dogs conducted by Köhler (2004) significant differences in heart rates were found between avalanche searches carried out at an altitude of 2600 m and alpine area searches at 600 m. During avalanche searches, heart rates were significantly higher, with the widest differences observed during the resting phases (see Figure 13.6). However, these differences in heart rate could not be clearly attributed to the influence of altitude because climatic factors also have been demonstrated to strongly influence heart rate.

In a study by Wust (2006), SAR dogs were then tested at three different altitudes. The dogs worked at nearly the same heart rates during the search operations

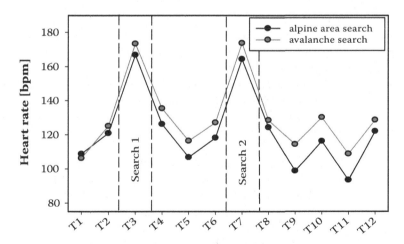

FIGURE 13.6 Heart rates of search and rescue dogs during a search work in alpine terrain in summer and an avalanche work in winter. T1 = measurement before first search run; T2 = immediately after first 20-minute search run; T3 = after 20-minute break; T4 = immediately after second 20-minute search run; T5 = after 60-minute recovery.

at all altitudes. In the resting phases before, between, and after the search runs, however, marked differences in heart rates were observed. Even at altitudes between 1400 and 1600 m, resting heart rates were higher compared to search activities at 400 to 600 m. This difference was even more marked at altitudes between 2400 and 2600 m. Accordingly, altitude exerts influence on the cardiovascular systems of dogs. However, this influence is detectable only in the resting phases. The dogs ran at their optimum heart rates during the search operations.

MUSCLE METABOLISM

Lactate concentration in the blood is considered an indicator of oxygen debt in muscles during a strain phase. Under strong physical strain, anaerobic energy gain will lead to increased lactate formation in the muscles. The blood lactate concentration during extended work shows the degree of muscular fatigue. Following strain, an increase in lactate levels by a factor of 5 to 20 can be detected in dogs. Exercised dogs form less lactate than unexercised ones because exercise enhances oxygen supply to muscles. No evidence indicates that mental stress can lead to increases in blood lactate levels (Hammel et al., 1977; Bjotvedt et al., 1984; Snow et al., 1988; Ilkiw et al., 1989; Burr et al., 1997; Murphy et al., 1997; Matwichuck et al., 1999; Engelhardt, 2000).

The optimum conditions under which rescue dogs work are also reflected in their muscle metabolism. Whereas dogs run at a level of anaerobic metabolism during short-term strain, e.g., greyhounds during sprint races, reaching average lactate values between 11 and 13 mmol/l, the values of the SAR dogs studied constantly remained within the aerobic range. The average lactate values were 2 mmol/l after operations in the alpine terrain lasting a total of 40 minutes and rubble search operations, each lasting 80 minutes. They were thus within the reference range of resting values stated in the literature.

Creatine kinase (CK) activity is considered an indicator of strain-related skeletal muscle damage. An increase in CK activity is the result of muscle trauma, unaccustomed physical strain, circulatory shock, or other factors causing muscle damage. A training effect is possible although CK activity decreases steadily in the course of exercise both in the resting phase and following physical strain. This is attributed to the abilities of the skeletal muscles of exercised dogs to provide more oxygen for metabolism (Sneddon et al., 1989; Aktas et al., 1993; Burr et al., 1997; Kolb and Seehawer, 2002).

The CK values measured after the two 20-minute avalanche search runs had risen slightly above the reference range and increased further after a 2-hour recovery phase. Compared to that, the CK values measured during alpine area searches of the same duration were always within the reference range. In the 3-day rubble search, on the other hand, a marked fatigue of the muscles was detectable after the daily 80-minute search runs. The highest CK values were measured at the end of study day 1, with a factor of 2.3 above the reference range. At the ends of the second and third study days, elevated CK values were measured as well, although they were considerably lower than on day 1. The higher CK values of study day 1 can be explained by the uncertain and cautious movements of the dogs on the rubble. Running tensely put their muscular systems under a more severe strain than the higher activity of the second study day or the accumulated total of 240 minutes of search time by the end of the third study day. Accordingly, a running activity of a total of 40 minutes during area search does not pose any noteworthy strain on the muscular systems of well exercised dogs. Even a 3-day rubble search is a feasible task for well exercised SAR dogs if they can recover at night. However, even after day 1, the dogs must be expected to exhibit a certain level of fatigue as well as stiffness and aching of the muscles. During avalanche search, a mere 40 minutes of activity resulted in muscular fatigue. This is why avalanche operations should be of considerably shorter duration than area or rubble searches. The decisive factor is the surface of the snow that influences the ability of dogs to walk on it without breaking through it.

STRESS

Stress is the body's specific reaction to unspecific internal and external stimuli. Cortisol concentration is generally used as an indicator of stress, with a distinction made between acute (intensive and short-term) and chronic (mild and long-term) stress. Any change in cortisol concentrations can be detected in blood, saliva, and feces (Palazzolo and Quadri, 1987; Kirschbaum and Hellhammer, 1989; Vincent and Michell, 1992; Rothuizen et al., 1993; Beerda et al., 1996, 1997, and 1998; Schatz and Palme, 2001).

Under conditions of physical and mental stress, cortisol release is increased. Salivary cortisol following mental stress may be increased by a factor of up to 10, with the maximum level occurring after 10 to 30 minutes. Following mental strain, salivary cortisol levels returned to normal within 60 minutes after the end of the stressor influence. Physical strain as well may trigger a considerable increase in cortisol secretion (Kirschbaum and Hellhammer, 1989; Vincent and Michell, 1992; Kolb, 1993; Beerda et al., 1997 and 1998; Kolb and Seehawer, 2002).

Salivary cortisol determination is a simple, noninvasive alternative method of measurement. Disturbance of the measuring results is only minor. Since sampling is relatively stress-free for the animals, sampling frequency can be increased. Determination of fecal cortisol levels involves analysis over time because cortisol yields information only after long-term stress exposure. The use of cortisol from the blood or the saliva, however, provides considerably more information about the temporal courses of stressors (Vincent et al., 1992; Beerda et al., 1996; Slotta-Bachmayr and Schwarzenberger, 2007).

Fecal cortisol levels in avalanche dogs are subject to very strong individual variations that may be explained by the dogs' personalities. Dogs that handle unfamiliar situations with a certain routine exhibit considerably lower levels of stress hormones than dogs that cannot cope with such situations. In addition, dogs need some time to adapt to unfamiliar situations. During SAR seminars and in animal shelters (Beerda at al., 1999), it takes dogs about 3 days to become accustomed to new situations. Additionally, a handler plays a major role. If a handler is nervous, for example, during a search mission, his or her nervous state transfers to the dog. In exercises with calm handlers, the stress hormone levels of the dogs were also considerably lower (Slotta-Bachmayr and Schwarzenberger, 2007).

During rubble searches, several points indicate increasing stress due to the repeated runs spanning 3 days. For instance, salivary cortisol reacted to each 20-minute search run by increasing. The increase during the first study day was within a range measured in the other two studies as well, and is typical of dogs capable of coping with stressors. On the second and third study days, however, considerably higher increases of salivary cortisol were measured after the fourth run. This higher increase in salivary cortisol levels in combination with simultaneous drops in the dogs' heart rates during the fourth search run indicates losses of motivation or mental fatigue of the dogs and a change from joyfully motivated eustress to an incipient, negatively felt distress.

The results of rubble search suggest that the change from eustress to distress occurs after 60 to 80 minutes of pure search time (see Chapter 7 by Helton for a cognitive approach to this phenomenon). One factor that must be considered, however, is that during the study a point was made to ensure that each search had a positive end. If a dog had already located a victim after a few minutes or was unable to find the victim, it was made to continue searching for the scheduled duration of 20 minutes and received another confirmation of success by finding another hidden person shortly before the end of the search. Thus, each search ended positively. This helped maintain the dogs' motivation to search throughout the entire 3 days.

CONSEQUENCES FOR ACTION AND TRAINING

The 3-day rubble search study provided impressive proof of the high efficiency of SAR dogs in searching for buried persons. A total of 236 search operations were analyzed. The dogs were successful within the first 5 minutes in 75% and within the first 10 minutes in 91% of the search operations. Only in two cases did the dogs fail to locate "victims" within 20 minutes; both operations took place on the third study day and were carried out by the two youngest dogs.

Generally, the young dogs (less than 4 years old) took longer to locate and pinpoint victims than the dogs aged 4 to 7 years. This age-related effect is attributable to the greater experience of the older dogs that are more accustomed to finding the exact positions of persons trapped underneath rubble. Younger dogs have less experience with the factors surrounding a search (unfamiliar locations, smells, dogs and persons, spending nights in unfamiliar places) and are more easily distracted. This observation substantiates the effects of level of experience and exercise of an individual dog on the success of a search operation. It takes some time for a dog's personality to mature, also. Moreover, it is easier for dogs with a certain level of experience to cope with the surroundings, and more experienced dogs are able to concentrate better on an actual operation.

The onset of signs of exhaustion earlier during alpine area and rubble searches and during avalanche searches is attributable both to the experimental set-up and the physical strain on the SAR dogs. During these search operations, the dogs were granted breaks of only 20 minutes between the two search runs. In contrast, during the 3-day rubble search, the dogs were allowed breaks of 60 minutes between search runs. The lengths of the search runs did not differ; each lasted 20 minutes. Accordingly, a break of 20 minutes is not enough for dogs to recover completely. This is underlined by the declines in activity during the ensuing search runs. A 60-minute break, on the other hand, allows dogs to regenerate to such an extent that no declines in activity were noticeable during the four search runs throughout an entire day.

During rubble searches, the dogs even exhibited markedly increased performances on day 2, and only on day 3—after 160 minutes of pure search time on the two previous days and two nights spent on the disaster terrain—did the SAR dogs exhibit distinct signs of exhaustion. These results show that employing care and consideration when using the dogs and granting them adequate breaks along with a habituation phase at the beginning of an extended search operation will help maintain their operational capability and efficiency even over several days. It was also shown, however, that search performance declined considerably during the course of the study, reaching a threshold value during the fourth search runs of days 2 and 3.

An overall result of these studies of alpine SAR dogs was that search operations in alpine rubble terrain and on avalanche terrain produced considerably more physical strain than simple area searches. The level of activity during the second search run in terrains comparable in terms of the search area size, altitude, and slope gradient was on average 5 to 10% lower during alpine area searches than during alpine rubble searches. This is attributable to the higher demands on the dogs' fine motor skills and adroitness when moving around alpine rubble terrain. The level of decline in activity during avalanche search fell between alpine rubble search and area search. The primary muscle-straining aspects of avalanche search were running in the snow and breaking through the snow cover.

Apart from terrain, canine performance is also significantly influenced by altitude and age. The young dogs (under 4 years) mainly undergo mental strain during SAR work. The middle-aged group (4 to 7 years) at their peak physical and mental performance cope best with the demands of SAR work. For dogs older than 7 years, SAR work is clearly physically straining. At altitudes between 2500 and 2700 m, the dogs older than 7 years experienced considerable declines in performance during the second search run. Activity in the first search run was still equivalent to levels of the other two groups but dropped dramatically in the second run. In addition, the older dogs' heart rates were consistently higher and they exhibited considerable delays in recovery after the second run (see Figure 13.7). It is therefore important to ensure short search times and adequately long breaks for older dogs during operations and exercises in alpine terrain. Paying no attention to the increased circulatory strain that occurs in older dogs at high altitudes presents risks of physical overload and, in extreme cases, circulatory collapse. Despite the greater physical strains on older SAR dogs, their searching efficiency equals that of young dogs; their experience makes up for their lower running speeds. However, the results also show that SAR dogs above age 7 need regular physical exercise to maintain their levels of performance.

In extreme climatic conditions such as very low ambient temperatures and heavy rain or snow in combination with strong winds, heat loss via body surface exceeds heat production by muscle activity and dogs begin to cool down and start to freeze. This intensifies the physical strain on them and at the same time reduces their working efficiency. This increased energy loss during inclement weather must be considered, particularly when selecting resting places for the dogs during recovery phases.

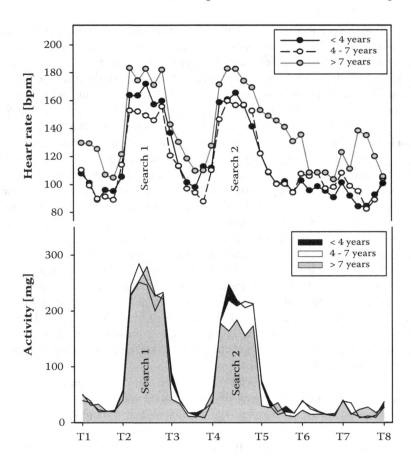

FIGURE 13.7 Heart rate and activity levels of search and rescue dogs during an alpine rubble search at altitude of 2700 m, comparison of three age groups (<4 years, 4 to 7 years, >7 years). T1 = measurement 15 minutes before first search run; T2 = immediately before first search run; T3 = immediately after first 20-minute search run; T4 = after 20-minute break; T5 = immediately after second 20-minute search run; T6 = after 20-minute recovery; T7 = after 40-minute recovery; T8 = after 60-minute recovery.

The body temperatures of avalanche dogs drop rapidly if they are kept in the open after a search operation. Their temperature returned to baseline within 20 minutes after the end of search runs. Body temperatures during summer search operations usually did not return to baseline levels until 40 minutes later. After resting for 60 minutes, the body temperatures of avalanche dogs dropped considerably below baseline, illustrating the degree of heat loss.

Apart from climatic factors, the quality of fur determines the degree of heat loss. The thicker the fur, the lower is the degree of heat loss. Fur traps air between individual hairs, thus providing a kind of insulation around the body. The denser and thicker a fur is, the stronger is its insulating effect. Short-haired dogs lose more heat via body surface than long-haired or stock-haired dogs. However, even thick fur does not protect the joints from cooling during extended resting phases. At outside temperatures in the plus range, however, this stronger ability to dissipate heat via the body surface is an advantage because short-haired dogs do not overheat as rapidly during search activities. In a study by Wust (2006), short-haired dogs were found to experience significantly lower rises in body temperature after search operations than long-haired or stock-haired dogs.

In particular during winter operations, it is important to ensure that the dogs are kept in a wind-protected place and may be provided with body warming bags to reduce cooling at low outside temperatures. In addition, in all search operations, but especially in avalanche search, the dogs should undergo warm-up procedures prior to starting search activities. This prepares the joints, ligaments, and muscles for the ensuing work and helps to prevent premature joint degeneration and pulled ligaments and muscles.

Dogs operate as very useful and efficient tools to find missing persons in various situations. Nevertheless this type of physical activity imposes a lot of stress and strain on the dogs and therefore adequate preparation and training are necessary before they are employed in this type of work.

REFERENCES

Aktas, M., Auguste, D., Lefebvre, H.P., Toutain, P.L., and Braun, J.P. (1993). Creatine kinase in the dog: a review. *Veterinary Research Communications, 17,* 353–369.

Almey, H. and Nicklin, S. (1996). How does your dog smell? A review of canine olfaction. *Journal of Defense Sciences, 1,* 345–352.

Beerda, B., Schidler, M.B., Bernadina, W., van Hooff, J.A., de Vries, H.W., and Mol, J.A. (1999). Chronic stress in dogs subjected to social and spatial restrictions. II. Hormonal and immunological responses. *Physiology and Behavior, 66,* 243–254.

Beerda, B., Schilder, M.B., Janssen, N.S., and Mol, J.A. (1996). The use of saliva cortisol, urinary cortisol, and catecholamine measurements for a noninvasive assessment of stress responses in dogs. *Hormones and Behavior, 30,* 272–279.

Beerda, B., Schilder, M.B., van Hooff, J.A., and de Vries, H.W., (1997). Manifestations of chronic and acute stress in dogs. *Applied Animal Behaviour Science, 52,* 307–319.

Beerda, B., Schilder, M.B., van Hooff, J.A., and de Vries, H.W., and Mol, J.A. (1998). Behavioural, saliva cortisol and heart rate response to different types of stimuli in dogs. *Applied Animal Behaviour Science, 58,* 365–381.

Bjotvedt, G., Weems, C.W., and Foley, K. (1984). Strenous exercise may cause health hazards for racing Greyhounds. *Veterinary Medicine and Small Animal Clinician, 79,* 1481–1487.

Burr, J.R., Reinhart, G.A., Swenson, R.A., Swaim, S.F., Vaughn, D.M., and Bradley, D.M. (1997). Serum biochemical values in sled dogs before and after competing in long-distance races. *Journal of the American Veterinary Medical Association, 211,* 175–179.

Engeland, W.C., Miller, P., and Gann, D.S. (1990). Pituitary–adrenal adrenomedullary responses to noise in awake dogs. *American Journal of Physiology, 258,* R672–R677.

Engelhardt, W. and Breeves, G. (2000). *Physiologie der Haustiere.* 1. Stuttgart: Enke, 482–489.

Ewert, J.P. (1998). *Neurobiologie des Verhaltens.* 1. Bern: Verlag Hans Huber, 140–153.

Franklin, D.L., Ellis, R.M., and Rushmer, R.F. (1959). Aortic blood flow in dogs during treadmill exercise. *Journal of Applied Physiology, 14,* 809–812.

Gazit, I. and Terkel, J. (2003). Explosives detection by sniffer dogs following strenuous physical activity. *Applied Animal Behaviour Science, 81,* 149–161.

Grandjean, D., Sergheraert, R., Valette, J., and Driss, F. (1998). Biological and nutritional consequences of work at high altitude in search and rescue dogs: the scientific expedition Chiens des Cimes-Licancabur, 1996. *Journal of Nutrition, 128,* 2694S–2697S.

Hammel, E.P., Kronfeld, D.S., Ganjam, V.K., and Dunlap, H.L. (1977). Metabolic responses to exhaustive exercise in racing sled dogs fed diets containing medium low or zero carbohydrate. *American Journal of Clinical Nutrition, 30,* 409–418.

Harmeyer, J. (2005). In Engelhardt, W. and Greves, G., Eds. *Physiologie der Haustiere,* 2. Stuttgart: Enke, 137–170.

Haupt, K.H. (1997). In Fenner, W.R., Ed. *Kleintierkrankheiten.* 2. Stuttgart: Enke, 45–61.

Hinchcliff, K.W., Olson, J., Crusberg, C., Kenyon, J., Long, R., Royle, W., Weber, W., and Burr, J. (1993). Serum biochemical-changes in dogs competing in a long distance sled race. *Journal of the American Veterinary Medical Association, 202,* 401–405.

Hornicke, H. (1987). Thermophysiologie. In Scheunert, A. and Trautmann, A., Eds. *Lehrbuch der Veterinärphysiologie.* Berlin: Paul Paray, 142–158.

Ilkiw, J.E., Davis, P.E., and Church, D.B. (1989). Hematologic, biochemical, blood-gas and acid-base values in greyhounds before and after exercise. *American Journal of Veterinary Research, 50,* 583–586.

Immelmann, K. (1982). *Wörterbuch der Verhaltensforschung.* 1. Berlin: Paul Paray, 230–232.

Jessen, C. (2005). In Engelhardt, W. and Breves, G., Eds. *Physiologie der Haustiere.* 2. Stuttgart: Enke, 446–469.

Kirschbaum, C. and Hellhammer, D.H. (1989). Salivary cortisol in psychobiological research: an overview. *Neuropsychobiology, 22,* 150–169.

Kohler, F. (2003). Vergleichende Untersuchungen zur Belastung von Lawinen- und Rettungshunden bei der Lauf und der Sucharbeit. Ludwig Maximilians Universität, München, dissertation.

Kolb, E. (1993). Anpassungsvorgänge in der Sekretion von Hormonen (Corticoliberin, ACTH, Cortisol) und im Stoffwechsel von Hunden bei Belastungen. *Monatshefe für Veterinamedizen 48,* 595–601.

Kolb, E. and Seehawer, J. (2002). Die Leistungsfähigkeit des Rennhundes und der Einfluss der Anwendung von Vitaminen. Tierärztliche *Umschau 57,* 317–325.

Ladewig, J. (1994) In Docke, F., Ed. *Veterinärmedizinische Endokrinologie.* 3. Stuttgart: Gustav Fischer, 379–399.

Mackintosh, I.C., Dormehl, I.C., Van Gelder, A.L., and Du Plessis, M. (1983). Blood volume, heart rate, and left ventricular ejection fraction changes in dogs before and after exercise during endurance training. *American Journal of Veterinary Research, 44,* 1960–1962.

Matwichuk, C.L., Taylor, S.M., Shmon, C.L., Kass, P.H., and Shelton, G.D. (1999). Changes in rectal temperature and hematologic, biochemical, blood gas, and acid–base values in healthy Labrador Retrievers before and after strenuous exercise. *American Journal of Veterinary Research, 60,* 88–92.

Murphy, M.G., Conroy, S., and Lowe, J.A. (1997). Aspects of exercise physiology in the dog. *Irish Veterinary Journal, 50,* 65–69.

Musch, T.I., Haudet, G.C., Ordway, G.A., Longhurst, J.C., and Mitchell, J.H. (1985). Arterial blood gases and acid–base status of dogs during graded dynamic exercise. *Journal of Applied Physiology, 61,* 1914–1919.

Palazzolo, D. and Quadri, S. (1987). The effects of aging on the circadian rhythm of serum cortisol in the dog. *Experimental Gerontology, 22,* 379–387.

Phillips, C.J., Coppinger, R.P., and Schimel, D.S. (1981). Hyperthermia in running sled dogs. *Journal of Applied Physiology, 51,* 135–142.

Ready, A.E. and Morgan, G. (1984). The physiological response of Siberian Husky dogs to exercise: effect of interval training. *Canadian Veterinary Journal, 25,* 86–91.

Rooney, N.J., Bradshaw, J.W.S., and Almey, H. (2004). Attributes of specialist search dogs: a questionnaire survey of UK dog handlers and trainers. *Journal of Forensic Science, 49,* 1–7.

Rothuizen, J., Reul, J.M., Van Sluijs, F.J., Mol, J.A., Rjinberk, A., and De Kloet, E.R. (1993). Increased neuroendocrine reactivity and decreased brain mineralocorticoid receptor-binding capacity in aged dogs. *Endocrinology, 132,* 161–168.

Sanders, M., White, F., and Bloor, C. (1977). Cardiovascular responses of dogs and pigs exposed to a similar physiological stress. *Comparative Biochemistry and Physiology, 58A,* 365–370.

Schatz, S. and Palme, R. (2001). Measurement of faecal cortisol metabolites in cats and dogs: a non-invasive method for evaluating adrenocortical function. *Veterinary Research Communications, 25,* 271–287.

Schilder, M.B.H. (1992). Stress and welfare and its parameters in dogs. *Tijdschrift Dierg. 117,* 53S–54S.

Shivik, J.A., (2002). Odor-adsorptive clothing, environmental factors and search-dog ability. *Wildlife Society Bulletin, 30,* 721–727.

Slotta-Bachmayr, L. (2003). Lawinenhunde im Stress. In Slotta-Bachmayr, L., Ed. *Optimierung von Einsatz und Training bei Lawinenhunden,* Salsburg Land, 48–63.

Slotta-Bachmayr, L. and Schwarzenberger, F. (2007). Faecal cortisol metabolites as indicators of stress during training and search missions in avalanche dogs. *Veterinary Medicine Austria, 94,* 110–117.

Smulyan, H., Cuddy, R.P., Vincent, W.A., Kashemsant, U., and Eich, R.H. (1965). Initial haemodynamic response to mild exercise in trained dogs. *Journal of Applied Physiology, 20,* 437–442.

Sneddon, J., Minnaar, P.P., Grosskopf, J.K.W., and Groeneveld, H.T. (1989). Physiological and blood biochemical responses to submaximal treadmill exercise in Canaan dogs before, during and after training. *Journal of the South African Veterinary Association, 60,* 87–91.

Snow, D.H., Harris, R.C., and Stuttard, E. (1988). Changes in haematology and plasma biochemistry during maximal exercise in greyhounds. *Veterinary Record, 123,* 487–489.

Sporri, H., (1987). In Scheunert, A. and Trautmann, A., Eds. *Lehrbuch der Veterinärphysiologie,* 7. Berlin: Paul Paray, 217–218.

Strasser, A., Hochleithner, M., and Bubna-Littitz, H. (1993). Stress in dogs used for searching for drugs. *Wien Tierärztl Mschr, 80,* 352–355.

Vincent, I.C. and Leahy, R.A. (1997). Real-time non-invasive measurement of heart rate in working dogs: a technique with potential applications in the objective assessment of welfare problems. *Veterinary Journal, 153,* 179–183.

Vincent, I.C. and Michell, A.R. (1992). Comparison of cortisol in saliva and plasma of dogs. *Research in Veterinary Science, 53,* 342–345.

Wilhelm, S.F. (2007). Belastung von Rettungshunden während einer dreitägigen Trümmersuche auf einem Katastrophenübungsgelände. Ludwig Maximilians Universität, München, dissertation.

Wust, C. (2006). Einfluss der Höhenlage und Geländebeschaffenheit auf die leistungsphysiologischen Parameter von alpinen Rettungshunden. Ludwig Maximilians Universität, München, dissertation.

Young, D.R., Mosher, R., Erve, P., and Spector, H. (1959). Body temperature and heat exchange during treadmill running in dogs. *Journal of Applied Physiology, 14,* 839–943.

14 Signs of Physiological Stress in Dogs Performing AAA/T Work

Dorit Haubenhofer

CONTENTS

INTRODUCTION

The first part of this chapter is devoted to general topics such as the nature of stress, its definition, symptoms, and measurement. We then enter the world of animal-assisted activities (AAA) and animal-assisted therapy (AAT). In Austria both fields are combined under the German "tiergestützte Therapie" umbrella term. After a short description of these terms, we outline a research program conducted by Haubenhofer and Kirchengast at the University of Vienna from 2003 to 2005. We investigated 13 dog handlers and 18 companion dogs working as teams. The handlers (owners) collected saliva samples on both control days without therapeutic work and days of therapeutic sessions. The concentration of the cortisol hormone in the samples was analyzed to evaluate the amounts of physiological stress caused by the therapy sessions.

We end this chapter by discussing results, and by delivering a look at other recent studies related to this topic. What still remains unanswered in all studies (including our own) is whether the dogs *perceived* feelings of stress emotionally. We may gather hormonal data to investigate physiological events, but the emotional states of dogs remain hidden to us. This last issue raises the specter of anthropomorphism discussed by Helton in Chapter 1.

DEFINITIONS OF STRESS

Stress is a term commonly heard in relation to both people and animals. If one reads through journals and clicks through Web sites, it seems that everyone and everything can become a victim of stress. Humans can, animals can, organs can, and even cells can. In a world that sometimes seems ruled by stress, basic questions still remain:

1. What is stress? What causes it and what are its symptoms?
2. Why is it important to write yet another piece about stress? More specifically, why are the topics of stress detection and stress abatement for dogs important?

Several models have been proposed to explain what stress is. Each model's orientation depends on the scientific field and culture in which it was developed. Traditionally stress researchers focused on physiological reactions, *stress responses*, or the environmental determinants of stress, *stressors*. Physiologists and physiologically oriented psychologists focused on physiological stress responses. In 1915, Walter B. Cannon defined stress as a nonspecific reaction of the body to any kind of demand. He characterized a stress response as a subaction of an animal's fight-or-flight mechanism. Hans Selye (1957) elaborated on this perspective with many studies during the second half of the 20th century. He interpreted stress as an animal's response to harmful stimuli or environmental stressors, and defined stress as a nonspecific corporal (physical) response to any harmful stimulation from the environment an animal encounters. He developed a general model of physiological reactions to stress called the general adaptation syndrome (GAS) based on three principles. GAS:

1. Is a defence response independent of the nature of the stressor
2. Consists of three phases: alarm, resistance, and exhaustion
3. May lead to long-term adaptation diseases if it lasts too long or is too strong

Early physiological perspectives of Selye and others defined stress as a dependent variable, an animal's response to external stimulation, similar to the y output of a linear mathematical function, $y = mx + b$. Other stress researchers devoted more effort to manipulating environmental variables that were labeled *a priori* as stressors, for example, thermal stress (extreme cold or heat), confinement, crowding, noise, shock (electrical), sleep deprivation, time pressure, and whole-body vibration (Hancock, 1984).

From this perspective the focus of stress was on the independent variable or x input in the linear $y = mx + b$ mathematical function. Although the research on stressors and their concomitant physiological responses such as GAS was informative, certain issues became apparent. First, some of the so-called stressors (stress inducers) were subjectively reported by people as pleasant in certain situations (Poulton, 1976). Second, studies revealed that physiological and psychological (self-reported) responses often do not correlate highly (Schacter and Singer, 1962). Animals, including people, are dynamic systems that interact with situations; they do not merely endure situations.

Today Selye's views are considered outdated. Modern researchers believe that stress is specific to situations and stimuli, and subject to an individual's response. Most current theories still adopt the idea that stress is a disruption caused by a harmful influence or stressor (Hill-Rice, 2000), but these theories also make room for psychological experiences rather than external stimuli alone (Holmes and Rahe, 1967; Masuda and Holmes, 1967). A stressor can be a change or event in a person's life (for example marriage, loss of a beloved person, pregnancy, vacation, divorce, retirement, etc.). This newer perspective suggests that stress experiences are dynamic and that stress is therefore the outcome of an interaction between an organism and its environment or context (Lazarus, 1966). In this view, stress does not exist as an event but instead results from a transaction of a person with his or her environment. This is called the *transactional* approach to stress. From this perspective, stress is not directly measurable like a single independent or dependent variable. Instead, it is a summary of a total functional relationship of stressors (x variables) and stress or coping responses (y variables) unique to every animal. In essence, every individual animal has a unique functional relationship between external stressors, including psychological life stressors, and stress responses that are contingent on the animal's appraisals, evaluations, or interpretations of a situation.

In conclusion, not one of these theories is ultimately true or false. Each theory describes a different facet of the same physiological processes that occur in any mammalian body when confronted by a challenging situation. To understand these processes, we must understand that two different kinds of stress reactions exist. The first is a quick adaptation to a situation that suddenly arises; this is the fight-or-flight syndrome. The second reaction is a somewhat slower adaptation to a longer lasting situation (GAS).

ROLE OF CORTISOL

The first physiological stress response happens in a matter of seconds and is caused by the release of adrenalin and noradrenalin hormones from the adrenal glands. These hormonal changes instantly ready the body for fight or flight: blood pressure and pulse increase, the central blood vessels dilate and the peripheral blood vessels become narrow, muscles strain, blood circulation in the brain increases, and a whole set of hormones are released. This serves to provide an optimal supply of blood to the skeletal muscles and maximal input of energy to prepare the muscles for action. A typical example of such a situation is walking down a dark and narrow street of a city and suddenly noticing a figure lingering in a partially hidden door frame (Von Faber and Haid, 1995). Your body reacts with a fast response to prepare you and your body's muscles for combat or a hasty retreat, hence, the flight-or-fight response phrase.

The GAS on the other hand, tries to protect the body against longer lasting strains by releasing increased amounts certain hormones, the best studied of which is cortisol. The underlying reason for the GAS response may be a series of harmful conditions like heat, cold, hunger, thirst, radiation, infection, or corporal (bodily) or emotional injury. Cortisol is a hormone produced in the adrenal glands and is absolutely essential for survival (Griffin and Ojeda, 1996; Hadley, 1996). It is considered a major indicator of altered physiological states in response to stressful stimulation in most mammals including humans and dogs (Haubenhofer and Kirchengast, 2006 and 2007). Cortisol has many ways to help the body to survive long periods of stress and regulate its reactions to inflammation (Griffin and Ojeda, 1996; Hadley 1996), for example, by triggering catabolic and anabolic mechanisms, influencing the metabolism of carbohydrates, proteins, and fats, and promoting immunosuppressive effects. In this last case, cortisol inhibits overreaction of the immune system to inflammations, injuries, or foreign bodies.

Many studies have already investigated the secretion patterns of cortisol in humans and dogs. An overview of these studies appears in our recent paper (Haubenhofer and Kirchengast, 2007). As an excerpt, adult, healthy humans secrete 8 to 25 mg of cortisol in typical daily rhythmic patterns. Secretion increases a few hours before awaking and reaches its maximum approximately 1 hour after awaking. During the rest of the day, cortisol secretion decreases again and reaches its minimum between late evening and night. Adult, healthy dogs secrete about one-tenth the human amounts of cortisol. Scientists are not sure whether dogs exhibit daily rhythmic patterns of cortisol secretion. This hypothesis has been both supported (Kolevská, Brunclík, and Svoboda, 2003; Palazzolo and Quadri, 1987; Rijnberk, der Kinderen, and Thijssen, 1968) and refuted (Kemppainen and Sartin, 1984; Koyama, Omata, and Saito, 2003; Takahashi et al., 1981).

Measuring cortisol is a tricky matter. Because it is found in certain body fluids, the sampling or measurement process must not lead to a sampling-caused cortisol release that would bias the results. The requirement is to gather bodily fluids without causing stress or injury to the person or dog. Blood sampling, which is uncomfortable, may serve as a stressor and therefore lead to biased results by artificially increasing cortisol (Beerda et al., 1996). Fortunately, cortisol is also present in saliva and saliva cortisol levels correlate to concentrations of cortisol in blood (Beerda et al., 1996; Kirschbaum and Hellhammer, 1989; Vincent and Michell, 1992). Many studies have focused on the use of salivary cortisol concentrations to evaluate stress of humans (Fujiwara et al., 2004; Kunz-Ebrecht et al., 2003; Kunz-Ebrecht, Kirschbaum, and Steptoe, 2004; Schlotz et al., 2004; Yang et al., 2001) and dogs (Beerda et al., 1997 and 2000). It is true that even saliva sampling can create stressful situations, but as Kobelt et al. (2003) note, no changes in saliva cortisol concentration can be detected within the first 4 minutes of handling a dog. If saliva samples can be collected in less than 4 minutes, data can be obtained without inducing much stress from handling.

STRESS: CAUSES AND SYMPTOMS

Stress reactions are individual and depend on the perception and handling of stimuli, information processing of outside influences, and individual factors like genetic disposition and past experiences (Feddersen-Petersen, 2004). Sleep deficits, corporal or mental exhaustion, hectic atmosphere, incorrect handling, and excessive attention are characteristic sources of stress in dogs (Nagel and Von Reinhart, 2003). Comparable situations can be found during therapeutic sessions. Humans working in health care environments sometimes show increased cortisol concentrations due to dealing with patients' diseases and deaths (Yang et al., 2001). Dogs, like people, exhibit a range of stress symptoms (Nagel and Reinhardt, 2003):

- Loss of appetite
- Disorders of the gastrointestinal tract, e.g., diarrhea or vomitus
- Hardened muscles or increased tonicity, shivering, and shaking, obvious erection or vibration of the vibrissae on the head and ruffling of neck hair
- Restlessness and nervousness, pacing, pulling and biting on their leashes, jumpy and agitated appearance, inability to relax, attentiveness to sounds, immoderate barking, whimpering, whining, snapping wildly without biting, chattering their teeth, opening their eyes widely and focusing on the cause of stress, panting, runny noses, sweaty paws, hanging tongues, profuse salivation, licking of snouts, lifting paws, and lowering bodies
- Overreaction, acting anxiously or aggressively during situations that would normally not trouble the animal; destroys objects
- Exhibiting calming signals
- Defecation and urination
- Riding up (males and females) and erection of parts of the penis (males); these behaviors are not always sexually motivated or signs of dominance
- Exaggerated personal hygiene (intensive licking that may produce wounds, particularly on legs, tail, and genitalia); unpleasant body smell, halitosis, and general unhealthy appearance
- Stereotypes of sound or movement over a long period without a reasonable explanation for the behavior (running in circles, pacing, hunting own tail, monotonous barking, excessive licking); displacement activities (behaviors that are untypical or unreasonable in a specific situation)
- Poor concentration, forgetfulness, passivity
- Allergies (to food, mites, fleas, pollens, grasses, insecticides); skin problems (eczema or pruritus)
- Sudden hair loss, dandruff, or other unhealthy hair conditions
- Changes of eye color or blood-shot eyes
- Hypersexuality or hyposexuality (stress can lead increased or reduced sexual drive); modified sexual cycles can occur in female dogs

At this point, the answers to the questions about the importance of stress in relation to therapy dogs posed at the beginning of this chapter become evident. Obviously, the health and welfare of our closest friends, domestic dogs, are important issues.

Dogs have been our working partners for a very long time. They were—and are—used for hunting, herding, guiding, searching, running, and playing, to name only a few activities (see Chapter 1 of this volume).

CREATING LINKS WITH ANIMAL-ASSISTED ACTIVITIES AND ANIMAL-ASSISTED THERAPY

A few decades ago, a new opportunity for dog use was developed, namely serving as co-therapists in animal-assisted activities (AAAs) and animal-assisted therapies (AATs). In both cases, dogs (or other species) are used to improve or restore a human's physical, mental, and/ or social health and wellbeing. This goal is accomplished by letting a person and an animal spend time together in positive interactions. The Delta Society, one of the largest international nonprofit organizations for fostering human–animal bonds (founded in 1977 in Portland, Oregon), has defined both AAA and AAT.

The society characterizes AAA as casual meet-and-greet activities—pets visit people. It is not a special therapy program designed for a particular client and does not involve special treatment goals for each service. AAA is flexible and is usually not conducted under the supervision of a skilled physician or physiotherapist. AAT, on the other hand, strives for special therapeutic goals, is supervised by medical professionals, and activities are documented and evaluated (http://www.deltasociety.org). Although the idea of using animals to improve human physical and mental health is certainly not new, its (semi)professional implementation did not take place until the 19th century; its scientific analysis was not undertaken until the 20th century.

Compared to other areas of canine work, such as herding and guarding, AAA and AAT are relatively new fields. For most traditional activities, special dogs have been bred throughout the centuries. For example, sight hounds were bred for quick pursuit of prey in open areas, terriers for subterranean combat, collies for herding sheep, and mastiffs for guarding homes. AAA and AAT do not require specialized breeding. Dog breeds developed for other purposes such as running or defense are now used in both situations. However, the work may lead to certain problems for the dogs, e.g., stress. To investigate this possibility, Haubenhofer and Kirchengast investigated whether AAA/T work can cause physiological stress in dogs, indicated by increases in GAS.

Answers to this question have already been published in scientific papers (Haubenhofer and Kirchengast, 2006 and 2007). Some of the results presented in these articles will be discussed later in this chapter along with some yet unpublished material.

RESEARCH METHODS

PARTICIPATING TEAMS

Thirteen handlers and 18 companion dogs participated in the study (three handlers owned two dogs; one handler owned three). Twelve of the 13 dog handlers were females. Their ages ranged from 28 to 68 years when the samples were taken. The dogs were of different breeds since AAA/T requires no specific breed: one Bernese

Mountain dog, three Border Collies, one Bouvier des Flandres, one unknown cross-breed, one Dachshund crossbreed, three Golden Retrievers, one Great Pyrenees, three Labrador Retrievers, one Pointer crossbreed, one Polski Owczarek Podhalanski, and two Pulis. The dogs ranged in age from 2 to 9 years. Fifteen were females (four neutered), and three were males (one neutered). None of the non-neutered females were in estrus (heat) during the evaluation and no dogs suffered from pathological allergies, skin diseases, vomitus, diarrhea, or other chronic diseases. All teams completed their training at the same training center, although in different years, and had different levels of experience.

AUSTRIA SETTINGS

The study was performed in Austria where the term *tiergestützte Therapie* applies jointly to AAA and AAT. Therapeutic sessions combined elements of both techniques, may or may not have been supervised by professionals, and may or may not have included therapeutic goals for sessions, but all activities were documented and evaluated. The sessions were held in different locations such as primary schools, hospitals, rehabilitation centers, and homes for the elderly. It may be important to add that none of the visiting teams were new to the facilities they visited.

The designs of the sessions depended on the wishes and abilities of the clients and the goals to be accomplished. Children's therapy time was spent actively (e.g., playing with dogs). Elderly interactions were calmer and involved talking to and hugging the dogs. Usually, sessions lasted 1 to 3 hours. Due to the short activity periods, most sessions did not include breaks or rest periods. Other types of sessions also represented in this study were animal-assisted services that were integral parts of the handlers' occupations. The handlers were joined by their dogs at their own working places and followed the therapeutic program as part of their jobs as physiotherapists, ergo-therapists (called occupational therapists in North America), social workers, and teachers. These sessions typically included many breaks for the dogs, because the dogs could not be involved in all parts of their handlers' work routines. The handlers worked normal 8-hour days. Some of the dog–handler teams worked only in short sessions visiting institutions, some completed normal long workdays, and some worked short and long sessions. All teams completed 9 to 50 therapeutic sessions over a sampling period of 3 consecutive months.

STUDY PROTOCOL

The study was designed to minimize its influence on the teams' daily routines. A self-administered questionnaire was used to obtain demographic information about handlers and dogs. Saliva samples of both humans and dogs were collected with Salivettes—little plastic tubes containing cotton swabs like those used by dentists. A swab is inserted into the mouth cheek of a human or a dog and left there until it is saturated with saliva. It is then returned to the plastic tube and conserved until analyzed. More detailed descriptions of the techniques can be found in Haubenhofer and Kirchengast (2006 and 2007). In addition, a self-administered form was used to detail each sample taken: date, time, and place of sampling; actions preceding sampling; possible special activities preceding sampling.

Handlers were taught how to apply the cotton swabs of the Salivettes first to themselves and then to their dogs. Handlers owning more than one dog were instructed to fix an order of sampling among the dogs and to retain this order for the duration of the study. Every Salivette was labeled for only one sampling point. Sampling periods began with 3 control days (no therapy sessions) during which samples were taken three times daily at 8 a.m., and 2 and 8 p.m. Control days included all typical daily routines including non-AAA/T work. During a following period of 3 consecutive months, saliva samples were taken immediately before and after a therapeutic session while the teams were still at the session location, independent of time of day.

The 3 consecutive months were followed by another 3 control days of daily routine without therapy sessions. For evaluation, cortisol data from control days were grouped by the three sampling times (8 a.m, 2 p.m., and 8 p.m.). Data from therapy days were also separated into three groups (therapy sessions before noon, from noon to 2 p.m., and after 2 p.m.).

RESULTS

CORTISOL LEVELS OF HANDLERS AND OWNERS ON CONTROL AND THERAPY DAYS

Six hundred fifty-five valid samples were collected from the handlers (90% between 4.03 nmol and 39.8 nmol salivary cortisol per liter); 554 valid samples were collected from dogs (90% samples between 0.3 nmol and 11.3 nmol salivary cortisol per liter). Interestingly, cortisol levels were significantly higher on therapy days than on control days for both handlers and dogs (Table 14.1). It is unlikely that these changes in cortisol levels were caused by daily variation patterns of cortisol secretion.

All valid data from control days were matched against all valid data from therapy days. Thus, daily variation patterns were included in both sets of data. Control days were characterized as free from therapeutic sessions, not free from daily stressors. In a nonlaboratory environment as in this study, stressful situations cannot be avoided. The main difference in the lives of handlers and dogs on control days and therapy days was therefore not the absence or existence of stressors, but the absence

TABLE 14.1
Cortisol Levels on Control and Therapy Days

	Handlers			Dogs		
	Median	Percentile 25	Percentile 75	Median	Percentile 25	Percentile 75
Control days	12.43	6.59	21.92	1.72	1.19	2.51
Before therapy	16.11	10.53	25.57	2.06	1.17	4.25
After therapy	12.68	7.62	18.29	2.18	1.03	4.64

or existence of therapeutic sessions. We therefore concluded that therapeutic work is a source of increased cortisol levels in both handlers and dogs.

Further analyses showed that dog handlers and dogs had different cortisol level distributions before and after therapeutic sessions. Dog handlers showed significantly higher cortisol levels before therapeutic sessions than afterward (Table 14.1). Cortisol levels after therapeutic sessions were higher than the levels averaged from all valid control data, independent of times of day of therapeutic sessions or their durations. We concluded that therapeutic work caused a certain amount of anticipatory stress for the handlers. Situations of worry as stressor are supported by Schlotz et al. (2004) and Kirschbaum (1991).

Dogs, however, showed significantly higher cortisol levels after therapeutic sessions than before, if the sessions took place before 2 p.m. Dogs in therapeutic sessions later than 2 p.m. showed, like their owners, significantly higher cortisol levels before than after the sessions. We concluded that situations related to therapy sessions led to increased cortisol levels. The half-life of cortisol lies between 70 and 110 minutes (Kirschbaum, 1991). If extratherapeutic factors alone caused the increases, the dogs' cortisol levels should have dropped to lower levels before the end of the sessions. We could not explain the differences in results before and after 2 p.m. Future studies will have to answer this question.

INFLUENCE OF DURATION OF THERAPY ON CORTISOL LEVELS

Cortisol levels of handlers and dogs varied for different durations of therapeutic sessions. Sampling sizes and statistical results are shown in Table 14.2. Cortisol levels of handlers increased directly proportional to the durations of sessions. This may be explained by the fact that the samples before the start of long therapeutic sessions were taken always during early mornings because the therapeutic work was part of the handlers' jobs. Yet, this analysis appears unlikely because short sessions were sometimes conducted during the morning hours. It seems more likely that longer sessions provided greater sources of arousal for the handlers than short sessions.

TABLE 14.2
Cortisol Levels for Different Durations of Therapy Sessions

	Handlers			Dogs		
Duration	Median	Percentile 25	Percentile 75	Median	Percentile 25	Percentile 75
1 to 2 hours	12.9	7.07	19.61	1.93	0.88	4.17
2 to 3 hours	13.36	9.27	17.64	2.48	1.29	15.53
5 to 6 hours	29.24	21.4	43.37	2.62	1.91	3.44
6 to 7 hours	40.58	22.54	73.58	1.76	1.27	2.95
7 to 8 hours	56.13	29.89	81.82	2.58	1.45	4.46

Reasons for this arousal may be manifold, e.g., different types of therapeutic work involved or lacked longer breaks.

In dogs, cortisol levels were significantly related to length of session, with levels higher in shorter sessions than in longer ones. Highest cortisol levels were caused by therapeutic sessions of about 3 hours' duration independent of the type of session. It is unlikely that the novelty of the situation caused these increases as neither therapeutic work nor the locations were new to any of the dogs. Apparently, the dogs were affected by the type of therapy, which was closely connected to the durations of sessions. Short sessions typically included visits to a certain location with no major breaks, and lasted 1 to 3 hours. A steady increase of cortisol secretion may be typical for this kind of therapeutic work. Long sessions, lasting up to 8 hours, included many breaks. The time that the dogs actually spent in therapeutic work was no longer than for dogs in short sessions. In longer sessions, the work was more spaced out and gave the dogs opportunities to rest and recover.

INFLUENCE OF FREQUENCY OF THERAPEUTIC WORK ON CORTISOL LEVELS

Within the sampling period, the teams completed 9 to 50 therapeutic sessions. Cortisol levels in dogs collected on therapy days increased directly proportional to the number of therapeutic sessions held. Furthermore, cortisol levels did not exceed about 3 nmol/L in dogs who participated in fewer than about 25 sessions within the sampling period. Dogs who worked more sessions showed steady increases in their cortisol levels. No comparable trend could be found among the handlers. Table 14.3 summarizes sample sizes and cortisol levels for all groups. These results suggest that several days of rest after each therapeutic session may be necessary for dogs to

TABLE 14.3
Summary of Sample Sizes and Cortisol Levels for All Groups

	Handlers			Dogs		
Settings	Median	Percentile 25	Percentile 75	Median	Percentile 25	Percentile 75
9	12.32	7.33	18.03	1.47	1.04	–
10	10.49	8.02	17.71	2.16	1.23	2.74
13	13.92	9.28	20.03	0.28	0.09	0.99
14	7.79	3.9	13.61	1.86	1.36	2.49
16	13.56	10.15	21.2	1.82	0.8	2.95
17	8.81	4.35	14.23	0.28	0.04	1.56
19	27.21	19.78	49.83	1.79	1.37	2.17
25	40.25	27.31	62.09	3.15	2.19	6.75
30	17.36	11.88	23.96	3.31	1.97	7.53
50	11.33	5.18	15.46	8.9	2.78	22.72

Percentile 75 for 9 settings missing due to lack of valid data.

prevent extreme physiological stress. The number of sessions from which cortisol levels started to rise steadily was about 25 within 3 months. This is equivalent to two therapeutic sessions per week. Further investigation should be made to test whether changing schedules would lead to other results in the cortisol levels.

COMPARISON OF SALIVARY CORTISOL LEVELS AND CLIENT GROUPS

The next question was whether the type of client affected levels of salivary cortisol. Eleven teams worked with old and/or bedridden people, three with disabled adults, and five with children (Table 14.4). The results indicate that work with disabled grown-ups was least stressful for handlers, followed by work with old and/or bedridden people and children (kindergarten and primary school). For dogs, work with disabled adults was least stressful, followed by work with children and then work with old and/or bedridden people.

We also compared the data collected during therapeutic settings from humans and dogs that worked at different places. Four teams worked both with old and/or bedridden people and with children, and one team worked with old and/or bedridden people, disabled adults, and with children. For the handlers, none of the differences were statistically significant. Among the animals, only one showed statistically significant differences in its salivary cortisol levels during therapeutic settings depending on whether it worked with old and/or bedridden people or with children. We examined the salivary cortisol levels of this dog and found out that working with children in a primary school was more straining for the animal than working with old and/or bedridden people in a hospital (see Figure 14.1).

We may explain these results by examining the typical procedures of a therapeutic setting. Work with disabled and old people was conducted mostly in groups (if the clients were mobile enough to meet in public rooms) or as individual therapy for bed-ridden clients. The dogs were primarily brushed, petted, and fed. Exercises were done to improve the clients' motor functions, fine motor activity, communication, concentration, and memory. A very important part of working with old people is talking and listening to them. A dog would lie on a table while clients sat around it or moved from client to client. The dogs had to remain calm and not move quickly.

TABLE 14.4
Salivary Cortisol Levels and Client Groups

Type of Patient	Handlers			Dogs		
	Median	Percentile 25	Percentile 75	Median	Percentile 25	Percentile 75
Old/bedridden	14.59	9.09	23.37	2.62	1.49	5.43
Disabled	7.57	2.51	12.26	0.29	0.04	1.59
Children	17.04	11.51	25.89	1.96	1.28	2.76

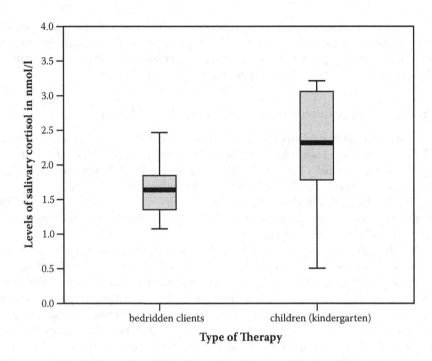

FIGURE 14.1 Levels of salivary cortisol (mol/L) from one dog that showed statistically significant differences in its results depending on the type of therapy it did. Outliners and extreme values excluded.

Several handlers said that they did not like working with disabled people, especially disabled children because they feared the depressing and distressing surroundings. However, those who worked with disabled clients mentioned that the therapeutic settings were anything but scary and noted the friendly surroundings and mostly happy moods of the clients and nursing staffs. Contrary to institutions for elderly people, such places are always bright and merry and emanate much vitality.

Usually the teams performed precise exercises to increase communication and motor functions (climb over dog, crawl under it, walk with it on a leash, brush it, feed it, do tricks and stunts with the dog, play with the children), and the dogs were urged to move around. What we can derive from these statements is that the work with elderly people was not physically demanding but was often mentally straining for the handlers. It is difficult for dogs to remain calm and still for longer periods, particularly for young dogs. This may have led to the increased levels of salivary cortisol among both handlers and dogs compared to data derived from the control days. The increased levels of salivary cortisol from working with children can be explained by the higher levels of physical action. Although it may have been fun for most of the teams, action and noise are still exhausting and stressful. Work with disabled clients seems to be a mixture of the positive effects of both types of therapies. It involved action, fun, and creativity, but not to the extent of work with children.

None of the dog handlers and only one animal that visited different kinds of clients showed different results. The animal was a female, sterilized Golden Retriever born in 1996 and thus the oldest dog of the group. She exhibited higher levels of salivary cortisol during therapy in a school than in a hospital with bedridden clients. Since this was a single case in our survey, we can only guess that the age of the dog was related to the results. The dog owner stated that she thought the dog was more stressed by therapy in the school than in the hospital and explained this by several symptoms of old age like articular problems.

Decreases in vitality of mammals during senescence (old age) are well known (Hofecker et al., 1981). Thus we may assume that young dogs are better adapted for working with lively children because they are still lively. Older dogs that do not like to jump and run may be better suited for calm types of therapy like sessions with elderly people. But this is only a supposition and surely something that should be investigated more precisely. Working with disabled clients seems appropriate for both younger and older dogs, depending on the design of the therapeutic units.

COMPARISON OF SALIVARY CORTISOL LEVELS AND TRAINING

We separated the teams by year of training (year of joining the organization) at their training centers and compared their concentrations of saliva cortisol on days of therapeutic settings (Table 14.5). The handlers showed no trend toward higher levels of salivary cortisol among those who joined the organization earlier, but such a trend was noted among the dogs. After we excluded year 2005 (represented by only one dog), we found statistically significant higher levels for 2000 than 2002, and 2000 than 2003.

Such variations could not be found when we analyzed the concentrations of cortisol of all the age classes of the dogs and we assume the dogs' ages were not connected to their levels of salivary cortisol. This leads us again to our hypothesis that this kind of therapy is stressful for dogs—steady increases of cortisol levels depending on the number of years and settings they had already worked during animal-supported therapy.

TABLE 14.5
Salivary Cortisol and Training Levels

	Year of Joining	Median	Percentile 25	Percentile 75
Handlers	1993	16.05	9.75	22.74
	2000	12.62	6.16	15.93
	2002	16.62	11.12	27.38
	2003	7.54	3.2	12.35
Dogs	2000	3.17	1.71	15.83
	2002	1.94	1.16	3.09
	2003	1.69	0.38	3.08
	2005	1.28	0.003	2.45

The lack of similar relations between years of therapeutic work and levels of cortisol secretion among handlers may have resulted from various factors. First, one year differs for dogs and humans relative to total lifespan. Second, these handlers work voluntarily and find some kind of pleasure and emotional satisfaction from it. Third, real species differences may exist; humans may be better able to regenerate or experience less stress than dogs in therapeutic work. This should certainly be a topic of further investigation: why do dogs' cortisol levels increase over their therapeutic careers and human levels do not.

DISCUSSION AND CONCLUSION

Although it can be said that therapeutic sessions caused physiological stress in both humans and dogs, it cannot be said that this increase is necessarily negative. Whether physiological stress as indicated by increased cortisol is perceived emotionally remains unanswered by this research.

Further research is needed to determine whether teams doing such work experience negative arousal that may be detrimental to their health or excitement generated by their therapeutic activities. Unfortunately, the correlation between physiological indicators of stress and subjective reports is weak in relation to humans (Schacter and Singer, 1962). Furthermore we must keep in mind that the general levels of salivary cortisol were low in the therapy dogs of our study compared to dogs used in other studies. Thus we must refine our statement insofar as the dogs exhibited higher cortisol concentrations compared to their own results from the control days, but ranged around the same baseline area of the dogs used in the study by Beerda et al. (1998).

The study surely has a list of limitations, such as the impossibility of examining subjectively felt stress by collecting cortisol; the participation of only one male dog handler; the collection of samples at other times of day compared to samples collected on control days; the small sample size; and the failure to consider domiciles, sexes, ages of humans and for dogs, sterilization status and breed.

Further investigations are highly recommended. AAA/T is an area in which future canine ergonomic researchers could produce large impacts. More studies should be designed to investigate the effects of scheduling (frequency of therapeutic work, duration of therapeutic work, and breaks) on emotions and cortisol concentrations in handlers and dogs. These studies should be compared to the already existing ergonomic literature covering work–rest schedules and recommended work session durations.

In a survey published by Kunz-Ebrecht et al. (2004) mean levels of cortisol in humans directly after awakening in the morning reached 18.9 ± 10.5 nmol/L and increased to 28.0 ± 13.6 nmol/L 30 minutes later. During the rest of the day, levels of about 7.59 ± 2.6 nmol/L were typical. These data represented employed women of higher socioeconomic status, comparable to those in our study. In their study evaluating fear in dogs, King et al. (2003) sampled salivary cortisol levels of 108 dogs of various breed, sex, and experience before and after four tests thought to cause fear. The average concentration of saliva cortisol before the tests was 2.8 ± 3.8 nmol/L and the mean concentration after the tests was 4.7 ± 5.6 nmol/L, indicating a tendency of, but no significance for, concentrations to be higher after the tests ($p \leq 0.07$).

In another paper (Beerda et al., 1998), dogs of different breed, sex, and age (selected randomly) were evaluated for behavioral, saliva cortisol, and heart rate responses to different types of stimuli. The scientists measured a mean basal cortisol level of 6.0 nmol/L, then exposed the dogs to stressful situations such like pulling them on a leash, releasing a paper-filled bag from the ceiling, sudden noise, and electric shock. The leash pulling stimulus caused mean levels of saliva cortisol of 16.7 ± 12.1 nmol/L (due to one dog that showed extreme levels of 100.7 and 69.4 nmol/L), loud noise increased cortisol up to 20.4 ± 4.5 nmol/L, the falling bag caused 18.7 ± 6.1 nmol/L, and the electric shock result was 15.5 ± 4.6 nmol/L. Other stimuli tested (forcing a dog onto the floor, opening an umbrella in front of a dog) caused no significantly increased cortisol secretion. The scientists attribute this to the predictability of stimuli used. Electric shocks, sound blasts, and falling bags are sudden, nonsocial stimuli and therefore caused increased cortisol secretion.

As mentioned, sex and sterilization did not appear as influencing factors. This does not, however, automatically mean that these parameters have no effects; it only means that possible effects did not become visible in this study. Svartberg (2002) tested the relationship between personality and performance in working dog trails. He showed that breed and sex differences disappear in dogs that achieve certain levels of training. Only dogs with a special kind of personality and amount of training reach the highest levels of performance in working dog trails. We hypothesize that the same is true for dogs working in animal-supported therapy.

Only dogs that meet the extremely high demands for personality (temperament) and training can become therapy dogs. These dogs should show a restricted range of personality and behavior, regardless of sex, age, or breed. Similar principles could apply to the handlers as well. These concerns for screening and selection of dogs and handlers for AAA/T work are shared in common with the larger personnel selection literature in industrial (or occupational) psychology and human resources.

Although both humans and dogs showed increased levels of salivary cortisol on therapeutic days, all handlers emphasized that they and their dogs liked the work. The handlers claimed that their dogs entered the therapeutic settings in a happy mood and that they would instantly stop their work if their dogs started to show signs of distress. Values surely play an important role in work-related processes and outcomes. The basic assumption is that a person will be happier, more motivated, satisfied, and committed when his or her values are congruent with those emphasized at a place of employment and that people prefer vocations that allow them to work in line with their values (Berings et al., 2004). If animal-supported therapy is what these people want to do, they may feel happily busy and not stressed. The same may be true for the dogs. They may be stressed physiologically by therapy, but as long as they enjoy their jobs, live in harmony with their handlers, and are not obliged to do what they do not like, they will not interpret negatively and will enjoy their lives.

Furthermore, one should not forget that, based on the theories of Lazarus (1966), positive stimuli in life may as well cause changes in an individual's physiology comparable to those caused by unpleasant experiences, but with the great difference that they are not interpreted negatively. Increases of cortisol levels are caused by stimuli sensed positively and negatively. Thus it is important to compare measured hormone levels to subjective perceptions or at least the behaviors of the same individual to circumvent false conclusions about dangers to wellbeing.

BEYOND THE BOX

Although the first analysis of the question whether working in animal-supported therapy could present a health hazard for the animals occurred in the late 1980s, only the tip of the iceberg has been investigated. Iannuzzi and Rowan (1991) published a paper delineating the potential for abuse associated with fatigue and burnout for animals that live in institutions. Heimlich's paper (2001) deals with the consequences of animal-supported therapy on her own companion dog that works in therapy. She claims that after 8 weeks of therapeutic work, the animal already started to show first signs of stress: excessive panting, frequent urination, and ear and urinary tract infections. The dog was treated with antibiotics and sent back to the therapy program, during which it appeared tired. Further veterinary investigations determined that the dog suffered from Cushing's syndrome, also known as canine hyperadreno-cortisism (HAC), a hormonal disorder that results in chronic elevation of circulating blood cortisol concentrations and often results from chronic stress (Tilley and Smith, 1997). After the diagnosis, the dog was removed from the therapeutic program, but will suffer from the illness for the rest of its life. Although one cannot say certainly that the dog developed the illness because of therapeutic activities, the possibility cannot be excluded. Alternative conditions can cause Cushing's syndrome, for example, adrenal cortex tumors, prolonged administration of cortisone drugs, or (most commonly) pituitary gland abnormality.

We want to stress that Heimlich's case was a very extreme example demonstrating both the consequences of chronic stress on a dog and also the great responsibilities placed on a handler to ensure the health and wellbeing of her canine co-worker. Subjecting dogs to veterinary investigations before starting therapeutic work and ending work as soon as symptoms of stress appear can prevent health problems for dogs.

At the Tenth International Conference of IAHAIO in Glasgow in October 2004, Ferrara et al. presented a poster from a study at the University of Rome, Italy evaluating the welfare of dogs during animal-supported therapy. They observed nine therapy dogs of different breeds (seven Golden Retrievers, one Golden Labrador crossbreed, one other crossbreed) 173 hours before, during, and after therapy sessions by focal-animal and all-occurrences sampling methods. They constructed an echogram from data consisting of 136 behavioral patterns grouped in 15 categories. They concluded that the nine dogs observed did not show stressed behavior or stereotypes due to anxiety or hard work during animal-supported therapy and contrarily even showed affiliative and playful behavior more frequently during work than before or after it. They also found that the dogs looked to their handlers more often than to other humans present during therapy sessions, implying very good relationships of dog and handler. The study has not yet been published.

Wellbeing is not simply the absence of negative emotions. It is also (and even predominantly) the presence of positive emotions (Boissy et al., 2007). Based on the nature of emotional experience, we have no way to know whether animals experience emotions similar to humans. We can never see the world from a dog's perspective or even know for certain whether a dog has a point of view (a subjective life—consciousness). We should never forget that this principle also applies to people. Subjective emotions of humans are certainly private. However, the behaviors, structures, and brain chemistries of dogs and humans are similar. It is therefore likely that dogs feel somewhat as we do and experience negative and positive emotions. At least this perspective, seeing dogs as subjects with lives, is a pragmatic view for working with them and is useful in predicting their behavior in work settings.

REFERENCES

Beerda, B., Schilder, M.B., Janssen, N.S., and Mol, J.A. (1996). The use of saliva cortisol, urinary cortisol and catecholamine measurements for a noninvasive assessment of stress responses in dogs. *Hormones and Behavior, 30,* 272–279.

Beerda, B., Schilder, M.B., Van Hooff, J.A., and De Vries, H.W. (1997). Manifestations of chronic and acute stress in dogs. *Applied Animal Behaviour Science, 52,* 307–319.

Beerda, B., Schilder, M.B., Van Hooff, J.A., De Vries, H.W., and Mol, J.A. (1998). Behavioural, saliva cortisol and heart rate responses to different types of stimuli in dogs. *Applied Animal Behaviour Science, 58,* 365–381.

Beerda, B., Schilder, M.B., Van Hooff, J.A., De Vries, H.W., and Mol, J.A. (2000). Behavioural and hormonal indicators of enduring environmental stress in dogs. *Animal Welfare, 9,* 49–62.

Berings, D., De Fruyt, F., and Bouwen, R. (2004). Work values and personality traits as predictors of enterprising and social vocational interests. *Personality and Individual Differences 36,* 349–364.

Boissy, A., Manteuffel, G., Bak Jensen, M., Oppermann Moe, R., Spruijt, B., Keeling, L.J., Winckler, C., Forkman, B., Dimitrov, I., Langbein, J., Bakken, M., Veissier, I., and Aubert, A., (2007). Assessment of positive emotions in animals to improve their welfare. *Physiology and Behavior 92,* 375–397.

Cannon, W.B. (1915). *Bodily Changes in Pain, Hunger, Fear, and Rage.* Boston: Bradford.

Feddersen-Petersen, D. (2004). *Hundepsychologie: Sozialverhalten und Wesen; Emotionen und Individualität* 4th ed. Stuttgart: Franck Kosmos.

Ferrara, M., Natoli, E., and Fantini, C. (2004). Dog welfare during animal-assisted activities and animal-assisted therapy. Poster presented at Tenth International Conference of IAHAIO, October, Glasgow, not published.

Fujiwara, K., Tsukishima, E., Kasai, S., Masuchi, A., Tsutsumi, A., Kawakami, N., Miyake, H., and Kishi, R. (2004). Urinary catecholamines and salivary cortisol on workdays and days off in relation to job strain among female health care providers. *Scandinavian Journal of Work, Environment and Health, 30,* 129–138.

Griffin, J.E. and Ojeda, S.R. (1996). *Textbook of Endocrine Physiology,* 3rd ed. New York: Oxford University Press.

Hadley, M.E. (1996). *Endocrinology,* 4th ed. New York: Prentice Hall.

Hancock, P.A. (1984). Environmental stressors. In Warm, J.S., Ed. *Sustained Attention in Human Performance.* Chichester, John Wiley & Sons, 103–142.

Haubenhofer, D.K. and Kirchengast, S. (2006). Physiological arousal for companion dogs working with their owners in animal-assisted activities and animal-assisted therapy. *Journal of Applied Animal Welfare Science 9,* 165–172.

Haubenhofer, D.K. and Kirchengast, S. (2007). Dog handlers' and dogs' emotional and cortisol secretion responses associated with animal-assisted therapy sessions. *Society and Animals 15,* 127–150.

Heimlich, K. (2001). Animal-assisted therapy and the severely disabled child: a quantitative study. *Journal of Rehabilitation 67*, 48–54.

Hill-Rice, V., Ed. (2000). *Handbook of Stress, Coping and Health: Implications for Nursing Research, Theory, and Practice*. London: Sage.

Hofecker, G., Niedermüller, H., and Skalicky, M. (1981) Der altersbedingte Leitungsabfall und seine Beeinflussung im Tierexperiment. *Aktuelle Gerontologie 11*, 188–194.

Holmes, T. and Rahe, R. (1967). The social readjustment rating scale. *Journal of Psychosomatic Research 12*, 213–233.

Iannuzzi, D. and Rowan, A.N. (1991). Ethical issues in animal-assisted therapy programs. *Anthrozoös 4*, 154–163.

Kemppainen, R.J. and Sartin, J.L. (1984). Evidence for episodic but not circadian activity in plasma concentrations of adrenocorticotrophin, cortisol and thyroxine in dogs. *Journal of Endocrinology, 103*, 219–226.

King, T., Hemsworth, P.H., and Coleman, G.J. (2003). Fear of novel and startling stimuli in domestic dogs. *Applied Animal Behaviour Science 82*, 45–64.

Kirschbaum, C. (1991). *Cortisolmessungen im Speichel: eine Methode der Biologischen Psychologie*. Bern: Hans Huber.

Kirschbaum, C. and Hellhammer, D.H. (1989). Salivary cortisol in psychobiological research: an overview. *Neuropsychobiology 22*, 150–169.

Kobelt, A.J., Hemsworth, P.H., Barnett, J.L., and Butler, K.L. (2003). Sources of sampling variation in saliva cortisol in dogs. *Research in Veterinary Science, 75*, 157–161.

Kolevská, J., Brunclík, V., and Svoboda, M. (2003). Circadian rhythm of cortisol secretion in dogs of different daily activities. *Acta Veterinaria Brno, 72*, 599–605.

Koyama, T., Omata, Y., and Saito, A. (2003). Changes in salivary cortisol concentrations during a 24-hour period in dogs. *Hormone and Metabolic Research, 35*, 355–357.

Kunz-Ebrecht, S.R., Kirschbaum, C., Marmot, M., and Steptoe, A. (2003). Differences in cortisol awakening response on work days and weekends in women and men from the Whitehall II cohort. *Psychoneuroendocrinology, 29*, 516–528.

Kunz-Ebrecht, S.R., Kirschbaum, C., and Steptoe, A. (2004). Work stress, socioeconomic status and neuroendocrine activation over the working day. *Social Science and Medicine 58*, 1523–1530.

Lazarus, R.S. (1966). *Psychological Stress and the Coping Process*. New York: McGraw Hill.

Masuda, M. and Holmes, T.H. (1967). Magnitude estimations of social readjustments. *Journal of Psychsomatic Research 11*, 219–225.

Nagel, M. and van Reinhardt, C. (2003). *Stress bei Hunden*, 2nd Ed. Grassau: Animal Learning.

Palazzolo, D.L. and Quadri, S.K. (1987). Plasma thyroxine and cortisol under basal conditions and during cold stress in the aging dog. *Proceedings of the Society for Experimental Biology and Medicine, 185*, 305–311.

Poulton, E.C. (1976). Arousing environmental stresses can improve performance, whatever people say. *Aviation, Space, and Environmental Medicine, 47*, 1193–1204.

Rijnberk, A., der Kinderen, P.J., and Thijssen, J.H. (1968). Investigations on the adrenocortical function of normal dogs. *Journal of Endocrinology, 41*, 387–395.

Schacter, S. and Singer, J.E. (1962). Cognitive, social and physiological determinants of emotional state. *Psychological Review, 69*, 379–399.

Schlotz, W., Hellhammer, J., Schulz, P., and Stone, A. (2004). Perceived work overload and chronic worrying predict weekend-weekday differences in the cortisol awakening response. *Psychosomatic Medicine, 66*, 207–214.

Selye, H. (1957). *Stress beherrscht unser Leben*. Düsseldorf: Econ Verlag.

Svartberg, K. (2002). Shyness–boldness predicts performance in working dogs. *Applied Animal Behaviour Science 79*, 157–174.

Takahashi, Y., Ebihara, S., Nakamura, Y., and Takahashi, K. (1981). A model of human sleep-related growth hormone secretion in dogs: effects of 3, 6, and 12 hours of forced wakefulness on plasma growth hormone, cortisol, and sleep stages. *Endocrinology, 109,* 262–272.

Tilley, L.P. and Smith, F. (1997). *The Five-Minute Veterinary Consult: Canine and Feline.* Philadelphia: Lippincott Williams and Wilkins.

Vincent, I.C. and Michell, A.R. (1992). Comparison of cortisol concentrations in saliva and plasma of dogs. *Research in Veterinary Science, 53,* 342–345.

Von Faber, H. and Haid, H. (1995). *Endokrinologie,* 4. Stuttgart: Eugen Ulmer.

Yang, Y., Koh, D., Ng, V., Lee, F.C., Chan, G., Dong, F., and Chia, S.E. (2001). Salivary cortisol levels and work-related stress among emergency department nurses. *Journal for Occupational and Environmental Medicine, 43,* 1011–1018.

15 Benefits of Animal Contact and Assistance Dogs for Individuals with Disabilities

Natalie Sachs-Ericsson and Nancy Hansen Merbitz

CONTENTS

INTRODUCTION

After the passage of the Americans with Disabilities Act (ADA, 1990), the use of assistance dogs trained and placed for the purpose of reducing the impacts of disabling conditions has continually increased. The health and functioning of people with disabilities involves many factors. The World Health Organization (WHO) has provided a model within which to consider past, current, and future factors that obstruct or further their full inclusion in society. In this chapter, research on the benefits of assistance dogs is surveyed and critiqued and areas of potential research are elaborated. Where research is lacking, the elegant structure of the WHO model also serves well to highlight potential relationships among these factors, including the benefits of ADs, and these issues also will be discussed.

Different terms are used to refer to dogs in relation to their training or purpose (Eames and Eames, 1997a). Family pets often are referred to as companion animals. *Assistance dog* (AD) is a general term for a dog specially trained to assist individuals with disabilities. The most common type of AD is the *guide dog* (GD), an animal trained to help blind or vision-impaired individuals. This chapter will focus primarily on two less-known types of ADs: *service dogs* (SDs), who assist people with mobility impairments, and *hearing dogs* (HDs), who assist individuals who are deaf or hard of hearing. It is important to note that an increasing number of dogs are trained to assist individuals with a range of disabilities including seizure disorders, Parkinson's disease, heart disease, and psychiatric disorders. Sometimes these dogs are referred to by their functions (e.g., seizure response dog); sometimes they are simply called service dogs.

Typical conditions of individuals who obtain SDs include spinal cord injuries, multiple sclerosis, muscular dystrophy, cerebral palsy, polio and post-polio syndrome, and acquired brain injuries. SDs generally serve two main functions: (1) enhancing an individual's mobility and (2) retrieving objects. SDs are taught such tasks as pulling wheelchairs, opening doors, turning light switches on and off, getting the phone, and picking up objects (Lane, McNicholas, and Collis, 1998). For ambulatory people with mobility impairments, SDs may assist with bracing as a person stands up and help them balance during ambulation. In these ways, SDs reduce the expenditures of time and physical exertion by patients and caregivers, allowing them more efficient uses of resources (Cusack and Smith, 1984).

Individuals who are deaf or hard of hearing often are not aware of important sounds. This circumstance may adversely affect the individual across several domains of functioning. HDs are trained to alert individuals to such sounds as a knock at the door, the ring of a telephone, smoke and fire alarms, a microwave oven, an alarm clock, an intruder, a baby crying, or someone calling their names (Mowry, Carnahan, and Watson, 1994).

The first AD training program in the United States, founded in 1929, was Seeing Eye, Inc., which trained GDs to aid blind individuals. The use of specially trained dogs to assist individuals who had disabilities other than blindness started about 30 years ago. In the mid-to-late 1970's programs were initiated to train SDs for individuals with mobility impairments and to train HDs for individuals who were deaf or hard of hearing (Bergin, 1981 and 2000). Informal surveys suggest that 10,000 to 16,000 individuals currently have ADs (Beck, 2000; Duncan, 1998; Hines, 1991). Approximately 50 regional programs train SDs and HDs. A list of AD providers who belong to Assistance Dogs International (ADI) can be found at http://www.adionline. org. Most AD training programs are nonprofit organizations that provide dogs at no charge to individuals. The cost of training one SD has been estimated at $12,000 to $20,000 (Duncan, 1998). Because of funding limitations, individuals often wait two or more years to obtain dogs.

A growing body of research shows that contacts with animals and pet ownership (companion animals) may be beneficial to humans (Jennings, 1997; Sachs-Ericsson, Hansen, and Fitzgerald, 2002). These studies have examined the benefits of companion animals in general population samples and in samples at high risk for psychosocial or health problems, such as elderly or widowed persons and individuals with medical problems.

Prominent in many of the theories examining the impacts of dogs on humans is an emphasis on the human–animal bond. Katcher (1983) identified four elements of the human–animal bond: safety, intimacy, kinship, and constancy. Friedmann (1990) postulated that pet ownership leads to benefits by: (1) improving fitness by providing a stimulus for exercise, (2) decreasing anxiety by providing a source of physical contact, and (3) decreasing loneliness by providing companionship. McNicholas and Collis (2000) suggest that the presence of a dog is beneficial for two reasons: (1) increased social interactions with other people, and (2) the relationship of the individual and the dog (Collis and McNicholas, 1998); the relationship shows similarities to human-to-human relations (Bonas, McNicholas, and Collis; 2000). Garrity and Stallones (1998) conclude that the positive impact of dogs on humans is consistent with the literature on the benefits associated with human social support. In a recent review on dogs' contribution to human health, the author concluded that the evidence collectively suggests that dogs have prophylactic and therapeutic value for people (Wells 2007). AD impacts are likely to include and extend the benefits of simply having a companion animal.

A variety of methodological designs have been used to investigate these effects. Because some designs are methodologically stronger than others (Campbell and Stanley, 1966), it is important to take into consideration the designs used to evaluate research findings. Retrospective, no-control group designs rely on individuals' memories about perceptions of life changes after obtaining ADs and typically do not use standardized measures of change. Since these studies include no comparison groups, reported changes may be due to other factors such as general adaptations over time that might also have occurred without ADs. Cross-sectional studies examine differences among individuals who have ADs and those who do not; comparison groups typically are comprised of individuals on waiting lists to obtain ADs. However, systematic differences may exist among individuals who obtained ADs sooner and those still on waiting lists and may account for any observed differences. Longitudinal studies with appropriate comparison groups and examination of within-group changes over time show the most promise for identifying benefits attributable to ADs, but this type of research is resource-intensive.

SUMMARY AND CRITIQUE OF RESEARCH

ORGANIZING FRAMEWORK FOR EXAMINING AD BENEFITS

The purpose of placing trained ADs is to reduce the impact of disabling conditions on the day-to-day lives of individuals across a variety of domains (health, mobility, mood, social interaction, employment) and situations (at home, in stores, at friends' houses, in workplaces; Sachs-Ericsson, Hansen, and Fitzgerald, (2002). Outcomes of AD use and the planning of future research may be considered within WHO's biopsychosocial model of functioning and disability, whose elements encompass the range of life areas potentially affected by disabling conditions.

This model, known as the International Classification of Functioning, Disability and Health (ICF), describes outcomes in disability and functioning proceeding from interactions of health conditions (diseases, disorders and injuries) and contextual factors (WHO, 2002). The ICF invites the study of functioning and disability from three perspectives: body, individual, and society. Health-related domains are classified via lists of (1) body functions and structures and (2) domains of activity and participation. Functioning may be facilitated or hindered by features of the environment and by a person's unique set of attributes. At body level, functioning is considered in terms of the integrity of functions (movement, hearing, etc.) and structures (limbs, ears, etc.). The *impairment* term is reserved for problems at this level. At activity level, the performance of whole-person activities (communication, eating, ambulation, etc.) is considered. *Activity limitation* is the phrase describing problems at this level. At the level of participation, involvement in normative life situations such as education, homemaking, employment, and parenting is considered. Problems at this level are termed *participation restrictions.*

Internal and external contextual factors including physical and social environments and personal attributes may facilitate or hinder functioning at any of these levels. Internal contextual factors include but are not limited to "... gender, age, coping styles, social background, education, profession, past and current experience, overall behavior pattern, character, and other factors that influence how disability is experienced by the individual." External contextual factors include such variables as accessibility, availability of resources, and attitudes and behaviors of people encountered by an individual with a disability.

The ICF is intended to serve as a standardized method for describing functioning and disability within populations of interest, with the ultimate goal of improving healthcare policies internationally. Within that goal, it is intended for use in documenting outcomes of interventions and policy changes. It provides an excellent framework for examining the benefits of ADs. Moreover, beyond its formal use as a system of classification and rating, the ICF is likely to find broader use in shaping thought and discussion about function and disability.

BODY: EFFECTS

Research has shown that individuals may experience immediate physiological consequences from simply touching an animal, particularly a dog (Jennings, 1997).

Several studies have found that simply stroking a dog decreases physiological arousal by lowering blood pressure (Katcher, 1985; Katcher et al., 1983; Vormbrock and Grossberg, 1988), decreasing heart rate and slowing respiration (Lynch et al., 1974), and increasing finger temperature, another index of lower sympathetic arousal (Schuelke et al., 1991).

Consistent with the theories of the human–animal bond, the effect of lowering blood pressure has been found to be greater when petting a dog with whom a relationship has been established (Astrup, Gantt, and Stephens, 1979; Baun et al., 1984; Schuelke et al., 1991). Charnetski and colleagues found indices of immune system functioning (assessed by secretory immunoglobulin A) levels to be higher in the experimental groups in which participants petted dogs compared to other groups (Charnetski et al., 2004).

If merely touching a dog can have such a remarkable impact, what are the long-term effects of pet ownership? Several large general population studies have found a positive relation between pet ownership and health. Pet owners (mostly of dogs) exhibited lower blood pressure, lower triglyceride levels, and lower cholesterol (Anderson, Reid, and Jennings, 1992) and required fewer doctor visits and less medication use (Headey, 1999). Serpell's (1991) longitudinal study found that new dog owners showed decreased health complaints and improvements in general health. In addition, among individuals with low social support (Garrity et al., 1989) dog ownership was associated with better physical health. Among individuals at increased risk for health and psychological problems (such as elderly or recently widowed persons), research has shown that dog ownership may reduce health problems, physician visits, and medication use (Akiyama, Holtzman, and Britz, 1986; Siegel, 1990). Dog ownership and associated physical activity have been successfully included in weight loss treatments (Kushner et. al 2006).

Lifestyle differences between pet owners and nonowners may account for differences in health status. Surveys indicate that dog owners engage in more physical activity than nonowners (Raina et al., 1999; Heady, 1999; Serpell, 1991). In a recent longitudinal population study (Ham and Epping, 2006), dog ownership positively affected physical activity. Variabilities in activity related to dog size, with ownership of larger dogs associated with more frequent dog walking. In a controlled crossover study conducted in Japan (Motooka et al., 2006), researchers compared changes in autonomic nervous system activity in healthy senior individuals who walked with and without dogs. The authors concluded that walking a dog has potentially greater health benefits for senior citizens than walking without a dog.

Dog ownership may provide a buffer between stressful life events and the subsequent utilization of physicians (Siegel, 1990). Longitudinal studies show that pet ownership may enhance survival in individuals with such serious illnesses as myocardial infarction and severe angina (Friedmann et al., 1980; Friedmann and Thomas, 1995).

However, not all studies of pet ownership have shown clear evidence of enhanced health status. Lago et al. (1989) found no differences in health functioning in pet owning and nonowning elderly rural residents. Robb and Stegman's (1983) retrospective study of elderly veterans found no association of health and dog ownership. The authors of another study of elderly persons (Thorpe et al., 2006) suggest that a benefit from dog ownership was seen only in dog walkers and the benefit was similar to that associated with any walking activity. In a recent review of dog ownership and physical activity, Cutt and colleagues (2007) concluded that while evidence suggests that dog ownership produces considerable health benefit and an important form of social support that encourages dog owners to walk, evidence about the physical, environmental, and policy-related factors that affect dog owners who walk with their dogs is limited.

ADs and General Physical Health

Some research specifically examined the effects of ADs on health. Mowry, Carnahan, and Watson (1994) conducted a large retrospective study of individuals partnered with HDs. Among the individuals who reported that their health was problematic (69%), most reported that HDs helped their health (86%). However, no specific details were provided as to how health was improved. Fairman (1998) and Fairman and Huebner (2001) conducted a large retrospective study of individuals who obtained SDs. Most participants indicated that their SDs assisted them with health maintenance (59.4%) including physical fitness, nutrition, and decreasing health risk behaviors. Some (18%) reported that their SDs assisted them with oral hygiene including getting supplies and performing procedures. Lane, McNicholas, and Collis (1998) conducted a retrospective survey of individuals partnered with SDs from a program in England. Almost half (47%) reported that their health improved despite the fact that most had degenerative illnesses. However, in a cross-sectional study, Marks (1993) found no differences in the health ratings of individuals who had SDs and individuals on the waiting list.

ADs and Psychological Functioning

Current ICF classifies mental functioning including mood regulation, cognitive abilities, and personality on the list of body functions. The placement of particular aspects of psychological functioning under the body domain has not gone unquestioned (Merbitz, 2006). For purposes of this chapter, the research on effects of companion animals and ADs on mood and anxiety is summarized here in the section on the body domain. Results of studies of effects on attitudes, feelings of safety, and so forth, are summarized under personal contextual factors.

Valentine, Kiddoo, and LaFleur (1993) retrospectively surveyed individuals partnered with SDs and HDs. SD participants reported better control of anxiety (70%), feeling less depressed (70%), and feeling less irritable (70%). Guest et al. (2006) found a number of significant reductions over time in measures of tension, anxiety, and depression between the period before placement of the HD and the period after placement; moreover, no comparable differences were revealed for the year-long waiting period prior to placement. In Mowry, Carnahan, and Watson's (1994) large study, most HD participants reported prior problems with depression and tension, and most reported that the HD helped a lot in improving their moods and enabling them to relax. Most of Lane et al.'s (1998) participants also reported that they relaxed more after they obtained SDs.

In a study by Collins and colleagues (2006) comparing individuals with ADs to similar individuals without dogs, they concluded that among participants with progressive conditions, those with service dogs demonstrated significantly higher positive affect scores than comparison group participants. Among those with clinical depression, service dog partners scored significantly higher in positive affect. They concluded that select individuals may experience psychosocial benefits from partnering with SDs.

ACTIVITY: EFFECTS AT LEVEL OF INDIVIDUAL PERFORMANCE

The impact of pet ownership and ADs on activity is typically discussed in terms of its implications for activities of daily living (ADLs). For example, in a large sample of elderly, nondisabled adults. More deterioration of ADLs was among the non-pet owners than among the pet owners over a 1-year period (Raina et al., 1999).

For deaf or hard-of-hearing individuals, sound awareness is of crucial importance in many daily activities. Several studies showed that HDs increase individuals' awareness of different sounds. Among participants in Mowry, Carnahan, and Watson's study (1994) more than three-quarters indicated that their HDs performed satisfactorily in alerting, for example, to door knocks, clock alarms, smoke alarms, phones, name called, baby crying, oven timer, tea kettle, and burglar alert. Alerting to sounds was cited most often as "a good thing about having a hearing dog."

Hart, Zasloff, and Benfatto (1996) conducted a cross-sectional study of individuals partnered with HDs and individuals on a waiting list to obtain HDs from the San Francisco SPCA. Respondents reported that their HDs fulfilled their primary expectations of alerting them to sounds including doorbells, smoke alarms, and alarm clocks. Researchers asked the participants to retrospectively rate their awareness to each of these sounds before and after obtaining an HD. Respondents reported that they were more aware of these sounds after obtaining HDs. Moreover, the waiting list participants reported less awareness of these sounds. In a recent study (Guest et al., 2006) HD recipients reported significant reductions in hearing-related problems. No comparable differences were noted for the year-long waiting period prior to dog placements.

Fairman and Huebner's (2001) respondents identified the ADLs for which their SDs assisted them. These included getting around the community (84%), getting around the house (78.2%), obtaining communication devices (72%), dressing (48%), grooming (44%), emergency responses (43%), bathing (20%), feeding self (18%), and toileting (18%). In Roth's (1992) retrospective survey, respondents reported that their SDs assisted them with eating (22%), dressing (20%), retrieving phones (20%), and grooming (17.4%). When asked to list the most important task that their SDs performed, respondents in several studies identified retrieval of objects (Roth, 1992; Lane et al., 1998; Marks, 1993).

Rintala (2006) followed for 6 to 7 months 18 individuals who received SDs and 15 participants who on a wait list to receive SDs. Rintala asked her SD participants about the tasks with which their service dog assisted them. The most frequent routine task routinely performed by the SDs was item retrieval. Other tasks reported by at least 50% of the participants were carrying items by mouth, barking in emergencies, opening and/or closing external and interior doors, carrying items in doggy backpacks, and pushing buttons to activate automatic door openers. Nonetheless, on the motor FIM total score, a standardized measure assessing abilities to perform ADLs, no significant changes in independence were found in the SD and control groups.

Rintala (2006) also followed 10 individuals for 6 to 7 months. Four received HDs and 6 were on a list to receive HDs. The participants who received dogs reported that alerting to telephone or alarm clock ringing was the most frequently performed task that produced the greatest impact on their daily activities.

ATTENDANT CARE

A growing amount of literature focusing on family burden shows that those who care for individuals with physical or health disabilities may suffer higher rates of depression and poorer quality of life. Specifically, increased rates of depressive symptoms and disorders among caregivers were readily apparent in an extensive review conducted by Dilworth-Anderson, Williams, and Gibson (2002). It is common for family members who live with disabled individuals to provide care in some way (Wallstein, 2000). Both needing care and providing care are often associated with depressive symptoms and frustration (Dilworth-Anderson et al., 2002; Llacer et al., 2002; Ramos, 2004). Thus, the potential for an AD to lessen family burden is an important consideration for individuals with disabilities and for family functioning as a whole.

The ability to handle ADLs by using an assistance dog can reduce the need for hours of caregiving by others. Thus, a reduction in paid and unpaid attendant care is an important, indirect measure of an AD's impact on ability to perform ADLs. In a cross-sectional study comparing individuals with SDs to those on a waiting list, Hackett (1994) found no significant difference between groups in the use of paid assistants, but it should be noted that no assessment was made of unpaid assistants. Fairman and Huebner's (2001) participants retrospectively reported the amounts of care needed from paid and unpaid assistants before and after obtaining their SDs. They reported using 2 fewer hours of paid assistance each week and 6 fewer hours of unpaid assistance each week. Unpaid assistants often are family members. The reduction in demand on family for assistance is likely to have an important impact on decreasing family burden.

Rintala's longitudinal study (2006) of individuals partnered with SDs found a significant reduction in the amount of assistance needed from other persons. The AD group averaged 4.5 hours of paid assistance daily at baseline and only 2.9 hours of paid assistance at follow-up. Rintala also specified the activities for which those who had SDs needed less human assistance. All participants with SDs but one indicated that the dogs made a difference in the lives of family members, friends, and/or attendants. Among Rintala's participants, the proportion receiving help from another person decreased from baseline to follow-up. Help included retrieving items (38.9% to 0%), pulling towels from a rack and placing them in a hamper (38% to 11%,), carrying a bag (27.8% to 5.6%), picking items from store shelves (44.4% to 22.2%), helping pull linens from beds and/or pulling clean linens into place (61.1% to 44.4%), and activating emergency call devices (22.2% to 5.68%). Thus, in considering the potential of ADs to reduce attendant care, it should be noted that these animals affect the environment in which individuals live by increasing the number of independently performed activities— another example of the recursive nature of variables related to functioning and disability.

PARTICIPATION: EFFECTS AT SOCIETY AND LIFE SITUATION LEVELS

In the WHO model, participation is a multifaceted category. An individual's participation in social roles and life situations in a community rests upon contextual factors (environmental and personal) and the person's functioning at the body and activity levels. Thus, this category expresses a fully biopsychosocial perspective on functioning and disability. Participation depends upon both social and physical access.

Fairman and Huebner (2001) asked participants whether their SDs assisted them in a variety of activities relevant to participation in social roles. Respondents indicated that they participated in more activities with their SDs (80%). Their SDs assisted them with shopping (76%), cleaning (55%), clothing care (40%), household maintenance (33%), care of others (22%), leaving their homes (77%), and using community resources (72.5%). Roth's (1992) respondents reported that their SDs assisted them with general shopping (69%), grocery shopping (50%), and banking (28.3%). Neither study identified the specific mechanisms by which the SDs facilitated these tasks. In the study by Guest and colleagues (2006), recipients of HDs subsequently reported significant improvements in social involvement and independence, compared to the year prior to receiving their dogs. Hart and colleagues (1996) found that among participants with HDs, 75.6% reported the HDs positively influenced their interactions with the hearing community.

Employment

Among Roth's (1992) sample, 50% reported that they were employed (part or full time) and 82.6% of these participants reported that their SDs assisted them at work. Fairman and Huebner's (2001) participants indicated that their SDs assisted them in work (46%) and school activities (22%). However, the type of assistance was not specifically assessed. Among Mowry et al.'s (1994) participants, 44.3% were employed, and among them 62.6% took their HDs to work. Participants reported that at work their HDs assisted by alerting them to sounds similar to those at home. Although 23% of Marks' (1993) respondents reported that their SDs were essential for them to work or attend school, she did not find a difference in employment status of individuals partnered with SDs (40.6%) and those on a waiting list (32.1%) to be statistically different.

Hart, Zasloff, and Benfatto (1996) found that 57% of individuals partnered with HDs were employed compared to 46% of the waiting list sample, although this difference was not statistically significant. In a cross-sectional study, Hackett (1994) found no significant difference in the employment status of individuals partnered with SDs and those on a waiting list, although she did not report the percentage of individuals in each group who were employed. Rintala (2000) conducted a longitudinal study (with no comparison group) of individuals before and 6 months after obtaining SDs. Using the Craig Handicap Assessment and Reporting Technique (Whiteneck et al., 1992), participants reported positive changes in productive use of time; however, this effect did not reach statistical significance. Rintala performed a second longitudinal study (6-month follow-up) for the Veterans' Administration in 2006. She included 42 participants. Half received SDs and the remainder were on a control wait list. Of those who completed the study, 18 participants obtained SDs and 15 participants did not (control group). Based on the Short Form Craig Handicap Assessment and Reporting Technique (SF-CHART) Rintala found no differences in levels of employment of the SD and control groups. In a longitudinal study conducted by Fitzpatrick (2007) she found employment levels increases over time for the SD group (N = 12), suggesting that the SDs were instrumental in getting people back to work. However, no differences were found across groups over time with respect to improvements in functioning as measured by CHART.

CONTEXTUAL FACTORS

The literature related to contextual factors is considered in two sections. In the first, we consider the potentially recursive relationships between attributes of a person and the presence of a companion animal or AD. In the second section, we consider the evidence indicating whether companion animals or ADs bring about changes in the external environment including behaviors of other people and access to resources that in turn could facilitate activity and participation.

INTERNAL CONTEXTUAL FACTORS: ATTRIBUTES OF PERSON

These factors include, among others, the internal strategies that a person uses for coping with life challenges (reminders about previous successes), attitudes and beliefs (e.g., about disability), prior learning history (e.g., experiences of vulnerability and threat when traveling with a wheelchair), education, knowledge and skills (knowing which restaurants are accessible, knowing how to communicate with healthcare providers, development of skills in social interactions), and so forth.

Evidence indicates that individuals partnered with ADs feel safer and thus feel more comfortable when using community resources and traveling away from home. Even in studies of nondisabled individuals (Serpell, 1991), new dog owners reported significant reductions in fear. In Mowry et al.'s (1994) large study, most (95.3%) participants reported problems in security or safety in their homes and communities and 91% reported that their HDs helped with these concerns.

In Pang's (1999) retrospective study of individuals partnered with HDs, only 33.3% felt comfortable traveling away from home before they had HDs, whereas 85.7% felt comfortable traveling away from home after obtaining their HDs. Only 9.5% of Pang's participants felt safe in their environments before obtaining HDs; 95.2% agreed that they felt safe after receiving their HDs. Most (91%) of Fairman's respondents (personal communication, February 2001) and 92% of participants in the study by Valentine, Kiddoo, and LaFleur, (1993) indicated that they felt safer after obtaining ADs.

An extensive but uneven body of literature cites the effects of pets on psychological well being. It is often impossible to distinguish effects of pet ownership from pre-existing differences (personal attributes) that may lead people toward or away from acquiring a pet. Results from studies addressing this issue have been equivocal, with a minority of investigators even concluding that pet owners have more psychological problems than nonowners (Cameron and Mattson, 1972; Cameron et al., 1966; Guttman, 1981). In contrast, a number of studies indicated a positive relation between pet ownership and psychological well-being (Akiyama et al., 1986: Serpell, 1991). Additionally, several studies show no relationship between psychological variables and pet ownership (Friedmann et al., 1984; Wilson, 1991; Kidd and Martinez, 1980; Lawton, Moss, and Moles, 1984; St. Yves et al., 1990; Johnson and Rule, 1991; Watson and Weinstein, 1993).

The ability to become attached to a dog may be related to the decision to have a companion animal or AD. Studies found individual differences in the extent to which individuals are attached to their pets, and these differences in attachment may affect the psychological impact of a pet on an individual (Brown, Shaw, and Kirkland, 1972; Ory and Goldberg, 1983; Joubert, 1987; Garity et al., 1989; Miller, Staats, and Partlo, 1992).

Retrospective reports have suggested that ADs exert positive psychological impacts on most individuals with whom they are partnered. Fairman and Huebner's (2001) respondents reported that they were more in control of their lives (83%), more independent (88%), felt better about themselves (75%), were better able to manage stress (77%), and had increased confidence (81%). Lane et al.'s (1998) participants reported that since obtaining SDs they worried less about their health (52%). Roth's (1992) participants reported that they took more risks (54.3%), felt more in control of their lives (78.3%), and were more accepting of their disabilities (45.7%). Moreover, while only 26.1% reported experiencing high life satisfaction before obtaining SDs, 71.1% reported high life satisfaction after obtaining SDs.

Valentine, Kiddoo, and LaFleur (1993) retrospectively surveyed individuals partnered with SDs or HDs. SD participants reported experiencing higher self-esteem (80%), more confidence (70%), more assertiveness (80%), more contentment (80%), more capabilities (100%), less loneliness (90%), and greater senses of belonging (36%). The HD respondents reported being more independent (79%) and feeling healthier (79%). In addition, Pang's (1999) respondents reported increases in confidence, acceptance of their disabilities, and quality of life after obtaining HDs.

Disability may contribute to deprivation in social interactions and thus fewer opportunities to develop social skills. Although Benshoff, Fried, and Roberto (1990) found no significant differences between disabled and nondisabled college students on an array of developmental skills (autonomy, independence, and planning), disabled students scored lower on skills related to interpersonal functioning. In Mowry, Carnahan, and Watson's (1994) large study, at least 60% or more of the HD participants reported having problems with social life, self-confidence, independence, loneliness, and companionship. Most reported that HDs helped "a lot" in problem areas. Participants emphasized the dog's affection and their bonds with the dogs; many said they no longer felt lonely. When asked about "the good things about having a HD," companionship (29.7%) was the second most frequent response (after alerting to sounds).

Similarly, in their study of individuals with progressive disabilities, Collins and colleagues (2006) found that individuals partnered with SDs reported significantly less loneliness compared to similar individuals without dogs. In a cross-sectional study, Hart et al. (1996) found that those partnered with HDs reported experienced less loneliness than did those in the waiting list comparison group. Additionally, although the groups did not differ on the number of stressful life events experienced, individuals partnered with HDs reported experiencing less stress from these events. In addition to the companionships experienced with ADs, the decreases in loneliness may be related to changes in social interactions.

Cross-sectional studies of individuals partnered with ADs that assessed changes in participants' self-concepts or self-esteem and used standardized measures have generally not found significant results. This may be due, in part, to factors such as the methodology employed, the specific measures used, sample size, and the length of time the participants had their ADs. A cross-sectional waiting list comparison study by Hackett (1994) found that levels of self-esteem measured by the Index of Self-Esteem (Hudson, 1982 and 1992) did not distinguish individuals partnered with SDs from individuals on the waiting list. However, the author pointed out that 43% of the respondents had their SDs for less than a year and had their disabilities on average for 15 years. Additionally, a cross-sectional study (Rushing, 1995) of self-concept (measured by the Tennessee Self-Concept Scale; Roid and Fitts, 1988) comparing individuals with quadriplegia who obtained SDs to those on waiting lists found no differences between the two groups. The author suggested that the lack of findings was due in part to the homogeneous nature of the participants. Moreover, they argue that the measure used to assess self-concept may not be sensitive enough to measure the subtle subjective and affective issues of this population. Finally, in another cross-sectional study using the same measure of self-concept (the TSCS), those having SDs scored higher in self-concept than the waiting list respondents but the difference was not significant (Marks, 1993).

A pre-test and post-test, 4-month longitudinal study (Donovan, 1995) examined differences among individuals who had recently received their SDs and individuals on a waiting list. The subjects were matched for age and gender. The groups were assessed on measures of depression (CES-D; Radloff, 1977), self-esteem (Coopersmith, 1959), acceptance of disability (ADPT; Yuker and Block, 1996), and quality of life (RAND; Sherbourne and Hays, 1991). No differences over time were found for any of the measures, except for pain, which was accounted for by pre-test differences. As the author noted, having a dog only for 4 months is likely too short a time for any substantive change to occur. Interestingly, both the SD and waiting list groups showed marginally significant ($p = 0.06$) increases in depressive symptoms over the study. In contrast, in her 6-month longitudinal study, Rintala (2000) found a decrease in depression (CES-D; Radloff, 1977), although it was not statistically significant. However, she found that the self-esteem (Rosenberg, 1979) of SD participants significantly increased over time. In the recent longitudinal study of Fitzpatrick (2007), no significant impact of service dogs with respect to psychosocial measures (affect, self-esteem, loneliness) was noted.

Contextual Factors: Environment

An AD is an external contextual factor—a resource within an individual's environment that has the potential to affect several of the ICF's domains. We reviewed the evidence for these effects on body, activity and participation, and personal contextual factors. Here, we consider the potential impact of an AD on other features of the environment that in turn may facilitate an individual's functioning. These are indirect effects of having an AD. We first consider the evidence for positive effects, then discuss potential disadvantages.

Social Environment

The behavior of other people forms the social environment of an individual. Studies of nondisabled populations consistently found that the presence of a companion dog increases social interactions (Mugford and M'Comisky, 1975; Messent, 1983; Sanders and Robins, 1991). Many well designed studies have found this effect to be quite robust, regardless of age, gender, or manner of dress (McNicholas and Collis, 2000). In addition to the other benefits that an AD may bring to an individual, an impact on social interaction may have broad implications for several areas of life. As an example of the social environment affecting participation, Belgrave and Walker (1991) found that social support predicted vocational functioning among disabled African-Americans.

A person accompanied by a dog appears more approachable. In a laboratory study, when shown pictures of nondisabled individuals, research participants gave higher ratings of happiness and safety to individuals pictured with dogs than they rated the same individuals without dogs (Rossbach and Wilson, 1992). Similar results were found in studies of disabled individuals when accompanied by their SDs. In several observational studies of adults and children with disabilities, the presence of a dog was found to increase communication and friendly contacts with strangers (Eddy, Hart, and Boltz, 1988; Hart, Hart, and Bergin, 1987; Mader, Hart, and Bergin, 1989). Thus, the presence of an AD may bring about changes in the attributions and behavior of people in a social environment.

Numerous studies have shown that having a social support network is related to psychological well being and healthy functioning (Uchino, Cacioppo, and Kiecolt-Glaser, 1996; Vinokur and Van Ryn, 1993). While individuals with ADs may experience increases in social acknowledgment from strangers, do the dogs produce long-term social benefits? Individuals partnered with SDs noted retrospectively that not only did their social interactions increase, but their numbers of friends increased as well. Fairman and Huebner's (2001) respondents reported that after they obtained SDs, more people approached them (100%), their social interactions increased (87%), and their numbers of friends increased (59%). Lane et al.'s (1998) participants reported that more people stopped to talk to them after they obtained SDs (92%), they had more friends (75%), and better social lives (34%).

Roth's (1992) respondents reported that their SDs facilitated social interactions (76.1%) and led people to be more friendly (85.8%). Valentine, Kiddoo, and LaFleur,

(1993) also reported that SD and HD recipients experienced positive changes in social functioning. SD participants reported more friendliness from strangers (80%), more contacts with friends (60%), and better relations with family (60%). Those with HDs reported experiencing more friendliness from strangers (50%) and better relations with family (29%).

Most (68%) of participants in Mowry and colleagues' (1994) study reported that their social lives were problematic before they acquired HDs. Among those participants, 86% reported that their social lives improved after they got dogs. Only 26.6% of Pang's (1999) respondents reported being socially accepted before obtaining HDs; however, 81% reported being socially accepted after obtaining HDs. While three-quarters of Hart et al.'s (1996) respondents reported that their HDs made positive changes in their interactions with the hearing community, only 26.6% of the waiting list participants anticipated this type of impact on their lives. Thus, some individuals may underestimate the social benefits they will receive from ADs.

ENVIRONMENT: ACCESSIBILITY AND SAFETY

In Mowry et al.'s (1994) large study, most (95.3%) participants reported fewer security and safety problems in their homes and communities after receiving their HDs.

ENVIRONMENT: FINANCIAL RESOURCES

These resources are likely to have complex and recursive relationships with ownership of an AD. Lack of financial resources can delay or prevent the acquisition of an AD, whereas receiving an AD may eventually add to financial resources through the facilitation of education and employment (see the section on participation). The care of an AD, however, also involves expense. If insurance policies change in the future to cover ADs (as some cover other assistive technologies), the result will be a change in resources for many individuals and for organizations training ADs.

ENVIRONMENT: DISADVANTAGES OF ASSISTANCE DOGS

Individuals considering obtaining ADs should review the potential disadvantages including financial costs, the responsibilities of caring for an animal, problem behaviors of an AD, and challenges to public access when accompanied by their ADs.

COSTS

Costs associated with pet ownership include food and routine veterinary care; veterinarian costs can be substantial. The minimum annual cost of owning a healthy dog including food and an annual veterinarian visit was estimated at $400 (Masullo, 2000). Among Fairman and Huebner's (2001) SD respondents, 75% reported spending less than $1000 on the care and feeding of their SDs annually. Additionally, 32% reported getting some government assistance for care of their SDs. Among Mowry et al.'s (1994) HD respondents, only 7.9% noted that the financial burdens of dogs created problems.

AD BEHAVIOR PROBLEMS

Dog ownership always presents responsibilities (Albert and Bulcroft, 1987) and even the best trained AD does not always perform tasks reliably. Moreover, ongoing practice is required to maintain an AD's skill. Using an open-ended format, Roth's (1992) respondents listed problems with their SDs including taking longer to complete some tasks (17.4%), people petting the dog (17.7%), and difficulty maneuvering through small spaces (13.2%). One-fourth of Fairman's (personal communication, February 2001) respondents reported problems maintaining their dogs including clipping nails and bathing. However, her respondents reported rarely having any of the other problem behaviors assessed, such as misbehaving at home (1.5%) or in public (2.5%).

Mowry et al.'s (1994), HD participants cited problems such as pulling on a leash (23%), barking too much (19%), toileting indoors (8%), and growling and biting (0.3%). In an open-ended format, respondents were asked to list "bad things" about having HDs; 13% did not list problems and 31.6% stated that they had no problems. Among those who described problems, 23.4% listed taking care of the dog including feeding, grooming, and walking.

Rintala's SD participants mentioned that their dogs needed a lot of attention. Five of the 18 participants who obtained SDs mentioned tasks the dogs were supposedly trained to do but did *not* do well. Several dogs had problems with obedience, at least in certain situations. Among Hart, Zasloff, and Benfatto's (1995) participants, half reported problems with their HDs, but only 13% of the waiting-list respondents anticipated having problems. The problems reported included aggression to people or other dogs, barking, and destructiveness. The authors pointed out that the lack of awareness of potential problems may lead to subsequent placement difficulties and even failures.

ACCESS PROBLEMS

The ADA (1990) grants individuals with disabilities public access (the right to go wherever the public is allowed) when accompanied by dogs specially trained to assist them. Several anecdotal reports concern denial of public access (Duncan, 1996). Some such interactions are brief, requiring minimal education or comments to explain. However, some such problems can be confrontational and distressing. Mowry and colleagues (1994) found that over one-third (36.6%) of their HD participants experienced problems with access. Among Fairman and Huebner's (2001) SD participants, 46% reported problems with access at some point. Roth (1992) asked participants to identify problems with their SDs and 15.2% reported problems with access. Moreover, 11 of Rintala's 18 SD participants said they experienced unwanted attention in public, including a challenge about bringing a dog into a restaurant.

From a different perspective, ensuring appropriate accommodations for an AD can be challenging in some situations; for example, a hospital setting may present legitimate concerns as to whether the presence of a dog may compromise the health and safety of staff or patients (Houghtalen and Doody, 1995).

PLACEMENT FAILURES

Some partner teams do not work out. This can be a difficult experience for an individual who may have become attached to his or her dog. In a study of placement success rates of 75 newly trained HDs (Mowry et al., 1994), the overall success rate was only 59.3%. Anecdotal reports from AD training schools identified various reasons why placements have not worked including problem behaviors of dogs that were not apparent during training, poor matches of a dog's temperament and skills and an individual's needs, and interference or resentment of family members toward the AD.

EMOTIONAL DISTRESS ASSOCIATED WITH ADs

One of the most difficult problems that individuals partnered with ADs must face is the death of a dog. The loss may be similar to the loss of a beloved family member (Arkow, 1993; Cusack, 1988). It can be months or years before another AD can be integrated into the individual's life (Ptak, 1994; Eames and Eames, 1997a). No investigations of short- or long-term effects of the loss of an AD were identified.

CONCLUSIONS AND FUTURE DIRECTIONS

Through clinical observation, anecdotal reports, and retrospective and cross-sectional studies, preliminary support was found for the conclusion that ADs exert positive impacts on individuals' health, psychological functioning and well being, social interactions, performance of activities, and participation in various roles at home and in the community. Researchers have, however, relied predominantly on retrospective and cross-sectional studies, subjective ratings, and one-time interviews with individuals after they received their dogs. Such results provide less convincing support for efficacy than might be provided by large-scale longitudinal studies that compare measures before and after obtaining ADs and include appropriate comparison groups.

Longitudinal studies present numerous problems in addition to the considerable cost involved. Random assignment to receive or not receive an AD may be impossible to accomplish because of the costs and ethical issues involved. Participants on wait lists are likely not randomly assigned to wait or to receive a dog, and some data suggests that the most disabled individuals receive priority in obtaining ADs. If individuals receiving dogs have greater difficulties (e.g., higher percentage of progressive illnesses) than control groups, positive changes over time related to AD ownership is difficult to determine.

AD outcome research may always be challenged by problems of selection bias. To attempt to control and minimize selection problems, comparison groups should include individuals matched on relevant variables such as age, gender, education, and disability. Comparison groups should include individuals who have applied and have been accepted to receive ADs (waiting list individuals) as well as individuals who have not applied for ADs, matched on relevant characteristics. Longitudinal studies should be conducted for a sufficient period to effectively evaluate the efficacy of the ADs. When an individual obtains an AD, the person and dog must learn to work together as a team and undergo an initial adjustment period. Based on anecdotal accounts, it is unlikely that any substantial benefits will occur in less than a year.

Indeed, initially obtaining an AD may be stressful and even impair functioning. Initially, an individual must adapt to a dramatically new life style; the individual and the AD face a lot of work and practice to work successfully as a team. To date, scientifically sound longitudinal studies of individuals with ADs have produced inconsistent findings about the benefits of ADs. This is in stark contrast to the direct reports from most AD users who cite major positive changes across the domains outlined in this chapter.

In conducting future research on the benefits of ADs it should be considered that the benefits may be multidimensional, affecting different areas of functioning, and thus requiring diverse strategies and targets of measurement. Areas of investigation should initially include psychosocial and health variables identified by previous studies to improve after obtaining companion animals or ADs. Researchers should select measures that are sensitive to change based on other intervention studies of individuals with disabilities. They should avoid parameters that are more appropriately characterized as stable personality characteristics or trait measures.

The use of standardized measures is encouraged. However, the use of participants' direct reports about the frequency of relevant activities recalled within a recent period (e.g., past 24 hours) is also encouraged (Merbitz, 1996). Ideally this frequency data would be sampled regularly during a longitudinal study to increase the reliability of measurement and sensitivity to change. A related measurement strategy would be to regularly sample participants' estimates of time required for completion of various ADLs on a given day.

Along with reports of frequency and time expenditure, information should be included about the presence and nature of assistance used (human, AD, assistive device). In the course of the study, medication use, physician visits, and health functioning also should be monitored for changes. However, because individuals who obtain ADs often have progressive illnesses along with co-morbid health problems, there is likely to be much variability across participants on outcomes related to health functioning. Observation of direct evidence of improvement in health functioning may be unlikely unless the sample sizes are large.

It would be of interest to examine individual differences in outcomes among the individuals who have ADs. This could include psychosocial variables, health variables, adherence to AD training program requirements, and aspects of the human–animal bond that could be examined as predictors of differential outcomes. Future studies could also examine how specific training of a dog may affect the long-term success of a team.

The availability of well-trained ADs is low and the demand is high. Moreover, the training of ADs is expensive. However, it is possible that the use of ADs by individuals with disabilities may save money. Individuals partnered with ADs may be more likely than nonusers to live and work independently. They may have fewer needs for hired caretakers, have better psychological and health functioning resulting in lowered health care costs, and require less assistance from social agencies. Careful documentation of these effects may open the way for third-party reimbursement for ADs. Identification of any areas of functioning that receive less than satisfactory benefits can point the way to improvements in training, placement, and follow-up procedures used by AD agencies. Accountability should be promoted as should recognition of the real and potential benefits of well-trained ADs.

REFERENCES

Akiyama, H., Holtzman, J., and Britz, W. (1986). Pet ownership and health status during bereavement. *Omega, 17*, 187–193.
Albert, A. and Bulcroft, K. (1987). Pets and urban life. *Anthrozoos, 1*, 9–25.
Allen, K.M. (1996). Response to Eames and Eames, *Disabilities Studies Quarterly 16*, 23–25.
Allen, K.M. and Blascovich, J. (1996). The value of a service dogs for people with severe ambulatory disabilities. *Journal of American Medical Association, 275*, 1001–1006.
Americans with Disabilities Act (ADA) of 1990. Public Law 101-336. *Federal Register,* Washington, D.C.
Anderson, W.P., Reid, C.M., and Jennings, G.L. (1992). Pet ownership and risk factors for cardiovascular disease. *Medical Journals of Australia, 157*, 298–301.
Arkow, P. (1993). Pets for the visually and hearing impaired. In Arkow, P., Ed. *Pet Therapy: A Study and Resource Guide for the Use of Companion Animals in Selected Therapies.* Stratford, NJ: Phil Arkow, 37–39.
Astrup, C.W., Gantt, W.H., and Stephens, J.H. (1979). Differential effect of person in the dog and in the human. *Journal of Pavlovian Biological Science, 14*, 104–107.
Baun, M.M., Bergstrom, N., Langston, N.F., and Thoma, L. (1984). Physiological effects of human/companion animal bonding. *Nursing Research, 33*, 126–129.
Beck, A.M. (2000). The use of animals to benefit humans: animal assisted therapy. In Fine, A., Ed. *Handbook on Animal-Assisted Therapy.* San Diego: Academic Press, 21–40.
Belgrave, F.Z. and Walker, S. (1991). Predictors of employment outcome of black persons with disabilities. *Rehabilitation Psychology, 36*, 111–119.
Bell, C. (1990). Hearing dogs are a relatively new service dog, *Alert*, Spring, 1–2.
Benshoff, J.J., Fried, J.H., and Roberto, K.A. (1990). Developmental skill attainment among college students with disabilities. *Rehabilitation Counseling Bulletin, 34*, 44–52.
Bergin, B. (1981). Companion animals for the handicapped. In Fogle, B. and Edney, A., Ed. *Interrelations between People and Pets.* Springfield, IL: Charles C Thomas, 191–236.
Bergin, B. (2000). World's first service dog team: Kerry and Abdul nearing the end of a 16-year partnership. *The Assistance Dog Institute Quest, 9*, 1.
Bonas, S., McNicholas, J., and Collis, G. (2000). Pets in the network of family relationships: an empirical study. In Podberscek, A.L. et al., Eds. *Companion Animals and Us.* Cambridge: Cambridge University Press, 209–236.
Bradburn, N.M. (1969). *The Structure of Psychological Well Being.* Chicago: Aldine.
Brown, L.T., Shaw, T.G., and Kirkland, K.D. (1972). Affection for people as a function of affection for dogs. *Psychological Reports, 31*, 957–958.
Cameron, P., Conrad, C., Kirkpatrick, D.D., and Bateen, R.J. (1966). Pet ownership and sex as determinants of stated affect toward others and estimates of others' regard of self. *Psychological Reports, 19*, 884–886.
Cameron, P. and Mattson, M. (1972). Psychological correlates of pet ownership. *Psychological Reports, 30*, 286.

Campbell, D. and Stanley, J. (1966). *Experimental and Quasi-Experimental Design for Research*. New York: Houghton Mifflin.

Charnetski, C.J., Riggers, S., and Brennan, F.X. (2004). Effect of petting a dog on immune system function. *Psychological Reports, 95*, 1087–1091.

Collins, D.M., Fitzgerald, S.G., Sachs-Ericsson, N., Scherer, M., Cooper, R.A., and Boninger, M.L. (2006). Psychosocial well-being and community participation of service dog partners, *Disability and Rehabilitation: Assistive Technology, 1*, 41–48,

Collis, G.M. and McNicholas, J. (1998). A theoretical basis for health benefits of pet ownership. In Wilson, C.C. and Turner, D.C., Eds. *Companion Animals in Human Health*. Thousand Oaks, CA: Sage, 105–122.

Coopersmith, S. (1959). A method for determining types of self-esteem. *Journal of Abnormal and Social Psychology, 59*, 87–94.

Crewe, N.M. and Dijkers, M. (1995). Functional assessment. In Cushman, L. and Scherer, M., Eds. *Psychological Assessment in Medical Rehabilitation*. Washington: American Psychological Association, 101–144.

Cusack, O. (1988). The death of a pet. In Cusack, O., Ed. *Pets and Mental Health*. New York: Hawthorne Press, 181–195.

Cusack, O. and Smith, E. (1984). *Pets and the Elderly: The Therapeutic Bond*. New York: Hawthorne Press.

Cutt, H., Giles-Corti, B., Knuiman, M., and Burke V. (2007). Dog ownership, health and physical activity: a critical review of the literature. In *Health and Place*. Amsterdam: Elsevier.

Dilworth-Anderson, P., Williams, I.C., and Gibson, B.E. (2002). Issues of race, ethnicity, and culture in caregiving research: a 20-year review (1980–2000). *Gerontologist, 42*, 237–272.

Donovan, W.P. (1995). The psychological impact of service dogs on their physically disabled owners. *Dissertation Abstracts International, 55*, 3010.

Duncan, S. (1996). Access denied. Now what? *Alert, 7*, 1.

Duncan, S. (1998). The importance of training standards and policy for service animals. In Wilson, C.C. and Turner, D.C., Eds. *Companion Animals in Human Health*. Thousand Oaks, CA: Sage, 251–266.

Eames, E. and Eames, T. (1994). *A Guide to Guide Dog Schools*. Fresno, CA: Eames and Eames.

Eames, E. and Eames, T. (1996). Economic consequences of partnership with service dogs. *Disabilities Studies Quarterly, 16*, 19–23.

Eames, E. and Eames, T. (1997a). *Partners in Independence*. New York: Howell.

Eames, E. and Eames, T. (1997b). Additional comments on the reported impact of service dogs on the lives of people were severe ambulatory difficulties. *Disabilities Studies Quarterly*, Winter, 22.

Eames, E. and Eames, T. (1998a). Team talk: where have all the doggies gone? *Off Lead*, February, 8–10.

Eames, E. and Eames, T. (1998b). Where have all the trainers gone? *American Pet Dog Trainers*, July–August, 9.

Eddy, J., Hart L.A., and Boltz, R.P. (1988). The effects of service dogs on social acknowledgments of people in wheelchairs. *Journal of Psychology, 122*, 39–45.

Fairman, S. (1998). Service dogs: a resource for occupational therapists to improve function. Master's thesis, Eastern Kentucky University, Richmond.

Fairman, S. and Huebner, R. (2001) Service dogs: a compensatory resource to improve function. *Occupational Therapy in Health Care, 13*, 41–52.

Fitzgerald, S. (2006). Effects of assistant dogs on persons with mobility or hearing impairments (4-12-16): Final Report to Veterans' Administration.

Friedmann, E. (1990). The value of pets for health and recovery. *Proceedings of Waltham Symposium, 20*, 9–17.

Friedmann, E., Honori, A., Lynch, J.J., and Thomas, S.A. (1980). Animal companion and one-year survival of patients after discharge from coronary care unit. *Public Health Reports, 95*, 307–312.

Friedmann, E., Katcher, A.H., Eaton, M., and Berger, B. (1984). Pet ownership and psychological status. In Anderson, R.K. et al., Eds. *The Pet Connection*. Minneapolis: University of Minnesota Press, 300–308.

Friedmann, E. and Thomas, S.A. (1995). Pet ownership, social support, and one-year survival after acute myocardial infarction in the Cardiac Arrhythmia Suppression Trial (CAST). *American Journal of Cardiology, 76*, 1213–1217.

Froling, J. (1995a). Vocational Rehabilitation funding possible. *Partners' Forum, 2*, 16.

Froling, J. (1995b). SSI allowance for assistance dogs. *Partners' Forum, 2*, 6.

Garrity, T.F. and Stallones, L. (1998). Effects of pet contact on human well being. In Wilson, C.C. and Turner, D.C., Eds. *Companion Animals in Human Health*. Thousand Oaks, CA: Sage, 3–22.

Garrity, T.F., Stallones, L., Marx, M.B., and Johnson, T.P. (1989). Pet ownership and attachment as supportive factors in the health of the elderly. *Anthrozoos, 3*, 35–44.

Guest, C.M., Collis, G.M., and McNicholas, J. (2006). Hearing dogs: a longitudinal study of social and psychological effects on deaf and hard-of-hearing recipients. *Journal of Deaf Studies and Deaf Education, 11*, 252–261.

Guttman, G. (1981). The psychological determinants of keeping pets. In Fogle, B., Ed. *Interrelations between People and Pets*. Springfield, IL: Charles C Thomas, 89–98.

Hackett, D. (1994). Levels of self-esteem in owners of service dogs. Master's thesis, California State University, Long Beach.

Ham, S.A. and Epping, J. (2006). Dog walking and physical activity in the United States. *Prevention of Chronic Disorders, 3*, A47.

Hart, L.A, Hart, B.L, and Bergin, B. (1987). Socializing effects of Service dogs for people with disabilities. *Anthrozoos, 1*, 41–44.

Hart, L.A., Zasloff, R.L., and Benfatto, A.M. (1995). The pleasures and problems of hearing dog ownership. *Psychological Reports, 77*, 969–970.

Hart, L.A., Zasloff, R.L., and Benfatto, A.M. (1996). The socializing role of hearing dogs. *Applied Animal Behaviour Science, 47*, 7–15.

Headey, B. (1999). Health benefits and health costs savings due to pets: preliminary estimates from an Australian national survey. *Social Indicators Research, 47*, 233–243.

Hines, L. (1991). National hearing dog center survey results. *Alert, 2*, 4.

Houghtalen, R.P. and Doody, J. (1995). After the ADA: service dogs on inpatient psychiatric units. *Bulletin of American Academy of Psychiatric Law, 23*, 211–217.

Hudson, W.W. (1982). *The Clinical Measurement Package: A Field Manual*. Homewood, IL: Dorsey.

Hudson, W.W. (1992). *The Walmyr Assessment Scales Scoring Manual*. Tempe, AZ: Walmyr.

Jennings, L.B. (1997). Potential benefits of pet ownership in health promotion. *Journal of Holistic Nursing, 15*, 358–372.

Johnson, S.B. and Rule, W.R. (1991). Personality characteristics and self-esteem in pet owners and non-owners. *International Journal of Psychology, 26*, 241–252.

Joubert, C.C. (1987). Pet ownership, social interest and sociability. *Psychological Reports, 61*, 401–402.

Katcher, A.H. (1983). Health and the living environment. In Katcher, A.H. and Beck, A.M., Eds. *New Perspectives on Our Lives with Companion Animals*. Philadelphia: University of Pennsylvania Press.

Katcher, A.H. (1985). Physiologic and behavioral responses to companion animals. *The Veterinary Clinics of North America: Small Animal Practice, 15*, 403–410.

Katcher, A.H., Friedmann, E., Beck, A.M., and Lynch, J. (1983). Looking, talking, and blood pressure: the physiological consequences of interaction with the living environment. In Katcher, A.H. and Beck, A.M., Eds. *New Perspectives on Our Lives with Companion Animals.* Philadelphia: University of Pennsylvania Press, 351–362.

Kidd, A.H. and Martinez, R.L. (1980). Two personality characteristics in adult pet owners and non-owners. *Psychological Reports, 47,* 318.

Kushner, R.F., et al. (2006). The PPET Study: people and pets exercising together. *Obesity, 14,* 1762–1770.

Lago, D., Delaney, M., Miller, M., and Grill, C. (1989). Companion animals, attitudes toward pets, and health outcomes among the elderly: a long-term follow-up. *Anthrozoos, 3,* 25–34.

Lane, D.R., McNicholas, J., and Collis, G.M. (1998). Dogs for the disabled: benefits for the recipient and welfare of the dog. *Applied Animal Behavior Science, 59,* 49–60.

Lawton, M.P., Moos, M., and Moles. E. (1984). Pet ownership: a research note. *Gerontologist, 24,* 208–210.

Llacer, A., Zunzunegui, M.V., Gutierrez-Cuadra, P., Beland, F., and Zarit, S.H. (2002). Correlates of wellbeing of spousal and children carers of disabled people over 65 in Spain. *European Journal of Public Health, 12,* 3–9.

Lynch, J.J., Fregin, G.F., Mackie, J.B., and Monroe, R.R. (1974). Heart rate changes in the horse to human contact. *Psychophysiology, 11,* 472–478.

Mader, B., Hart, L.A., and Bergin, B. (1989). Social acknowledgements for children with disabilities: effects of service dogs. *Child Development, 60,* 1529–1534.

Marks, L. (1993). The effect of service dogs on the self-concept of the disabled. Master's thesis, Florida International University, Miami.

Masullo, R.A. (2000). The costly truth about cats and dogs: how much of a bite will a pet take out of your budget? *Sacramento Bee,* Sacramento, CA. http://www.sacbee.com/ourtown/pets/catsndogs.html.

McNicholas, J. and Collis, G.M. (2000). Dogs as a catalyst for social interactions: robustness of the effect. *British Journal of Psychology, 91,* 61–70.

Merbitz, C.T. (1996). Frequency measures of behavior for assistive technology and rehabilitation. *Assistive Technology, 8,* 121–130.

Merbitz, N.H. (2006). International Classification for Functioning Disability and Health: an orientation with practical applications. *SCI Psychosocial Process, 19.* Selected Abstracts from 2005 Annual Meeting of AASCIPSW.

Messent, P.R. (1983). Social facilitation of contact with other people by pet dogs. In Katcher, A.H. and Beck, A.M., Eds. *New Perspectives on Our Lives with Companion Animals.* Philadelphia: University of Pennsylvania Press, 36–46.

Miller, D., Staats, S., and Partlo, C. (1992). Discriminating positive and negative aspects of pet interaction: sex differences in the older population. *Social Indicators Research, 27,* 363–374.

Motooka, M., et al. (2006). Effect of dog walking on autonomic nervous activity in senior citizens. *Medical Journal of Australia, 184,* 60–63.

Mowry, R., Carnahan, S., and Watson, D. (1994). A national study of the training, selection and placement of hearing dogs. Technical Report, University of Arkansas, Fayetteville.

Mugford, R.A. and M'Comisky, J.G. (1975). Some recent work on the therapeutic value of cage birds with old people. In Anderson, R.S., Ed. *Pet Animals and Society.* London: Bailliere Tindall, 54–65.

Ory, M. and Goldberg, E. (1983). Pet ownership and life satisfaction in elderly women. In Katcher, A.H. and Beck, A.M., Eds. *New Perspectives on Our Lives with Companion Animals.* Philadelphia: University of Pennsylvania Press, 803–817.

Pang, P.E. (1999). Hearing dogs: enhancing human adaptability. *Dissertation Abstracts International, 60,* 1292.

Paulhus, D.L. (1983). Sphere specific measures of perceived control. *Journal of Personality and Social Psychology, 44,* 1253–1265.

Ptak, A. (1994). Requiem for a service dog. *Alert, 5,* 17–19.

Radloff, L.S. (1977). The CES-D scale: a self report depression scale for research in the general population. *Applied Psychological Measurement, 1,* 385–401.

Raina, P., Waltner-Toews, D., Bonnett, B., Woodward, C., and Abernathy, T. (1999). Influence of companion animals on the physical and psychological health of older people: an analysis of a one-year longitudinal study. *Journal of the American Geriatrics Society, 3,* 323–329.

Ramos, B.M. (2004). Culture, ethnicity, and caregiver stress among Puerto Ricans. *Journal of Applied Gerontology, 23,* 469–486.

Rintala, D. (2000). Effect of service dogs on community integration: a pilot study. Poster presentation at 14th Annual Conference of American Association of Spinal Cord Injury Psychologists and Social Workers, Las Vegas, NV. September 2000.

Rintala, D. (2006). Effects of assistant dogs on persons with mobility or hearing impairments (4-12-16): Final report to Veterans' Administration.

Robb, S.S. and Stegman, C.E. (1983). Companion animals and elderly people: a challenge for evaluators of social support. *Gerontologist, 23,* 277–282.

Roid, G.H. and Fitts, W.H. (1988). *Tennessee Self-Concept Scale.* Los Angeles: Western Psychological Services.

Rosenberg, M. (1979). *Conceiving the Self.* New York: Basic Books.

Rossbach, K.A. and Wilson, J.P. (1992). Does a dog's presence make a person appear more likable? *Anthrozoos, 5,* 40–51.

Roth, S. (1992). The effects of service dogs on the occupational performance and life satisfaction of individuals with spinal cord injuries. Master's thesis. Rush College, Chicago.

Rowan, A.N. (1996). Research and practice (editorial). *Anthrozoos, 9,* 2–3.

Rushing, C. (1995). The effect of service dogs on the self-concept of spinal injured adults. *Dissertation Abstracts International, 55,* 4133.

Sachs-Ericsson, N.J., Hansen, N., and Fitzgerald, S. (2002). Benefits of assistance dogs: a review. *Rehabilitation Psychology, 47,* 251–277.

Sanders, C.R. and Robins, D.M. (1991). Dogs and their people: pet facilitated interactions in a public setting. *Journal of Contemporary Ethnography, 20,* 3–25.

Schuelke, S.T., Trask, B., Wallace, C., Baun, M.M., Bergstrom, N., and McCabe, B. (1991). The physiological effects of the use of a companion dog as a cue to relation in diagnosed hypertensives. *Latham Letter,* Winter, 14–17.

Serpell, J. (1991). Beneficial aspects of pet ownership some aspects of human health and behaviour. *Journal of the Royal Society of Medicine, 84,* 717–720.

Sherbourne, C.D. and Hays, R. (1991). *RAND. User's Manual for the Medical Outcome Study (MOS) 36-Item Short Form.* Santa Monica, CA: Sherbourne.

Siegel, J.M. (1990). Stressful life events and use of physician services among the elderly: the moderating role of pet ownership. *Journal of Personality and Social Psychology, 58,* 1081–1086.

St-Yves, A., Freeston, M.H., Jacques, C., and Robitaille, C. (1990). Love of animals and interpersonal affectionate behavior. *Psychological Reports, 67,* 1067–1075.

Thorpe, R.J., et al. (2006). Dog ownership, walking behavior, and maintained mobility in late life. *Journal of the American Geriatric Society, 54,* 1419–1424.

Uchino, B.N., Cacioppo, J.T., and Kiecolt-Glaser, J.K. (1996). The relationship between social support and physiological processes: a review with emphasis on underlying mechanisms and implications for health. *Psychological Bulletin, 119,* 488–531.

Valentine, D.P., Kiddoo, M., and LaFleur, B. (1993). Psychosocial implications of service dog ownership for people who have mobility or hearing impairments. *Social Work in Health Care, 19,* 109–125.

Vinokur, A.D. and Van Ryn, M. (1993). Social support and undermining in close relationships: their independent effects on the mental health of unemployed persons. *Journal of Personality and Social Psychology, 65,* 350–359.

Vormbrock, J.K. and Grossberg, J.M. (1988). Cardiovascular effects of human–pet dog interactions. *Journal of Behavioral Medicine, 11,* 509–517.

Wallstein, S.S. (2000). Effects on caregiving, gender, and race on the health, mutuality, and social supports of older couples. *Journal of Aging Health, 12,* 90–111.

Watson, N.L. and Weinstein, M. (1993). Pet ownership relation to depression, anxiety and anger in working women. *Anthrozoos, 6,* 135–138.

Wells, D.L. (2007). Domestic dogs and human health: an overview. *British Journal of Health Psychology, 12,* 145–156.

Whiteneck, G., Charlifue, S.W., Gehart, K.A., Overholser, J.D., and Richardson, G.N. (1992). Quantifying handicap: a new measure of long term rehabilitation outcomes. *Archives of Physical Medicine Rehabilitation, 73,* 519–526.

Willer, B., Ottenbacher, K.J., and Coad, M.L. (1994). The community integration scale. *American Journal of Physical Medicine and Rehabilitation, 73,* 103–111.

Wilson, C. (1991) The pet as an anxiolytic intervention. *Journal of Nervous and Mental Disease, 179,* 482–489.

World Health Organization (2002). Towards a Common Language for Functioning, Disability and Health: ICF. Geneva. http://www.who.int/classifications/icf/site/beginners/bg.pdf.

Yuker, H.E. and Block, J. (1996) Research with attitudes toward disabled persons scales 1960–1985. Hofstra University, New York.

16 Conclusion
Working Dogs and the Future

William S. Helton

CONTENTS

CANINE ERGONOMICS

This book provides a first look at the new science of canine ergonomics—an integration of ergonomics and animal sciences. As scientists, we focus on impartial examination of canine abilities and limitations, but in reality we must work within a larger sociopolitical system. While working dogs clearly have limitations, they are the best options for many work contexts. The scientific community's recent reassessment of dogs' cognitive capabilities and increasing recognition of dogs as legitimate workers opens new opportunities for people with dogs to solve society's problems. We must balance the risk of excessive anthropomorphism, viewing dogs falsely as small people in furry suits, with excessive anthropocentrism, thinking only humans are smart enough to be workers or only human-designed machines are appropriate solutions to our problems. Our most reliable coworker has been, is, and for the foreseeable future will be, the domesticated dog. We are obligated to point out the fallacy of not investigating dogs as possibly the best solutions available.

With proper training, for example, dogs not only match the breast cancer detection accuracy, sensitivity, and specificity of mammography performed by trained radiologists, but dogs may actually exceed them (McCulloch et al., 2006). Dog cancer detection, because it is based on breath samples taken painlessly from patients and not on x-rays taken with often uncomfortable breast tissue compression, may be more attractive to patients. Dog detection presents no potential of radiation-induced side effects, no uncomfortable compression of breast tissue, and no need for patients to spend time traveling to a specialized clinic where they waste their time waiting to be examined. Breath samples for dog detection can be taken anywhere and sent to a remote dog laboratory. This is merely one example where dogs are largely dismissed in favor of expensive machines.

We must also increasingly keep in mind when comparing technologies the global impact of our choices. Dogs are, in some sense, natural sustainable products and may exert much smaller overall impacts on the environment than machines when material mining, processing, and the power needs of the machines are considered. Based on the catastrophic implications of global climate change, we are no longer free to pick nonsustainable options and will increasingly need to consider the larger ecological impacts of our choices. Future researchers must determine, for example, the carbon footprints of dogs versus machines in their comparisons of these two potential solutions. They must also determine to some degree the long-term impact of the choice between dogs and machines on human welfare and the environment.

Dogs will be used in increasing numbers in a variety of occupations and my hope is that this book may help foster this situation. While dog researchers are often enthusiastic supporters, we must not make the mistake of putting the animals on a pedestal and failing to recognize their limitations. A new ergonomic approach will be to augment dogs with technology. Instead of trying to entirely replace the biological with the artificial, we can use machines to enhance canine capabilities.

Our focus on dogs should not limit research on other animals. Despite dogs' unique qualities, other animals such as rats, seals, and dolphins may be more useful in some contexts. The chapters in this book provided a general introduction to what we know scientifically about canine ergonomics and provided some glimpses of where this new discipline is headed. In the rest of this concluding chapter, I will provide you with parting examples of new canine technologies on the horizon, with a particular emphasis on enhancing human–dog communications.

THE FUTURE: TEAM COMMUNICATIONS

Many of the mistakes occurring in canine operations are due to communication failures between dog and handler, not due to failures of the dogs. New communication devices will open new possibilities for dog workers. Three potential communication devices will be discussed: lexigram touch boards, touch screens, and tactile communication devices. None of these technologies are currently employed with working dogs, but they are suggestive of ideas that could be employed by canine ergonomic professionals.

LEXIGRAM TOUCH BOARDS

Comparative psychologists interested in language and communication in nonhuman animals have developed symbolic touch boards for great apes and dolphins. Duane Rumbaugh, for example, developed a lexigram keyboard to be used by Lana, a chimpanzee. This keyboard for primates consists of three panels of 384 noniconic arbitrary symbols (lexigrams). A primate, for example Lana the chimpanzee, can press the lexigrams to form strings of statements or requests such as "Lana want banana."

Recent research by Rossi and Ades (2008) demonstrates that dogs can use a simplified lexigram touch board as well.

Dogs are excellent at reading human gestures and nonverbal signs, such as pointing, head movements, and changes in eye gaze. Dogs are also able to match arbitrary spoken words with objects or actions. Although dogs are not language users per se, recent research by Kaminski et al. (2004) demonstrates that they are capable of reasoning by exclusion with words and fast-mapping words to objects. Essentially, Rico, the dog in Kaminski et al.'s study, already knew the associated names of a wide array of objects. When asked to retrieve a known object, for example a red ball, the dog could go into a room and properly retrieve the red ball. This is not surprising. More interestingly, when the dog was asked to retrieve an object with an unfamiliar name, it was able to retrieve that unknown object from a room where all the other objects in the room had associated names. These results suggest, but do not necessarily prove, dogs are able to reason by exclusion. Since all the other objects in the room had associated names, the single object in the room without an associated name must have been the newly named object. This ability of dogs has also been recently demonstrated by Aust and colleagues (in press) with pictures. Dogs are cognitive and their understanding of communicative signs may have been underestimated by cognitive scientists.

Dogs are typically seen as the receivers of communications in people–dog exchanges, but they are also capable of sending messages to people. Dogs express their motivational and emotional states through their vocalizations and physical movements (Miklosi et al., 2000; Molnar et al., 2006; Yin and McCowan, 2004). Dogs have not, however, been previously given a means to communicate information using arbitrary lexigrams and signs to people, but recent research by Rossi and Ades (2008) provides evidence that dogs, like chimpanzees and dolphins, are capable of using arbitrary signs to communicate with people. Inspired by the work of Rumbaugh and others in developing computer-controlled lexigram keyboards for great apes and dolphins, Rossi and Ades developed a keyboard for a female mixed breed dog named Sofia.

The dog lexigram keyboard was a 15 × 30 cm electronic keyboard consisting of 6 × 6 cm keys each marked by a lexigram (sign). When depressed, each key generated a recorded Portuguese word (the study was conducted in Brazil) that matched a key's lexigram. The words were walk, pet, toy, water, food, crate, and urine (in Portuguese: passear, carinho, brinquedo, agua, comida, casinha, xixi). The keyboard was placed on the floor to be freely accessible to Sofia. Training using the lexigram keyboard was initiated when she was 4 months old and the final test was conducted when she was 17 months old.

Sofia lived in the home of Rossi as a pet and was a participant in other research projects. In careful tests, Rossi and Ades (2008) provide convincing evidence that Sofia could use the keyboard to make meaningful and appropriate requests. In 87.4% of Sofia's keyboard communications in which the authors were able to video record Sofia's prior activities, such as gaze behavior and physical movements, her prior activities were predictive of which lexigram was selected by two other observers. In 97.2% of Sofia's keyboard communications in which the authors recorded Sofia's actions subsequent to a key press, her behavior was appropriate for the item pressed, e.g., if she pressed *toy* and received a toy, she appropriately played with the toy.

Furthermore, Rossi and Ades were careful to test Sofia's ability to map the words with the lexigrams by randomizing the positions of the keys on the board. Although rearranging the positions of keys affected Sofia's accuracy (as would rearranging the keys on a keyboard affect human accuracy), she still used the lexigrams more appropriately than by chance. Rossi and Ades (2008) cite an example of Sofia's keyboard activities: "Coming back from a walk, Sofia enters the room and goes immediately to the keyboard, pressing *water*. She licks her nose and stares at APR [Rossi], once again presses *water*, stares again at APR, looks in the direction of the water container, closely follows (nose almost touching it) the container while APR puts it on the floor."

Two other interesting points made by Rossi and Ades about Sophia's use of the keyboard were the generality of the lexigrams and some evidence Sophia grasped the communicative use of the keyboard. The toy and food lexigrams, for example, were induced by a diverse set of toys and food items instead of a single toy or food product. Also, Sophia seemed to grasp that the keyboard was a communicative device. She never used the keyboard when left alone with it and repeatedly pressed the appropriate key if the person present failed to quickly respond or comply with her request. Regardless of implications this may or may not have for questions of dog cognition, Sophia's appropriate use of the keyboard suggests this technique may be adapted to other working dog scenarios.

Two examples of how this technique could be employed with working dogs are as signaling devices in multiple target searches and as signaling devices for hearing assistance dogs (also known as signal dogs). Dogs are sometimes trained to search for more than one target type. Search and rescue dogs, for example, are sometimes trained to search for both live and dead people (cadavers). The dogs could be trained to associate a lexigram with a particular target type. When given the command to search, a dog could use the lexigram board as the overt response. By pressing a specific lexigram, the dog would give a clear signal to the handler of which target was detected: a cadaver or a living person. This technique could be used in other detection tasks as well. The lexigram board could also be useful to people with hearing assistance dogs. These animals are trained to signal hearing-impaired persons when certain environmental sounds occur. For example, when a door bell rings, a hearing dog runs to the hearing-impaired person, makes an overt response by touching the person, and induces the person to follow it to the door. With a lexigram board, the dog could, for example, press the appropriate *door* lexigram when the door bell rang and this information could be sent to a cell phone set to vibrate. Another approach could be that a trained dog would signal the person and when asked "what did you hear," the dog could press the appropriate lexigram. This opens intriguing possibilities—some will work and others will not, but giving a dog a clearer voice may facilitate working dog behavior.

VIDEO DISPLAY TOUCH SCREENS

Range and colleagues (2008) demonstrated that dogs can categorize pictures on a computer screen and interface with computers using touch screens. Range used a

15-inch TFT display equipped with an infrared touch frame (Carroll Touch, Round Rock, Texas; 32 vertical × 42 horizontal resolution). They set the distance between an array of light-emitting diodes and screen at 1 cm. When the dog's muscle passed the light diodes, a press was registered. The picture stimuli they used were presented at 150 × 111 pixels, producing 5.29 × 3.92 cm pictures on the monitor. The dogs in their study could reliably sort pictures showing dogs from pictures without dogs. Since the 1960s touch-sensitive screens have been used with a wide variety of birds in experimental studies of vision. Pictures are presented to the birds and the birds make overt responses by pecking on the pictures. These touch-screen applications have also been used with primates.

Dogs can perceive pictures presented on video terminals and screens. The mammalian visual system consists of two kinds of receptor cells that are sensitive to electromagnetic radiation (light): rods and cones. The rod receptors are more sensitive and adapted for low light conditions, and the cones are adapted for bright light and color detection. While dogs' visual systems appear to be specialized for low light conditions (dawn and dusk), dogs have dichromatic color vision and do not see only in black and white (Aguirre, 1978; Odom et al., 1983). Approximately 3% of canine photoreceptors are cones (Peichl, 1991). Human visual systems are trichromatic, with three types of cones sensitive to different wavelengths of light. About 5% of the photoreceptors in the human visual system are cones. Dogs' temporal resolution for their color-sensitive cone receptors is slightly higher (70 to 80 Hz) than in humans (50 to 60 Hz). While the degree of binocularity is lower in dogs than people, dogs do have binocular vision, although it may vary for different breeds (Aguirre, 1978). As long as these physiological differences are taken into consideration when designing displays (for example, dogs have problems discriminating shades of red and fine details and may be more sensitive to flickering), dogs will be able to use the displays.

A possible concern, for example, when using video displays for dogs is the perception of flicker. Pictures presented with most video display technologies are not stable; they are presented as a series of pictures moving very quickly. In motion pictures (movies), a series of still photographs are taken and when presented quickly enough, at a rate above what people can detect (the flicker fusion threshold), people perceive fused motion, not a series of still photographs.

This is also related to how many video displays work. Computer cathode ray tubes (CRTs) operate well over 60 Hz, modern ones reach around 100 Hz, so people rarely experience flickering. Modern LCD flat panels are even better because the backlight of the screen operates around 200 Hz. A dog's flicker fusion threshold is, however, much higher than that of a human, up to 80 Hz, so care must be taken when using video displays with dogs. Perception of flicker may reduce any illusion of smooth motion in a display and may be distracting. Despite this and other concerns, dogs can discern pictorial information in video displays (Aust et al., in press; Pongracz et al., 2003; Range et al., 2008). Range, moreover, reports that the dogs in her studies are motivated to work the video display apparatus; this suggests they do not find the work uncomfortable.

While no one has yet explored the use of touch-screen display terminals with working dogs, it is possible that these screens could be used to facilitate dog–handler communications. In a manner similar to the lexigram touch board discussed previously, dogs could with training use touch screens to interact with their handlers. The screens could also be used to convey information to the dogs since dogs can perceive pictures presented on video displays.

TACTILE COMMUNICATION DEVICES

One area requiring further research is the use of tactile communication devices by human–dog working teams. Simpler versions of these devices are currently used in dog training and in homes and businesses to restrict dogs to certain areas. These devices are often marketed as correction collars or electronic correction collars. A device consisting of electrodes and batteries housed in a unit is attached to a collar. With a remote control device, a handler can deliver shocks to a dog at various intensities and durations, or the electronic collar can be set to trigger a shock if the dog approaches a restricted area. These collars are often used as direct punishers (in psychological terminology, positive punishment) to deliver physical punishment after a dog behaves undesirably, for example, if a dog chases a squirrel. A shock is delivered to halt the behavior. These collars may also be used as negative reinforcers, in which case relief from a series of shocks is granted for appropriate behavior. This technique is similar to an approach often used by horse trainers in which behavior is rewarded by removing an aversive stimuli, for example, easing of the horse's bit to relieve pressure to the animal's mouth when it complies by stopping. The use of these training devices is controversial, with critics pointing out increases in dog fear and anxiety when these devices are used (Schilder and van der Borg, 2004). Despite their controversial nature, these devices lead to the possibility of using nonpainful tactile (touch) devices to communicate with dogs.

Touch or tactile sensing is not typically used as a communication channel by humans. We are much more accustomed to auditory language and visually recognizable gestures such as sign language to communicate with each other. We do, however, use touch to communicate. When you reach out to tap someone on his or her shoulder, you are using tactile communication. This form of communication has some advantages: it is silent, often invisible, and yet easily interpreted when the signals are not too complex. Communication signals can be transferred using a variety of techniques including vibration of mechanical devices, for example, the vibration modes of many cell phones and pagers.

Vibrating mechanical tactile communication devices (tactors) have been developed. They are electromechanical actuators that vibrate against the skin and have been used in a variety of applications, from covert military operations to helping astronauts and pilots orient under very low or very high gravitational forces. Companies have been developing a wide range of small wireless tactors that can be attached to vests or belts. These tactile systems composed of multiple tactors surrounding a person's body can be used to orient people to objects in their environment or let them know in which direction to move (Brill et al., 2004; Cholewiak, Brill, and Schwab, 2004). A vibration of the right side of the body can be used to instruct a person to look to the right or move to the right. In essence, movement and orientation can be controlled by sending appropriate signals to an array of tactors worn around the midsection or in a vest.

More interesting recent research has focused on the development of tactile icons (Brill et al., 2006) in which a series of vibrations through the array can be used to send simple commands. For example, a course of pulses around the body could instruct to turn around. Two pulses on the side of the body could warn a person to duck. This technique could probably also be used with dogs. With proper training and appropriate system design, dogs could be trained to interpret signals from a tactor harness.

Trainers currently use hand signals that could probably be translated into tactile icons. This opens a viable communication channel between dogs and their handlers. Dogs could be trained to take directional information from a harness. The device could also be used to guide a dog remotely through a building, in a manner similar to the technique employed by military teams (Brill et al., 2004). Dogs could also be given remote commands to sit or seize an intruder. When these devices are combined with the remote viewing video techniques discussed by Ferworn in Chapter 11, dog handlers will have a whole new array of possibilities. Dog sentries and search dogs may in the future operate remotely. Remote operation will not require invasive brain surgery or the implantation of electrodes into a dog's central nervous system in some macabre experiment; the simple requirements are appropriate technology and patient training. More details about tactors and tactile communication can be found at the Web site of the Tactile Research Group (currently http://eastcollege.poly.asu.edu/robgray/TRG/).

REFERENCES

Aguirre, G. (1978). Retinal degeneration in the dog: rod dysplasia. *Experimental Eye Research, 26*, 233–253.

Aust, U., Range, F., Steurer, M., and Huber, L. (2008). Inferential reasoning by exclusion in pigeons, dogs, and humans. *Animal Cognition*, in press.

Brill, J.C., Gilson, R.D., Mouloua, M., Hancock, P.A., and Terrence, P.I. (2004). Increasing situation awareness of dismounted soldiers via directional cueing. In Vincenzi, D.A. et al., Eds. *Human Performance, Situation Awareness, and Automation: Current Research and Trends*, Vol. 1. Mahwah, NJ: Lawrence A. Erlbaum, 130–132.

Brill, J.C., Terrence, P.I., Stafford, S., and Gilson, R.D. (2006). A wireless tactile communication system for conveying U.S. Army arm–hand signals. *Proceedings of 50th Annual Meeting of the Human Factors and Ergonomics Society*, San Francisco, CA.

Cholewiak, R.C., Brill, J.C., and Schwab, A. (2004). Vibrotactile localization on the abdomen: effects of place and space. *Perception and Psychophysics, 66*, 970–987.

Kaminski, J. Call, J., and Fischer, J. (2004). Word learning in a domestic dog: evidence for "fast mapping." *Science, 304*, 1682–1683.

McCulloch, M., Jezierski, T., Broffman, M., Hubbard, A., Turner, K., and Janecki, T. (2006). Diagnostic accuracy of canine scent detection in early- and late-stage lung and breast cancers. *Integrative Cancer Therapies, 5*, 30–39.

Miklósi, A.P.R., Topál J., and Csányi V. (2000). Intentional behaviour in dog–human communication: an experimental analysis of "showing" behaviour in the dog. *Animal Cognition, 3*, 159–166.

Molnár, C., Pongrácz, P., Dóka, A., and Miklósi, A. (2006). Can humans discriminate between dogs on the base of the acoustic parameters of barks? *Behavioral Processes, 73*, 76–83.

Odom, J.V., Bromberg, N.M., and Dawson, W.W. (1983). Canine visual acuity: retinal and cortical field potentials evoked by pattern stimulation. *American Journal of Physiology, 245,* 637–641.

Peichl, L. (1991). Catecholaminergic amacrine cells in the dog and wolf retina. *Visual Neuroscience, 7,* 575–587.

Pongraz, P., Miklosi, A., Doka, A., and Csanyi, V. (2003). Successful application of video projected human images for signaling to dogs. *Ethology, 109,* 809–821.

Range, F., Aust, U., Steurer, M., and Huber, L. (2008). Visual categorization of natural stimuli by domestic dogs. *Animal Cognition, 11,* 339–347.

Rossi, A.P. and Ades, C. (2008). A dog at the keyboard: using arbitrary signs to communicate requests. *Animal Cognition, 11,* 329-338.

Rumbaugh, D.M. (1977). *Language Learning by a Chimpanzee: The Lana Project.* New York: Academic Press.

Savage-Rumbaugh, E.S. (1990). Language acquisition in a nonhuman species: implications for the innateness debate. *Developmental Psychobiology, 23,* 599–620.

Schilder, M.B.H. and van der Borg, J.A.M. (2004). Training dogs with help of the shock collar: short and long term behavioural effects. *Applied Animal Behaviour Science, 85,* 319–334.

Yin, S. and McCowan, B. (2004). Barking in dogs: context specificity and individual specification. *Animal Behaviour, 68,* 343–355.

Index

Printed in the United States
by Baker & Taylor Publisher Services